O	In chi square, the observed frequency
r_s	A correlation coefficient for ranked data; named for Spearman
r	Pearson product-moment correlation coefficient
r^2	The coefficient of determination; the proportion of variance in variable Y that is associated with variance in variable X
S	The standard deviation of a sample; used to describe the sample
s	The standard deviation of a sample; used to estimate the standard deviation of a population
s^2	Variance of a sample; used to estimate the variance of a population
s_D	Standard deviation of the distribution of differences between correlated scores (direct-difference method)
$s_{\bar{D}}$	Standard error of the difference between correlated means (direct-difference method)
$s_{\bar{X}_1 - \bar{X}_2}$	Standard error of a difference between means
$s_{\bar{X}}$	Standard error of the mean for a sample
SS	Sum of squares; the sum of the squared deviations from the mean
T	A nonparametric statistic used to test the difference between two correlated samples (Wilcoxon matched-pairs signed-ranks test)
t	A ratio of the difference between means to the standard error of the difference between means; a sampling distribution of such ratios
t_α	Critical value of t; level of significance $= \alpha$
U	Nonparametric statistic used to determine whether two sets of data based on two independent samples came from the same population (Mann-Whitney U test)
UL	Upper limit of a confidence interval
X	A score
x	A deviation score
$\bar{X}$	The mean of a sample
$\bar{\bar{X}}$	The mean of a set of means
X_H	The upper limit of the highest score in a distribution
X_L	The lower limit of the lowest score in a distribution
Y'	The Y value predicted from some X value
$\bar{Y}$	The mean of the Y variable
z	A standard score; a score expressed in standard-deviation units
z_X	A z value for a score on variable X
z_Y	A z value for a score on variable Y

BASIC STATISTICS

Tales of Distributions
4th Edition

BASIC STATISTICS

Tales of Distributions

4th Edition

Chris Spatz
Hendrix College

James O. Johnston
Oglala Lakota College (Retired)

Brooks/Cole Publishing Company
Pacific Grove, California

MAI 338 1953

Consulting Editor
Roger E. Kirk, Baylor University

Brooks/Cole Publishing Company
A Division of Wadsworth, Inc.

Printed in the United States of America

10   9   8   7   6   5   4

Library of Congress Cataloging-in-Publication Data

Spatz, Chris, [date]–
    Basic statistics.

    Bibliography: p.
    Includes index.
    1. Statistics. I. Johnston, James O. II. Title.
QA276.16.S66   1988   519.5   88-14491
ISBN 0-534-09486-4

Sponsoring Editor: *Phil Curson*
Senior Editorial Assistant: *Amy Mayfield*
Production Editor: *Phyllis Larimore*
Production Assistants: *Dorothy Bell, Marie DuBois*
Manuscript Editor: *Carol Reitz*
Permissions Editor: *Carline Haga*
Interior and Cover Design: *Kelly Shoemaker*
Cover Photo: *Lee Hocker*
Art Coordinator: *Sue C. Howard*
Interior Illustration: *Maggie Stevens-Huft*
Typesetting: *Polyglot Compositors, Inc.*
Cover Printing: *The Lehigh Press, Inc.*
Printing and Binding: *Arcata Graphics, Fairfield*

To Thea and Carolyn

Even if our statistical appetite is far from keen, we all of us should like to know enough to understand, or to withstand, the statistics that are constantly being thrown at us in print or conversation — much of it pretty bad statistics. The only cure for bad statistics is apparently more and better statistics. All in all, it certainly appears that the rudiments of sound statistical sense are coming to be an essential of a liberal education.

— R. S. Woodworth

PREFACE

Basic Statistics: Tales of Distributions, Fourth Edition, is designed for a one-term, introductory course in statistics. In writing and revising this book, we have tried to be comprehensible and complete for the student who takes only one course and comprehensible and preparatory for the student who will take additional statistics courses.

Detailed directions and examples are given for each statistical procedure, but we have also concentrated heavily on conceptualization and interpretation of statistical results. In addition, we have included discussion and examples of experimental design in several chapters.

Our expectation for students who work through this book is that they will be able to

- solve statistical problems
- understand statistical reasoning
- tell the story of what a statistical analysis shows
- choose the proper statistical technique for simple experimental designs

Students who meet these expectations will be able to understand the statistical concepts in many journal articles, analyze data from their own research, and use their knowledge for years to come.

We think you will like this book. Most students find it a pleasant book to read because it is written in a conversational rather than a formal style. We have used examples and problems from many fields, including some that the pioneers in statistics worked on.

Besides the writing style, this book has a number of features that are designed to make statistics easier to learn. For example, the problems are an integral part of the text. The answers, complete with all necessary steps or explanations, are at the back of the book. Concepts that will be important in later chapters are identified as Clues to the Future and set off in boxes. The Error Detection boxes call attention to ways of catching and correcting mistakes. At the beginning of each chapter, a list of objectives provides orientation. This same list also serves as an effective summary of the chapter.

There are three glossaries: those of words and formulas are in Appendixes D and E; the symbol glossary is conveniently located on the inside covers of the book.

A careful examination of this fourth edition shows that the basic outline of previous editions remains intact. Except the introductory chapter, other chapters are now numbered one less than in previous editions because the arithmetic and algebra review material (formerly Chapter 2) has been moved to Appendix B. New problems from psychology, sociology, law, manufacturing, and health have been added. Additional information on the personal history and motivation of early statisticians has been added, particularly on Pearson, Galton, and Fisher. The explanation of $N - 1$ in the denominator of the sample standard deviation has been improved. The chapter on correlation and regression (Chapter 4) has been extensively revised, with new material on uses of r and the least-squares technique having been added. An explanation of the reason that .05 was adopted as the most popular α level is given. Also, the presentation of tests subsequent to ANOVA is shorter and simpler. Our belief in the importance and power of graphs can be found in many new places, including the 25 new or substantially revised figures and graphs. Contemporary problems have been brought up to date.

It is most satisfying to acknowledge that this book has benefited greatly from the resources of students, colleagues, and Hendrix College. Students at Hendrix used this material, gleefully but respectfully pointing out errors. Martha Chandler provided prompt custodial attention. Hendrix College awarded the first author a sabbatical and travel funds. Roger E. Kirk of Baylor University again made important suggestions and saved us from several errors. C. Deborah Laughton, Phil Curson, Phyllis Larimore, and the staff at Brooks/Cole contributed in many ways. We also acknowledge the help of the reviewers of this edition: Dennis Bonge, University of Arkansas; John Boswell, University of Missouri, St. Louis; Ric Brown, California State University, Fresno; Jerry Lackey, Stephen F. Austin State University; William McDaniel, Georgia College; Richard Sloop, Frostburg State College; Billy Smith, University of Central Arkansas; and Thomas Stonebraker, Greenville College.

In addition, we are grateful to the Literary Executor of the late Sir Ronald A. Fisher, F.R.S., to Dr. Frank Yates, F.R.S., and to Longman Group, Ltd., London, for permission to reprint Tables III, IV, and VII from their book *Statistical Tables for Biological, Agricultural and Medical Research* (Sixth Edition, 1974).

Our final and most important acknowledgment goes to our families, who have helped and supported us in many ways over the fifteen years of this project.

Both of us are teachers of long standing and parents too. We acknowledge the responsibility of educating the next generation—a task of the first order, in our opinion. We hope our enthusiasm for this task comes through in our writing.

Chris Spatz
James O. Johnston

CONTENTS

APPENDIXES

BASIC STATISTICS

Tales of Distributions
4th Edition

1 INTRODUCTION

Objectives for Chapter 1: After studying the text and working the problems in this chapter, you should be able to:

1. Distinguish between descriptive and inferential statistics
2. Define the words *population, sample, parameter, statistic,* and *variable* as these terms are used in statistics
3. Distinguish between quantitative and qualitative variables
4. Identify the lower and upper limits of a quantitative measurement
5. Identify four scales of measurement and give examples of each
6. Distinguish between statistics and experimental design
7. Define some experimental-design terms—*independent variable, dependent variable,* and *extraneous variable*—and identify these variables when you are given a description of an experiment
8. Describe the relationship of statistics to epistemology
9. Identify a few events in the history of statistics

This is a book about statistics, written for people who do not plan to be statisticians by two people who were not trained as statisticians. Your authors are psychologists who use statistics to analyze data from experiments (ours and others'). We have also used statistics to help lawyers establish their clients' cases more convincingly. In addition, we have observed that statistical reasoning and statistical techniques, which have proved so powerful in science, are being used in more and more occupations and are proving to be just as powerful there.

As professors, we teach students how to use statistics to analyze data. Our students learn in one term to identify the questions an experiment is asking, determine the statistical procedures that will answer those questions, carry out the procedures, and then tell in plain English the story of what the procedures have shown. The best way for *you* to learn all this (especially the part about telling the story) is to take an active approach to learning. Specialists in learning have demonstrated that active learning is superior to passive learning, and this book is based on that active learning approach. So get ready to *do* some statistics problems, to *think* about the meaning of your answers, and to *write* your explanations in words that lead your study partners to say some version of "This is clear." Your efforts will pay handsome dividends.

WHAT DO YOU MEAN, "STATISTICS"?

Before we can say much more about statistics, we need to discuss the term itself. The word *statistics* has two meanings. A **descriptive statistic**[1] is an index number that summarizes or describes a set of data. You are already familiar with at least one of these index numbers, the arithmetic average, called the **mean.** You have probably known how to compute a mean since elementary school—just add up the numbers and divide by the number of numbers. As you already know, the mean gives you one number that is somewhat typical of all the numbers. Thus, the mean is a descriptive statistic. You will learn about the other descriptive statistics starting with Chapter 2. The basic idea is simple: a descriptive statistic expresses a characteristic of a set of data with one number.

The *Oxford English Dictionary* says that the word *statistics* came into use more than 200 years ago during the 18th century. At that time statistics referred to a country's political characteristics that were quantifiable, characteristics like population, taxes, and area. Statistics meant "state numbers." Those tables and charts (descriptive statistics) turned out to be a most satisfactory method for understanding a country better, for comparing different countries, and for making projections about the future. Later, the word *statistics* was used to refer to tables and charts on trade (economics) and on natural phenomena (science).

Starting about 100 years ago, scientists began to develop a procedure that today is called **inferential statistics.** Like earlier statisticians, these scientists were trying to understand some phenomenon that had caught their interest. The situation was different, however. The descriptive statisticians could gather *all* the data on a particular topic, but these scientists were forced to work with *samples*. Sampling introduced chance errors and this added more uncertainty to the conclusions. Inferential statistics was developed as a way to measure the effects of chance errors (with an eye toward reducing them). A dictionary definition of inferential statistics is that it is a method of taking chance factors into account when you use samples to reach conclusions about populations.

Inferential statistics has proved to be a very powerful method. The results of most scientific experiments are based on inferential statistics. (See any issue of the journal *Science*.) Many other fields use inferential statistics, too. We have supported this statement by selecting examples and problems from a variety of disciplines for this text and its study guide.

Our first example of the use of inferential statistics comes from the field of psychology. Today there is a lot of evidence that uncompleted tasks are remembered better than completed tasks. This is known as the *Zeigarnik effect*. Let's go back to a time in Berlin when this phenomenon was about to be discovered by Bluma

---

[1] Words and phrases printed in boldface type are defined in Appendix D, Glossary of Words.

Zeigarnik.[2] Zeigarnik asked the participants in her experiment to do about 20 tasks such as to construct a box from cardboard, work a puzzle, and make a clay figure. For each participant, half the tasks were interrupted before completion. Later, when participants were asked to recall the tasks they had worked on, they listed more of the interrupted tasks (about seven) than the completed tasks (about four). However, as you well know, scores on tests are subject to a certain amount of chance or luck. So the question arises, Was the difference in recall due to the interruption or was it due to chance? One way to answer the question is to do the experiment again and again and observe whether the same results continue to occur. However, you can answer the question after only *one* experiment by using inferential statistics. The way inferential statistics works is that it tells you the probability that the "chance is responsible" answer is correct. If the probability is very low, the "chance is responsible" explanation can be discarded. An answer based on inferential statistics is a statement like "The difference between the means is due to interruption and not due to chance because chance would produce results like these only one time in a hundred." Probability, then, is at the heart of inferential statistics.

Here is a general overview of how statistics is used today. Researchers start with some phenomenon, event, or process that they want to understand better. They translate it into numbers, manipulate the numbers according to the rules of statistics, translate the numbers back into the phenomenon, and finally tell the story of their new understanding of the phenomenon, event, or process. You can understand why many textbooks refer to statistics as a tool.

CLUE TO THE FUTURE

> The first part of this book is devoted to descriptive statistics (Chapters 2–4) and the second part to inferential statistics (Chapters 5–12). The two parts are not independent, though, because to understand inferential statistics you must first understand descriptive statistics.

WHAT'S IN IT FOR ME?

That is a reasonable question. We have several answers; pick the ones that apply to you and add any others that are true in your own case. One thing that is in it for you is that, when you have successfully completed this course, you will understand how statistical analyses are used to *make decisions*. You will understand both the elegant beauty and the ugly warts of the process. H. G. Wells, a novelist who wrote about the future, said, "Statistical thinking will one day be as necessary for efficient citizenship as the ability to read and write." So chalk this course up under the heading "general education about decision making, to be used throughout life."

Another result of the successful completion of this course is that you will know how to use a tool that is basic to psychology, education, medicine, biology,

---

[2] A summary of this study can be found in Ellis (1938). The complete reference, and all others in the text, can be found in Appendix A, References.

sociology, forestry, economics, and several other disciplines. Each of these disciplines has its own interests, but when it comes to deciding whether there is any difference between two methods of therapy, three ways of teaching, five rates of administering hormones, two advertising campaigns, and so forth, they all rely on statistics. The ideas of statistics are multidisciplinary; because they are used by many, this course opens several doors for you at once.

More and more disciplines are finding that statistical methods are useful. In American history, for example, the authorship of 12 of the Federalist Papers was disputed for a number of years. (*The Federalist Papers,* a group of 85 short essays, were written under the pseudonym "Publius" and published in New York City newspapers in 1787 and 1788. Written by James Madison, Alexander Hamilton, and John Jay, they were designed to persuade the people of the state of New York to ratify the Constitution of the United States.) To determine authorship, each of the 12 papers was graded with a quantitative *value analysis* in which the importance of such values as national security, a comfortable life, justice, and equality was assessed. The value analysis scores were compared with value analyses of papers known to have been written by Madison and Hamilton. The authors of the study concluded that James Madison wrote all 12 papers (Rokeach et al., 1970). This result confirms the conclusion reached by historians who used more traditional methods.

Here is another example, this time from law. Rodrigo Partida was convicted of burglary in Hidalgo County, a border county in southern Texas. A grand jury rejected his motion for a new trial. Partida's attorney filed suit, claiming that the grand-jury selection process discriminated against Mexican-Americans. In the end (Castaneda v. Partida, 430 U. S. 482 [1976]), Justice Harry Blackmun of the U. S. Supreme Court, wrote, regarding the number of Mexican-Americans on the grand jury, "If the difference between the expected and the observed number is greater than two or three standard deviations, then the hypothesis that the jury drawing was random (is) suspect." In Partida's case the difference was approximately 12 standard deviations, and the Supreme Court ruled that Partida's attorney had presented *prima facie* evidence. (*Prima facie* evidence is so good that you win the case unless the other side rebuts the evidence, which in this case did not happen.)

If you are an independent sort—a person who distrusts authority or who prefers to find out things for yourself, statistics offers you a way to analyze the data that come from your own observations and experiments. We happen to think that experiments are lots of fun, and we recommend them to you. The best ones are those you think up yourself. Once you have designed an experiment and gathered the data, a statistical analysis is often the first step in the process that ends with you telling others exactly what your experiment has shown.

We hope that this course will encourage you or even teach you to ask questions. Consider the questions, one good and one better, asked of a restaurateur in Paris who served delicious rabbit pie. He served rabbit pie even when other restaurateurs could not obtain rabbits. A suspicious customer asked this restaurateur if he was stretching the rabbit supply by adding horse meat.

"Yes, a little," he replied.

"How much horse meat?"

"Oh, it's about 50–50," the restaurateur replied.

The suspicious customer, satisfied, began eating his pie. Another customer, one who had learned to be more thorough in asking questions about statistics, said, "50–50? What does *that* mean?"

"One rabbit and one horse" was the reply.

To end this section, we put into one sentence our answer to your question "What's in it for me?" *When you learn the basics of statistics you will understand data better, you will be able to communicate with others who use statistics, and you will be better able to persuade others.* (You will be able to present *prima facie* evidence.) In short, we think there will be lots of lasting value for you in this course.

SOME TERMINOLOGY

Like most courses, statistics introduces you to many new words. In statistics, most of the terms keep being used, time and time again. Your best move, when introduced to a new term, is to *stop, read* the definition carefully, and *memorize* it. As the term continues to be used, you will become more and more comfortable with it.

Populations, Samples, and Subsamples

A **population** consists of all members of some specified group.[3] A **sample** is a subset of a population. A **subsample** is a subset of a sample. The population is the thing of interest, is defined by the investigator, and includes all cases. The following are some populations:

Family income of college students in the fall of 1987
The proportion of salmon that actually return to their hatching stream
The IQs of females and the IQs of males
Depressed Alaskans
Racial attitudes of 9th- and 12th-grade students in the southern United States after school desegregation
The memory of human beings[4]

Investigators are always interested in a population. However, you can see from these examples that populations are often so large that not all the members can be measured. The investigator must often resort to measuring a sample that is small enough to be manageable but large enough to be representative of the population. A sample taken from the first population might include only 38 students. From the last population on the list, Zeigarnik used a sample of 164. Ways to assure yourself that samples are representative of their populations will be discussed at various points in this book.

---

[3] Actually, in statistics, a population consists of the *measurements* on the members and not the members themselves. Thus, for a group of fifth-graders, there might be a population of ages, a population of IQ scores, and a population of times for the 50-meter dash.

[4] We didn't pull these populations out of thin air; they are all populations that researchers have been interested in. Studies of these populations are described at various points in this book.

Going a step further, samples are often divided into subsamples and relation-ships among the subsamples are determined. The samples of 9th- and 12th-graders, for example, might be subdivided on the basis of gender or race. The investigator would then look for differences among the subsamples.

Resorting to the use of samples and subsamples introduces uncertainty into the conclusions because different samples from the same population nearly always differ from one another in some respects. Inferential statistics is used to determine whether or not such differences should be attributed to chance. You will meet this problem head-on in Chapter 6.

In research articles, authors usually identify the population only implicitly; they often do not announce how far they think their results will generalize. You should think about this, however. As an example, it seems a short, reasonable step to generalize the results of a drug therapy program with 14 teenage male schizo-phrenics to all teenage male schizophrenics. By knowing more, you may be comfortable in generalizing to all teenage schizophrenics or even to all schizo-phrenics. As another example, many researchers are willing to generalize from an albino rat study to "all mammals." The reason for doing the study is to get the generalization.

Parameters and Statistics

A **parameter** is some numerical characteristic of a population. An example is the mean IQ score of *all* fifth-grade pupils in the United States. A **statistic** is some numerical or nominal characteristic of a sample or subsample. Examples of statistics are the mean IQ score of 200 fifth-graders (a sample) and the mean IQ score of 100 male fifth-graders (a subsample). A parameter is constant; it does not change unless the population itself changes. There is only one number that is the mean of the population. Unfortunately, it often cannot be computed because the population is unmeasurable. So, a statistic is used as an estimate of the parameter, although, as we suggested before, a statistic tends to differ from one sample to another. If you have five samples from the same population, you will probably have five different sample means. Remember that parameters are constant; statistics are variable.

Variables

A **variable** is something that exists in more than one amount or in more than one form. The essence of *measurement* is the assignment of numbers on the basis of variation.

Height and gender are both variables. The variable height is measured using a standard scale of feet and inches. The notation 5′7″ is a numerical way to identify a group of persons who are similar in height. Of course, there are many other height groups, each with an identifying number. Gender is a variable. With gender, there are (usually) only two groups of people. Again, we may identify each group by assigning a number to it, usually 0 for females and 1 for males. All zeros are similar in gender. We will often refer to numbers like 5′7″ and 0 as *scores* or *test scores*. A score is simply the result of a measurement. So we may refer to an "age score" or "height score" or "anxiety score." Score just means number.

Quantitative Variables

Most of the variables you will work with in this text can be classified as **quantitative variables.** When a quantitative variable is measured, the scores tell you something about the amount or degree of the variable. At the very least, a larger score indicates more of the variable than a smaller score does.

Take a closer look at an example of a quantitative variable, IQ. IQ scores come only in whole numbers like 94 and 109. However, it is reasonable to think that of two persons who scored 103, one just barely made it and the other almost scored 104. Picture the variable IQ as the straight line in **Figure 1.1.**

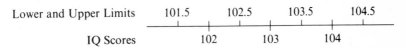

Figure 1.1   The lower and upper limits of IQ scores of 102, 103, and 104

Figure 1.1 shows that a score of 103 is used for a range of possible scores, the range from 102.5 to 103.5. The number 102.5 is the **lower limit** and 103.5 is the **upper limit** of the score 103. The idea is that IQ can take any fractional value between 102.5 and 103.5, but all scores in that range are rounded off to 103.

In a similar way, a score of 12 seconds is used to express all scores between 11.5 and 12.5 seconds; 11.5 seconds is the lower limit and 12.5 is the upper limit of a score of 12 seconds.

Sometimes scores are measured and expressed in tenths, hundredths, or thousandths. With a good stopwatch, for example, time can be measured to the nearest hundredth of a second. Scores like 3.82, 12.05, and 6.00 result. These scores also have lower and upper limits, and the same reasoning holds: each score represents a range of possible scores. The score 3.82 is the middle of a range of scores running from 3.815 to 3.825. Lower and upper limits of the scores 12.05 and 6.00 are 12.045 and 12.055 and 5.995 and 6.005, respectively.

To summarize, to find the lower and upper limits of any score, subtract half a unit from and add half a unit to the last digit on the right of the score. Thus, for the score 6.1 seconds, the lower limit is $6.1 - 0.05 = 6.05$ seconds and the upper limit is $6.1 + 0.05 = 6.15$ seconds.

The basic idea is that the limits of a score have one more decimal place than the score itself, and this last decimal place is 5. A good way to be sure you understand this is to draw a few number lines like the one in Figure 1.1, put scores like 3.0, 3.8, and 5.95 on them, and find the lower and upper limits.

<table>
<tr><td>**CLUE TO**
THE FUTURE</td><td>The lower and upper limits of a score on a quantitative variable will be important in Chapter 2, "The Organization of Data, Graphs, and Central Values."</td></tr>
</table>

Qualitative Variables

In addition to quantitative variables, sometimes you will deal with **qualitative variables.** With such variables, the scores (numbers) are simply names; they do not have quantitative meaning. For example, political affiliation is a qualitative variable. Terms like *Democrat, Republican, Independent,* and *Other* might be assigned the numbers 1, 2, 3, and 4, but these numbers clearly have no quantitative meaning. The same thing is true for the 0 and 1 used to identify gender. The numbers are simply used as substitutes for names.

PROBLEMS AND ANSWERS

At the beginning of this chapter, we urged you to take up the active learning approach. Did you? For instance, did you draw those lines and label them, as we suggested two paragraphs back? If not, try it now for the score 3.05. If you didn't get 3.045 as the lower limit, you need to reread that material. If you still have trouble, get some help from your instructor or from another student who understands it. Be active. Make it your policy throughout this course never to let anything slip by you.

Part of active learning is active reading. Read this text actively. You will know you are doing this if you reread a paragraph that you did not understand the first time, or wrinkle your brow in concentration, or say to yourself "Hmm, I really don't understand that; I'll ask my instructor about it tomorrow" or, best of all, nod to yourself, "Oh yes."

From time to time we will interrupt you with a set of problems so that you can practice what you have just been reading about. Working these problems correctly is also evidence of active reading. You will find the answers at the end of the book in Appendix F. Here are our suggestions for *efficient* learning.

1. Buy yourself a spiral notebook for statistics. Work all the problems for this course in it, since, for several problems, we refer you back to a problem you worked previously. When you make an error, don't erase it—circle it and work the problem correctly below. Seeing your error later serves as a reminder of what not to do on a test. If you find that *we* have made an error, write to us as a reminder of what not to do in the next edition.
2. Never, *never* look at an answer before you have worked the problem (or at least tried twice to work the problem).
3. When you come to a set of problems, work the first one and then immediately check your answer against the answer that we give. If you make an error, find out why you made it—faulty understanding, arithmetic error, or whatever.
4. Don't be satisfied with just getting the answer. Be sure you understand what the answer means and the process by which you got it.
5. When you finish a chapter, go back over the problems immediately, reminding yourself of the various techniques you have learned.
6. Use any blank spaces near the end of the book for your special notes and insights.

Now, here is an opportunity to see how actively you have been reading.

PROBLEMS

1. Give the lower and upper limits of the measurements below that are from quantitative variables. For the qualitative variables, write "qualitative."
   a. 8, gallons of gasoline
   b. 5, a category of daffodil in a flower show
   c. 10.00, a time required to run 100 meters
   d. 811.1, the identification of American poetry authors in the Dewey Decimal System
   e. 4.5, millions of dollars
   f. 2.95, tons of grain
2. Write a paragraph that gives the definitions of and the relationships among the following terms: *population, sample, parameter,* and *statistic.*
3. Classify each statement as being related to descriptive or inferential statistics.
   a. The population of the American colonies in 1776 was 2 million.
   b. Based on a sample, the pollster predicted that Demosthenes would win the election.
   c. Demosthenes won the election with 54.3 percent of the vote.
   d. The world population is projected to be 6.1 billion in the year 2000.
   e. The superior performance of those in Group 2 supports the hypothesis that moderate anxiety helps college students learn simple tasks.
   f. Based on the amount of C^{14} left in this sample, this wooden artifact is about 5000 years old.

SCALES OF MEASUREMENT

Numbers mean different things in different situations. Consider three answers that appear to be identical but are not:

"What number were you wearing in the race?"	"5"
"What place did you finish in?"	"5"
"How many minutes did it take you to finish?"	"5"

The 5's all look the same, but the situations they refer to are different (identification number, finish place, and time). As a result, each 5 carries a *different set* of information.

To illustrate this, consider another person whose answers to the same three questions were 10, 10, and 10. Taking the first question by itself, and knowing that the two people had scores of 5 and 10, what can you say? You can say that the first runner was different from the second, but *that is all.* (Think about this until you agree.) On the second question, with scores of 5 and 10, what can you say? You can say that the first runner was faster than the second and, of course, that they were different. Comparing the 5 and 10 on the third question, you can say that the first runner was twice as fast as the second runner (and, of course, was faster and different).

The point of this is to draw the distinction between the *thing* you are interested in and the *number* that stands for the thing. Much of your experience with numbers has been with pure numbers or with the measurement of things such as time, length, and amount. "Four is twice as much as two" is true for the pure numbers themselves and for time, length, and amount, but it is not true for finish places in a race. Fourth place is not twice anything in relation to second place—not twice as *slow* or twice as *far behind* another runner.

S. S. Stevens's 1946 article is probably the best known effort to focus attention on these distinctions. He identified four different kinds of *scales of measurement*, each of which carries a different set of information. Each scale uses numbers, but the information that can be inferred from the numbers differs. The four scales are nominal, ordinal, interval, and ratio. The first of these, the nominal scale, is not a scale at all in the usual sense. In the **nominal scale,** numbers are used simply as names and have no real quantitative value. It is the scale used for qualitative variables. Numerals on sports uniforms are an example; here, 45 is *different* from 32, but that is about all we can say. The person represented by 45 is not "more than" the person represented by 32, and certainly it would be meaningless to calculate a mean. Designating different colors, different psychological diagnoses, or different political parties by numbers produces nominal scales. With a nominal scale, you can even reassign the numbers and still maintain the original meaning, which is only that things with different numbers are different. Of course, all things that are alike have the same number.

A second kind of scale, the **ordinal scale,** has the characteristic of the nominal scale (different numbers mean different things) plus the characteristic of indicating "greater than" or "less than." In the ordinal scale, the object with the number 3 has less or more of something than the object with the number 5. Finish places in a race are an example of an ordinal scale. The runners finish in rank order, with 1 assigned to the winner, 2 to the runner-up, and so on. Here, 1 means less time than 2. Other examples of ordinal scales are house numbers, Government Service ranks like GS-5 and GS-7, and judgments such as "I am more anxious now" and "She is a better mathematician than he is."

The third kind of scale is the **interval scale,** which has the properties of both the ordinal and nominal scales, plus the additional property that intervals between the numbers are equal. "Equal interval" means that the distance between the things represented by 2 and 3 is the same as the distance between the things represented by 3 and 4. The Celsius thermometer is based on an interval scale. The difference in temperature between 10° and 20° is the same as the difference between 40° and 50°. The Celsius thermometer, like all interval scales, has an arbitrary zero point. On the Celsius thermometer, this zero point is the freezing point of water at sea level. Zero degrees on this scale does not mean the complete absence of heat; it is simply a convenient starting point. With interval data, there is one restriction: you may not make simple ratio statements. You may not say that 100° is twice as hot as 50° or that a person with an IQ of 60 is half as intelligent as a person with an IQ of 120.[5]

The fourth kind of scale, the **ratio scale,** has all the characteristics of the nominal, ordinal, and interval scales, plus one: it has a true zero point, which indicates a complete absence of the thing measured. On a ratio scale, zero means "none." Height, weight, and time are measured with ratio scales. Zero height, zero weight, and zero time mean that no amount of these variables is present. With a true zero point, you can make ratio statements like "16 kilograms is four times heavier than 4 kilograms."[6]

[5] Convert 100° C and 50° C to Fahrenheit, and suddenly the "twice as much" relationship disappears.

[6] Convert 16 and 4 kilograms to pounds (1 kg = 2.2 pounds), and the "four times heavier" relationship is maintained.

Having illustrated with examples the distinctions among these four scales—a practice common to most elementary statistics textbooks—we must tell you that it is sometimes difficult to classify the variables used in the social and behavioral sciences. Very often they appear to fall between the ordinal and interval scales. It may happen that a score provides more information than simply rank, but equal intervals cannot be proved. Intelligence test scores are an example. In such cases, researchers generally treat the data as if they were based on an interval scale.

The main reason this section on scales of measurement is important is that the kind of descriptive statistics you can compute on your numbers depends to some extent on the kind of scale of measurement the numbers represent. For example, it is not meaningful to compute a mean on nominal data such as the numbers on football players' jerseys. If the quarterback's number is 12 and a running back's number is 23, the mean of the two numbers (17.5) has no meaning at all. (See Gaito [1980], who tells when scales of measurement are important [measuring] and when they are not [inferential statistics].)

STATISTICS AND EXPERIMENTAL DESIGN

Here is a story that will help distinguish between statistics (applying straight logic) and experimental design (observing what actually happens). This is an excerpt from a delightful book by E. B. White, *The Trumpet of the Swan* (1970).

> The fifth-graders were having a lesson in arithmetic, and their teacher, Miss Annie Snug, greeted Sam with a question.
> "Sam, if a man can walk three miles in one hour, how many miles can he walk in four hours?"
> "It would depend on how tired he got after the first hour," replied Sam. The other pupils roared. Miss Snug rapped for order.
> "Sam is quite right," she said. "I never looked at the problem that way before. I always supposed that man could walk twelve miles in four hours, but Sam may be right: that man may not feel so spunky after the first hour. He may drag his feet. He may slow up."
> Albert Bigelow raised his hand. "My father knew a man who tried to walk twelve miles, and he died of heart failure," said Albert.
> "Goodness!" said the teacher. "I suppose *that* could happen, too."
> "Anything can happen in four hours," said Sam. "A man might develop a blister on his heel. Or he might find some berries growing along the road and stop to pick them. That would slow him up even if he wasn't tired or didn't have a blister."
> "It would indeed," agreed the teacher. "Well, children, I think we have all learned a great deal about arithmetic this morning, thanks to Sam Beaver."
> Everyone had learned how careful you have to be when dealing with figures.

Statistics involves the manipulation of numbers and the conclusions based on those manipulations (Miss Snug). Experimental design deals with all the things that influence the numbers you get (Sam and Albert). This text could have been a "pure" statistics book, from which you would learn to analyze numbers without knowing where they came from or what they referred to. You would learn about statistics, but such a book would be dull, dull, dull. On the other hand, to describe procedures for collecting numbers is to teach experimental design—and this book is for a statistics course. Our solution to this conflict is generally to side with Miss Snug but to include some aspects of experimental design throughout the book. We hope that the information on design that you pick up from this book will encourage you to try your hand at your own data gathering. We'll start our discussion of experimental design now.

Research is often conducted to discover cause-and-effect relationships. In a typical simple experiment, the experimenter is interested in the effect that one variable (called the **independent variable**) has on some other variable (called the **dependent variable**). In such experiments, a change in the independent variable is the presumed cause for a change in the dependent variable.

In an experiment, the experimenter chooses values for the independent variable, administers a different value of the independent variable to each group of subjects, and then measures the dependent variable for each subject. If the scores on the dependent variable differ as a result of differences in the independent variable, the experimenter may be able to conclude that there is a cause-and-effect relationship.

An example might help to clarify this important concept. Assume for the moment that you are a budding gourmet cook attempting to improve your roast beef gravy and you think that adding a little Worcestershire sauce will do the trick. So you give two dinner parties that feature roast beef. At one party, the gravy contains Worcestershire sauce. At the other party, it does not. At both parties, you count the number of favorable comments about the gravy. The independent variable in this example is Worcestershire sauce. The dependent variable is the guests' perceptions of the gravy as measured by favorable comments. The assumption is that differences in the number of favorable comments about the gravy are caused by the presence or absence of Worcestershire sauce.

Here is an exercise in identifying independent and dependent variables. Read each statement on the left of **Table 1.1** and answer for yourself, Which is the cause and which is the effect? With your own answer in mind, look at the right side of the table to see whether our independent variable is your cause and our dependent variable is your effect.

Table 1.1   Some cause-and-effect statements and their independent and dependent variables

Cause-and-effect statement	Independent variable	Dependent variable
Practice makes perfect.	Amount of practice	Degree of success
If at first you don't succeed, try, try again.	Number of tries	Success
You are what you eat.	What you eat	What you are
Too many cooks spoil the broth.	Number of cooks	Quality of broth
As the twig is bent, so grows the tree.	Early experience	Adult character

One of the problems with drawing cause-and-effect conclusions is that you must be sure that changes in the scores on the dependent variable are the result of changes in the independent variable and not the result of changes in some other variable(s). Variables other than the independent variable that can cause changes in the dependent variable are called **extraneous variables.** Many extraneous variables could be operating in the roast beef gravy example—and just one is enough to make it a mistake to draw a cause-and-effect conclusion. Some extraneous variables in that situation might be differences in the quality of beef on the two occasions, differences in the "party moods" of the two sets of guests, and differences in the quality and exact proportions of gravy ingredients other than Worcestershire sauce on the two occasions. These and other variables could affect the number of favorable comments about the gravy.

It is important, then, that experimenters be aware of and control extraneous variables that might influence their results. The simplest way to control an extraneous variable is to be sure that all subjects are equal on that variable. For example, if only well-rested subjects are used and the experiment is always conducted in the same room by the same experimenter, the extraneous variables of fatigue, environment, and experimenter are controlled.

Independent variables are often referred to as *treatments* because the experimenter frequently asks, "If I treat this group of subjects this way and treat another group another way, will there be a difference in their behavior?" The ways that the subjects are treated constitute the **levels** of the independent variable being studied, and experiments typically have two or more levels.

The experiment by Zeigarnik described on pages 2–3 can be used to illustrate these experimental-design terms. In her experiment, the independent variable was whether or not a task was interrupted. There were two levels of this variable: yes and no. The dependent variable was whether or not a task was recalled.

The design of the experiment used by Zeigarnik is called a *within-subject design* because each subject is tested under all levels of the independent variable. One of the nice characteristics of a within-subject design is that it controls many extraneous variables. Thus, in Zeigarnik's study, variables like the age of the subjects, how good their memories were, and their motivation were the same whether the task was interrupted or not.

The general question that experiments try to answer is, What effect does the independent variable have on the dependent variable? At various places in the following chapters, we will explain experiments and statistical analyses of their results using the terms *independent*, *dependent*, and *extraneous* variables. These explanations will usually assume that all extraneous variables were well-controlled; that is, you may assume that the experimenter knew how to design the experiment so that changes in the dependent variable can be correctly attributed to changes in the independent variable. However, we will present a few investigations (like the roast beef gravy example) that we hope you recognize as being so poorly designed that conclusions about the relationship between the independent variable and dependent variable cannot be drawn.

STATISTICS AND PHILOSOPHY

The previous section directed your thinking to the relationship between statistics and experimental design; this section will direct your thinking to statistics' place in the grand scheme of things.

The discipline of philosophy has the task of explaining the grand scheme of things and, as you know, there have been many schemes. For a scheme to be considered a grand one, it has to propose answers to questions of **epistemology**— that is, answers to questions about the nature of knowledge.

One of the big questions in epistemology is, How do we acquire knowledge? Answers such as *reason* and *experience* have gotten a lot of attention.[7] For those who emphasize the importance of reason, mathematics has been a continuing source of inspiration. In classical mathematics, you start with a set of axioms that are assumed to be true. Theorems are thought up and are then proved by giving axioms as reasons. Once a theorem is proved, it can be used as a reason in a proof of other theorems.

Since statistics has its foundations in mathematics, statistics is based on reason. As you go about the task of memorizing definitions to terms such as σ^2 and ΣX, calculating their values from data, and telling the story of what they mean, know that deep down what you are using is a technique in logical reasoning. Logical reasoning is rationalism, which is one answer to questions of epistemology. Experimental design is more complex—it is a combination of reasoning and experience (observation).

If we move from a formal description of philosophy to a more informal one, the task of many, many human beings can be described as *trying to understand*. Statistics has helped many in their search for better understanding and it is such people who have recommended (or demanded) that statistics be taught in school. A reasonable expectation is that you, too, will find statistics useful in your future efforts to understand.

STATISTICS: THEN AND NOW

The beginning point for statistics was with counting. That event, of course, was prehistory. The origin of the mean is almost as obscure. That statistic was in use by the early 1700s, but no one is credited with its discovery. A beginning point of graphs, however, is established; J. H. Lambert, a Swiss-German scientist and mathematician, and William Playfair, an English political economist, invented and improved graphs in the period 1765 to 1800 (Tufte, 1983).

In 1834 the Royal Statistical Society was formed in London. Just five years later, on November 27, 1839, at #15 Cornhill in Boston, a group of Americans founded the American Statistical Society. Less than three months later, for a reason that you can probably figure out even though 150 years have passed, the group changed its name to the American Statistical Association.

If you are willing to take 1839 as a beginning point, this 1989 textbook represents an introduction to the study of a 150-year-old discipline. Thus, your authors wish to dedicate this fourth edition of *Tales* as the American sesquicentennial (ses'kwuh centennial) edition.

[7] In philosophy, those who emphasize reason are rationalists and those who emphasize experience are empiricists.

According to Walker (1929), the first course in statistics given in a university in the United States was probably "Social Science and Statistics," taught at Columbia University in 1880. The professor was a political scientist and the course was offered in the Economics Department. In 1887 at the University of Pennsylvania, separate courses in statistics were offered by the departments of psychology and economics. By 1891, Clark University, the University of Michigan, and Yale had been added to the list of schools and anthropology had been added to the list of departments. Biology was added in 1899 (Harvard) and education in 1900 (Columbia).

You might be interested to know when statistics was first taught at your school and in what department. College catalogs are probably the most accessible source for this information.

Kirk (1984) divides present-day statisticians into four groups. Category 1— those who are able to understand material that is presented statistically; Category 2—those who are able to apply statistical techniques and interpret the results; Category 3—professional statisticians who help others apply statistical techniques to a particular problem; and Category 4—mathematical statisticians who develop new statistical techniques and discover new characteristics of old techniques.[8] If our hopes for this book are realized, by the end of this course you will be comfortable calling yourself a Category 2 statistician.

HOW TO USE THIS BOOK

At the beginning of this chapter, we mentioned the importance of your active participation in learning statistics. We have tried to organize this book in a way that will encourage active participation. Here are some of the features our students have found helpful.

Objectives

At the beginning of each chapter, there is a list of the skills that the chapter is designed to help you acquire. Read this list of objectives first to find out what you are to learn to do. Then thumb through the chapter and look at the headings. Next, study the chapter and work all the problems. Finally, reread the objectives to see if you have accomplished them.

Clues to the Future

Many times we will present an idea that will be used extensively in later chapters. You will find these ideas separated from the rest of the text in a box labeled "Clue to the Future." You have already met two of these "Clues" in this chapter. If you will pay particularly close attention to these concepts, it will help you later in the course.

---

[8] See the section, "Quantitative Methods in Psychology," in any edition of the journal *Psychological Bulletin.*

Error Detection

We have also boxed in, at various points in the book, some ways to detect errors. Some of these "Error Detection" tips will also help your understanding of statistical concepts. Since many of these checks can be made early in a statistical problem, they can prevent the frustrating experience of having to redo a ten-step problem because of an error in step two.

Figure and Table References

When we think it would be worthwhile for you to examine a figure or a table, we have put the word "Figure" or "Table" in boldface type. After your examination, it will be easy for you to find your place in the text again.

Problems and Answers

You have already been introduced to our method of presenting problems and answers. Some additional no-nos should be mentioned. Do not skip any assigned problems because we occasionally put new material in the problems or in the answers. Never start with the answer in front of you. Do not simply correct a wrong answer without discovering why you made the error.

Sometimes you will find minor differences between your answer and ours (in the last decimal place, for example). Most of these will probably be the result of rounding errors and do not deserve your worry and anxiety.

Many of the problems are conceptual questions that do not require you to do any arithmetic. Think these through and write your answers. The arithmetic associated with an experiment is of no value if you cannot tell what it means in English. Writing tells you whether you understand or not.

Transition Pages

At various points in the book, there will be major differences between the material you have finished and that which you are about to begin. At these points we have provided a "Transition Page," which points out those major differences. You should create a new category in your memory in preparation for the new kinds of problems.

Glossaries

We have compiled three glossaries that you can use to jog your memory, if necessary. Two of them are in the back of the book and one is printed on the inside cover of the book.

1. *A glossary of words (Appendix D)*. This is a glossary of statistical words and phrases used in the text. When these items are introduced in the text, they appear in boldface type.
2. *A glossary of formulas (Appendix E)*. The formulas for all the statistical techniques discussed in the text are printed here, in alphabetical order according to the name of the technique.

3. *A glossary of symbols* (inside cover of this book). Statistical symbols are defined here.

Arithmetic and Algebra Review

Appendix B gives you a review of arithmetic and algebra. It consists of a pretest (to see if you need to refresh your memory on any of the many techniques you have learned over the years) and a review (to provide that refresher). We recommend that you use this appendix as soon as you finish Chapter 1.

CONCLUDING THOUGHTS FOR THIS INTRODUCTORY CHAPTER

Statistics is the sort of subject that requires you to use old knowledge in the learning of new topics. You must know the material in Chapter 2 before you can learn the material in Chapter 3. You must understand both Chapters 2 and 3 before you can learn what is presented in Chapter 4, and so on throughout the book. Thus, skipping anything would be a serious mistake that would haunt you throughout the course.

Most students find that this book works well for them as a textbook in their statistics course. Those who keep this book find that it becomes a very useful reference book. In future courses and after leaving school, they find themselves pulling out their copy to look up some forgotten definition or vaguely recalled procedure. We hope that you not only learn from this book but that you, too, join that group of students who count this book as part of their personal library.

We wish to emphasize that this book is a fairly complete introduction to elementary statistics. There is more to the study of statistics—lots, lots more; but there is a limit to what you can do in one term. Some of you, however, will learn the material in this book and want to know more. For those students (may you live long and joyful lives!), we have provided references in footnotes.

We also recommend that you study and work the problems in the widely unassigned Chapter 13, the last chapter in the book. It is designed to be an overview/integrative chapter.

We find the study of statistics very satisfying. We hope you do, too.

PROBLEMS

4. Name the four kinds of scales identified by S. S. Stevens.
5. Give the properties of each of the four kinds of scales.
6. Identify which kind of scale is being used in each of the following cases.
   a. Geologists have a "hardness scale" for identifying different rocks, called the Moh scale. The hardest rock (diamond) has a value of 10 and will scratch all others. The second hardest will scratch all but the diamond, and so on, down to a rock such as talc, with a value of 1, which can be scratched by any other rock. (A fingernail, a truly handy field-test instrument, has a value of between 2 and 3.)
   b. Three different cubes are measured with a ruler and their volumes are computed to be 40, 65, and 75 cubic inches, respectively.
   c. Three different highways are identified by their numbers: 40, 65, and 75.

d. Republicans, Democrats, Independents, and Others are identified on the voters' list with the numbers 1, 2, 3, and 4.

e. The pages of a book are numbered 1 through 150.

f. The winner of the Miss America contest was Miss California; runners-up were Miss Ohio and Miss Pennsylvania.[9]

7. Identify the independent and dependent variables in the following statements.

a. The ability to solve new problems depends on previous experience.

b. The more you eat, the better you feel.

c. The more you eat, the worse you feel.

d. For some problems, the more anxious subjects solved the problems more quickly.

e. These plants are growing poorly since they do not get much sunlight.

f. Experience is the best teacher.

g. April showers bring May flowers.

8. For each of the three studies described below, identify the following:

a. the independent variable	**b.** the dependent variable
c. a controlled extraneous variable	**d.** a population of interest
e. a sample	**f.** a statistic that was calculated
g. a parameter of interest	**h.** a variable measured on a nominal scale
i. a variable measured on a ratio scale	**j.** Write a sentence that tells what the study found.

i. Theodore X. Barber has done many hypnosis experiments. (See Barber, 1976.) In one study, Barber hypnotized 25 people, gave them a series of suggestions, and recorded their responses. The suggestions were for things like arm rigidity, hallucinations, color blindness, and enhanced memory. The mean number of suggestions followed by this group was 4.8. For another group of 25, he simply solicited their help in achieving the best score they could (but there was no hypnosis used). This second group was given the same suggestions and their behavior was scored. For the nonhypnotized group the mean number was 5.1.

ii. Elizabeth Loftus has shown that our memory of an event changes and continues to change depending on the information we get *after* the event happens. She documents the problem that these changes produce in courts of law in *Eyewitness Testimony* (1979). In one study, subjects saw a film of a car accident. Shortly afterward, they answered some questions. Some of the subjects were asked how fast the car was going; others were asked how fast the car was going when it passed the barn. (There was no barn in the film.) A week later, when asked the question, "Did you see a barn?" 17 percent of those who had heard the question that mentioned the barn answered yes. Three percent of the other group said yes.

iii. Stanley Schachter believes that obese people's hunger is partly controlled by external (environmental) cues. The following description is similar to a study by Schachter and Gross (1968). After some preliminary data gathering with obese male students, a clock on the wall was correct (5:30 P.M.) for 20 volunteers, slow (4:30 P.M.) for 20 others, or fast (6:30 P.M.) for 20 more. The actual time, 5:30, was the usual dinner time for these students. While the participants filled out a final questionnaire, Wheat Thins were freely available. The weight of the crackers each student consumed was calculated; the means were: 4:30 group, 20 grams; 5:30 group, 30 grams; and 6:30 group, 40 grams.

---

[9] Contest winners have come most frequently from these three states, which have had 6, 5, and 5 winners, respectively.

2

THE ORGANIZATION OF DATA, GRAPHS, AND CENTRAL VALUES

Objectives for Chapter 2: After studying the text and working the problems in this chapter, you should be able to:

1. Arrange data into simple and grouped frequency distributions
2. Describe the characteristics and uses of frequency polygons, histograms, bar graphs, and line graphs, and be able to interpret them
3. Recognize some distributions by their shapes and name them
4. Find the mean, median, and mode of raw scores of a simple frequency distribution and of a grouped frequency distribution
5. Determine whether a central value measure is a statistic or a parameter and interpret it
6. Determine which central value measure is most appropriate for a set of data
7. Determine the direction of skew of a frequency distribution
8. Find the mean of a set of means

Now that the preliminaries are out of the way, you are ready to start on the basics of descriptive statistics. The starting point is a group of **raw scores** or measures, all obtained by administering the same test or procedure to a group of subjects. In an experiment, the scores are measurements of the dependent variable. Often this results in an unorganized mass of numbers such as those in **Table 2.1,** which lists the scores of 100 college students on a self-esteem inventory.

Table 2.1   Scores of 100 college students on a self-esteem inventory

40	40	39	36	41	23	25	34	25	29
23	59	42	55	29	30	39	29	34	30
40	32	38	42	55	25	36	29	36	53
51	41	38	25	23	34	44	44	44	30
55	30	34	46	44	39	34	42	39	49
47	55	58	44	41	36	36	47	39	32
34	39	41	53	42	27	47	42	46	27
51	39	51	39	41	47	39	34	40	49
32	39	41	42	65	53	46	34	51	51
34	41	38	30	41	32	40	38	27	29

In this chapter you will learn to (1) organize data like those in Table 2.1 into frequency distributions, (2) present the data graphically, and (3) calculate central values, interpret them, and determine whether they are statistics or parameters.

SIMPLE FREQUENCY DISTRIBUTIONS

Table 2.1, which is just a jumble of numbers, is not very interesting or informative. (Our guess is that you glanced at it and went quickly on.) With a little work, the self-esteem scores in Table 2.1 can be presented as **Table 2.2,** which you should look at now. Table 2.2 is a **simple frequency distribution**—an ordered arrangement that shows the frequency of each score. (We would guess that you spent more time on Table 2.2 and that you got more information from it than you did from Table 2.1.)

Look again at Table 2.2. The X column shows the scores and the f column shows how frequently each score appeared. The tally marks are used only to construct a rough draft of a frequency distribution. (Later we will describe a final form for formal presentation.) N is the number of scores and can be found by summing the f column.

Table 2.2 Simple frequency distribution of self-esteem scores, $N = 100$ (a rough draft)

X	Tally marks	f	X	Tally marks	f
65	/	1	43		0
64		0	42	卌 /	6
63		0	41	卌 ///	8
62		0	40	卌	5
61		0	39	卌 卌	10
60		0	38	////	4
59	/	1	37		0
58	/	1	36	卌	5
57		0	35		0
56		0	34	卌 ////	9
55	////	4	33		0
54		0	32	////	4
53	///	3	31		0
52		0	30	卌	5
51	卌	5	29	卌	5
50		0	28		0
49	//	2	27	///	3
48		0	26		0
47	////	4	25	////	4
46	///	3	24		0
45		0	23	///	3
44	卌	5			

A simple frequency distribution is a very common way to present the results of a study or experiment. Here are some reasons it is so useful; you will learn others later in the chapter. At just a glance, you can find the highest and lowest scores, and it is easy to find scores with zero frequencies. In addition, the general shape or form of the distribution can be ascertained (although you may find that a little practice is necessary for this). In the case of Table 2.2, the distribution is fat in the middle and skinny on both ends, since the most frequently occurring scores are grouped near the middle.

You will have a lot of opportunities to construct simple frequency distributions so here is a set of steps to follow:

1. Find the highest and lowest scores. In Table 2.1, the highest score is 65 and the lowest score is 23.
2. In column form, write in descending order all possible scores between the highest score (65) and the lowest score (23). The heading for this column is the letter X.
3. Start with the number in the upper left-hand corner of the unorganized scores (a score of 40 in Table 2.1), draw a line under it, and place a tally mark beside 40 in your frequency distribution.[1]
4. Continue this process through all the scores.
5. Count the number of tallies by each score and write that number beside the tallies in a column headed f. Add up the numbers in the f column to be sure they equal N. You have now constructed a simple frequency distribution.

A formal presentation of a simple frequency distribution does not include either the tally marks or the scores for which the frequency is zero. A final form would be used for presentation to other people such as colleagues, professors, supervisors, or editors. **Table 2.3** shows the data of Table 2.1 properly arranged for formal presentation.

Table 2.3   Simple frequency distribution of self-esteem scores for formal presentation, $N = 100$

Self-esteem scores (X)	f
65	1
59	1
58	1
55	4
53	3
51	5
49	2
47	4
46	3
44	5
42	6
41	8
40	5
39	10
38	4
36	5
34	9
32	4
30	5
29	5
27	3
25	4
23	3

[1] If you use the unorganized scores again, you will discover that underlining is better than crossing out.

PROBLEMS

1. The following numbers are the heights in inches of two groups of 18- to 24-year-old Americans. Choose the more interesting group and organize the 50 numbers into a simple frequency distribution, using the rough-draft format. For the group you choose, your result will be fairly representative of the whole population of 18- to 24-year-olds (*Statistical Abstract for the United States: 1986*, 1985).

WOMEN					MEN				
64	67	63	64	66	68	65	72	68	70
63	66	62	65	65	69	73	71	69	67
60	72	64	61	65	77	67	72	73	68
69	64	64	66	61	68	64	72	69	69
63	67	65	62	69	70	71	71	70	75
62	65	63	64	61	72	68	62	68	74
66	64	59	63	65	66	70	72	66	71
63	63	68	64	65	69	71	68	73	69
66	67	62	62	63	71	69	69	65	76
64	70	64	62	63	71	73	65	70	70

2. The frequency distribution you will construct for this problem is a little different. The "scores" are simply names, making this a frequency distribution of nominal data.

    A political science student covered a voting precinct on election day morning, noting the yard signs for the five candidates. Her plan was to find the relationship between yard signs and actual votes. The five candidates were Bolivar (B), Gandhi (G), Lenin (L), Mao (M), and Attila (A). Construct an appropriate frequency distribution.

G	B	M	M	M	G	G	L	B	G	A	B	B	G	G	A	L	M	M
B	G	G	A	G	L	M	B	B	L	M	G	G	M	G	L	G	B	B
A	L	G	G	B	G	B	M	L	M	G	A	B	G	L	G	M	B	

GROUPED FREQUENCY DISTRIBUTIONS

Most researchers would condense Table 2.3 even more. The result would be a **grouped frequency distribution,** and **Table 2.4** is a rough-draft example.[2] It is often necessary to group data when you want to construct a graph or to present your results in the form of a table. In the days before computers and cheap calculators, grouping also saved time by simplifying computations.

In a grouped frequency distribution, X values are grouped into equal-sized ranges called **class intervals.** In Table 2.4, the entire range of scores, from 65 to 23, has been reduced to 15 class intervals. Each interval covers three scores; the size of this interval (number of scores covered) is indicated by i. For Table 2.4, $i = 3$. The midpoint of each interval represents all scores in that interval. For example, nine students had scores of 33, 34, or 35. The midpoint of the class interval 33–35 is 34. All nine are represented by 34. There are no scores in the interval 60–62, but zero-

---

[2] As is the case with a simple frequency distribution, a grouped frequency distribution that is to be presented formally does not include tally marks.

Table 2.4 A grouped frequency distribution of self-esteem scores, $i = 3$ (rough-draft version)

Class interval	Midpoint (X)	Tally marks	f
63–65	64	/	1
60–62	61		0
57–59	58	//	2
54–56	55	////	4
51–53	52	₩ ///	8
48–50	49	//	2
45–47	46	₩ //	7
42–44	43	₩ ₩ /	11
39–41	40	₩ ₩ ₩ ₩ ///	23
36–38	37	₩ ////	9
33–35	34	₩ ////	9
30–32	31	₩ ////	9
27–29	28	₩ ///	8
24–26	25	////	4
21–23	22	///	3

frequency intervals are included in formal grouped frequency distributions if they are within the range of the distribution.

Class intervals have lower and upper limits, much like simple scores obtained by measuring a quantitative variable. A class interval of 33–35 has a lower limit of 32.5 and an upper limit of 35.5. Similarly, a class interval of 40–49 has a lower limit of 39.5 and an upper limit of 49.5. You will use these limits later when you calculate the median.

The only difference between grouped frequency distributions and simple frequency distributions is class intervals, which is our next topic.

Establishing Class Intervals

Three conventions are usually followed in establishing class intervals. We call them conventions because they are customs rather than hard-and-fast rules. There are two justifications for these conventions. First, they allow you to get maximum information from your data with minimum effort. Second, they provide some standardization of procedures, which aids in communication.

1. *The number of class intervals should be 10 to 20.* A primary purpose of grouping scores is to provide a clearer picture of trends in the data. For example, Table 2.4 shows that the frequencies are large near the center of the distribution but they get smaller and smaller as the ends of the distribution are approached. If the data are grouped into fewer than ten intervals, such trends are not as apparent. In **Table 2.5,** the same scores are grouped into only five class intervals. The concentration of frequencies in the center of the distribution is not nearly so apparent. On the other hand, the use of more than 20 class intervals puts you back into the simple frequency distribution situation—it is difficult to get an overall picture of the form of the distribution.

2. *Choose a convenient size for the class interval (i).* Three and five are often chosen for i because if i is odd, the midpoint of the interval will be a whole number.

Table 2.5   A grouped frequency
distribution of self-esteem scores ($i = 10$)
with too few intervals, $N = 100$

Class interval	Midpoint (X)	f
60–69	64.5	1
50–59	54.5	14
40–49	44.5	33
30–39	34.5	32
20–29	24.5	20

This midpoint is an important part of any graph of the data and whole numbers are
cleaner than decimal numbers in a graph. If an i of 5 produces more than 20 class
intervals, data groupers usually jump to an i of 10 or some multiple of 10. An
interval size of 25 is popular. You will occasionally find that $i = 2$ is necessary to
stay within 10 to 20 class intervals.

3. *Begin each class interval with a multiple of i.* For example, if the lowest score
is 44 and $i = 5$, the first class interval should be 40–44 because 40 is a multiple
of 5. This convention is violated more often than the other two. A violation that
seems to be justified occurs when $i = 5$. When the interval size is 5, it may be more
convenient to begin the interval such that multiples of 5 fall at the midpoint, since
multiples of 5 make graphs easier to read. For example, an interval 23–27 has 25
as its midpoint, whereas an interval 25–29 has 27 as its midpoint.

In addition to these three conventions, remember that the highest scores go at
the top of the distribution and the lowest scores at the bottom.

Converting Unorganized Data into a Grouped Frequency Distribution

Now that you know the conventions for establishing class intervals, we will go
through the steps for converting unorganized data like those in Table 2.1 into a
grouped frequency distribution like Table 2.4:

1. Find the highest and lowest scores. In Table 2.1, the highest score is 65 and
   the lowest score is 23.
2. Find the range the scores cover by subtracting the lower limit of the lowest
   score from the upper limit of the highest score. The use of lower and upper
   limits is necessary in order to include the full range of scores.[3]
3. Determine i by a trial-and-error procedure. Remember that there are to be 10
   to 20 class intervals and that the interval size should be convenient (3, 5, 10, or
   a multiple of 10). Dividing the range by a potential i value tells the number of
   class intervals that will result. For example, dividing the range of 43 by 5
   gives a quotient of 8.60. Thus, $i = 5$ produces 8.6 or 9 class intervals. That
   does not satisfy the rule calling for at least 10 intervals, but it is close and

---

[3] For a more complete explanation, with a picture, see the section in Chapter 3, "The Range."

might be acceptable. In most such cases, however, it is better to use a smaller i and get a larger number of intervals. Dividing the range by 3 ($43 \div 3$) gives 14.33 or 15 class intervals. It sometimes happens that this process results in an extra class interval. This occurs when the lowest score is such that extra scores must be added to the bottom of the distribution to start the interval with a multiple of i. For the data in Table 2.1, the most appropriate interval size is 3, resulting in 15 class intervals.

4. Begin the bottom interval with the lowest score, if it is a multiple of i. If the lowest score is not a multiple of i, begin the interval with the next lower number that is a multiple of i. In the data of Table 2.1, the lowest score, 23, is not a multiple of i so you must start the interval with 21. The lowest class interval, then, is 21–23. From there on, it's easy. Simply begin the next interval with the next number and end it such that it includes three score values (24–26). Look at the class intervals in Table 2.4. Notice that each interval begins with a number evenly divisible by 3.

5. The rest of the process is the same as for a simple frequency distribution. For each score in the unorganized data, put a tally mark beside its class interval and underline the score. Count the tally marks and put the number in the frequency column. Add up the frequency column to be sure that the sum equals N.

PROBLEMS

3. A sociology professor who was trying to decide how much statistics to present in her Introduction to Sociology class developed a 65-item test of statistical knowledge that covered concepts such as the median, grouped frequency distributions, and the standard deviation. She gave the test to one class of 50 students, and on the basis of the results she planned a course syllabus for that class and the other four sections being taught that year. Arrange the data into an appropriate rough-draft frequency distribution.

20	56	48	13	30	39	25	41	52	44
27	36	54	46	59	42	17	63	50	24
31	19	38	10	43	31	34	32	15	47
40	36	5	31	53	24	31	41	49	21
26	35	28	37	25	33	27	38	34	22

4. A psychology instructor read 60 related statements to a General Psychology class. He then asked the students to indicate which of the next 20 statements they had heard among the first 60. Due to the relationships among the concepts in the sentences, many seemed familar but, in fact, none of the 20 had been read before. The following scores represent the number (out of 20) that each student had "heard earlier." (See Bransford and Franks, 1971.) Arrange the scores into an appropriate rough-draft frequency distribution.

14	11	10	8	12	13	11	10	16	11
11	9	9	7	14	12	9	10	11	6
13	8	11	11	9	8	13	16	10	11
9	9	8	12	11	10	9	7	10	

5. For his project in a sociology class, a student decided to survey 99 randomly selected homes in Midwesternville. At each house he asked how many pairs of shoes the resident had. Arrange the data into an appropriate rough-draft frequency distribution.

10	7	14	15	22	18	5	12	8	17
23	5	3	7	6	14	11	13	20	19
5	16	7	9	29	4	5	9	6	8
11	14	6	20	11	4	8	13	17	7
5	6	9	5	8	25	19	6	2	4
8	5	7	10	6	11	21	4	7	3
9	32	10	16	12	6	33	18	15	19
17	6	10	12	7	5	20	9	13	4
13	10	3	11	28	7	9	18	5	12
5	17	8	23	4	16	8	15	19	

GRAPHIC PRESENTATION OF DATA

You have no doubt heard the saying *A picture is worth a thousand words.* We agree. And when it comes to numbers, a not-yet-well-known saying is *A graph is better than a thousand numbers.* Actually, as long as we are rewriting sayings, we would like to say *The more numbers you have, the more necessary graphics are.*

Pictures that present statistical data are called graphics. In 1983 Edward Tufte of Yale published *The Visual Display of Quantitative Information,* which demonstrates and celebrates the power that graphics have. He gives many examples of this power, including a reprint of "(perhaps) the best statistical graphic ever drawn."[4] Tufte says that, regardless of your field, when you construct a quality graphic, it improves your understanding of the phenomenon you are interested in. So if you find yourself somewhere on that path between confusion and understanding, you should try to construct a graphic. In the heartfelt words of a sophomore engineering student we know, "Graphs sure do help." To learn more, we recommend both Tufte's book and Wainer's article.

Also, a graphic is *very* helpful in educating and persuading others. After you understand, convincing others is usually the next step.

The most common graphic is a graph, composed of a horizontal axis (variously called the baseline, X axis, and **abscissa**) and a vertical axis (called the Y axis or **ordinate**). To the right and upward are both positive directions; to the left and downward are both negative directions.

---

[4] We'll give you just a hint about this graphic. It was drawn by a French engineer some time ago to convey a disastrous military campaign of Napoleon. Wainer (1984) was also impressed with this graphic, nominating it as the "World's Champion Graph."

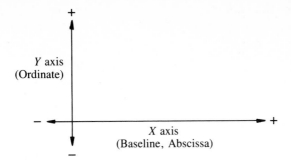

In this section, we will describe two kinds of graphs. The first kind is used to present frequency distributions like those you have been constructing. **Frequency polygons, histograms,** and **bar graphs** are examples of this first kind of graph. The second kind we will describe is the **line graph,** which is used to present the relationship between two different variables. We will have more for you on graphs in later chapters.

Frequency Distributions

The kind of variable you have measured determines whether you present your data with a frequency polygon, a histogram, or a bar graph. A frequency polygon or histogram is used for quantitative data, and a bar graph is used for qualitative data. It is not wrong to use a bar graph for quantitative data, but most researchers follow the rule just given. Qualitative data, however, *should not* be presented with a frequency polygon or a histogram. The self-esteem scores in Table 2.1 are an example of quantitative data; the data on yard signs in Problem 2 are an example of qualitative data.

The Frequency Polygon.   **Figure 2.1** shows a frequency polygon based on the grouped frequency distribution in Table 2.4. We will use it to demonstrate the characteristics of all frequency polygons. On the X axis we placed the midpoints of the class intervals and labeled the axis "Self-Esteem Scores." Notice that the midpoints are spaced at equal intervals, with the smallest midpoint at the left and the largest midpoint at the right. The Y axis is labeled "Frequency" and is also marked off into equal intervals.

Graphs are designed to "look right." Often they look right if the height of the figure (Y axis) is 60 percent to 75 percent of its length (X axis). Since the midpoints must be plotted along the X axis, you must divide the Y axis into units that will satisfy this rule. Usually this requires a little juggling on your part. Both Huff (1954) and Runyon (1981) offer excellent demonstrations of the misleading effects that can occur when this convention is violated.

The intersection of the X and Y axes is often the zero point for both variables. For the Y axis in Figure 2.1, this is indeed the case. The distance on the Y axis is the same from 0 to 2 as from 2 to 4, and so on. On the X axis, however, that is not the

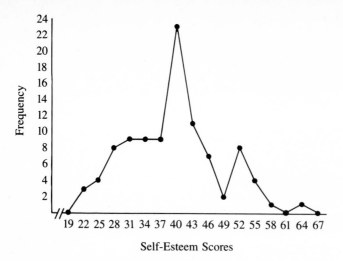

Figure 2.1   Frequency polygon of self-esteem scores of 100 college students

case. Here the scale jumps from 0 to 19 and then is divided into equal units of 3. It is conventional to indicate a break in the measuring scale by breaking the axis with slash marks between 0 and the lowest score used, as we did in Figure 2.1. It is also conventional to close a polygon at both ends by connecting the curve to the X axis.

Each point of the frequency polygon represents two numbers: the class midpoint directly below it on the X axis and the frequency of that class directly across from it on the Y axis. By looking at the points in Figure 2.1, you can readily see that three students are represented by the midpoint 22; nine students by each of the midpoints 31, 34, and 37; 23 students by the midpoint 40; and so on.

The major purpose of the frequency polygon is to present an overall view of the distribution of scores. Figure 2.1 makes it clear, for example, that the frequencies are greater for the lower scores than for the higher ones.

The Histogram.   **Figure 2.2** is a histogram constructed from the same data that were used for the frequency polygon of Figure 2.1. Researchers may choose either of these methods for a given distribution of quantitative data, but the frequency polygon is usually preferred for several reasons: it is easier to construct, gives a generally clearer picture of trends in the data, and can be used to compare two or more distributions on the same graph, making comparisons very easy. Frequencies, however, are easier to read from a histogram.

Actually, the two figures are very similar. They differ only in that the histogram is made by raising bars from the X axis to the appropriate frequencies instead of plotting points above the midpoints. The width of a bar is from the lower to the upper limit of its class interval. Notice that there is no space between the bars.

The Bar Graph.   A bar graph is used to present the frequencies of the categories of a qualitative variable. A conventional bar graph looks exactly like a histogram

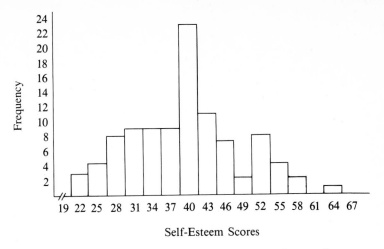

Figure 2.2   Histogram of self-esteem scores of 100 college students

except that there are spaces between the bars. The space is the conventional signal that the variable being graphed is a qualitative one. Since the variable that is being graphed consists of names that have no inherent order, any order on the *X* axis is appropriate; alphabetizing is common (but often not best).

We decided that a bar graph of college majors would be interesting to you, so we found data for 1981–82 college graduates in the United States. We constructed a conventional bar graph with the names of the majors in alphabetical order on the *X* axis and the frequencies on the *Y* axis. It *looked* terrible. In addition, it was awkward to use, since you had to turn the graph on its side to read the majors.

We returned to our drawing board, thinking of the admonition in Tufte's (1983) epilogue: "It is better to violate any principle than to place graceless or inelegant marks on paper." The result was **Figure 2.3**. The variable being graphed (majors) is on the *Y* axis, allowing names to be written horizontally so they are easy to read. Frequency is on the *X* axis. We arranged the majors from most frequent to least frequent.[5] In addition, we wrote whole numbers at the end of each bar to indicate thousands of graduates. All in all, we were very pleased with the result. (In several of our early drafts, we used numbers with decimals to indicate the number of graduates. The decimals, however, detracted from the goal of the graph, which is to present an overall picture of the quantitative relationships.)

Like histograms, bar graphs are not very satisfactory for presenting two or more variables. Imagine adding 1971–72 data to Figure 2.3. The result would be a cluttered graph. However, by using a little creative thought, the two sets of data can be combined, with the result that both variables are presented with elegance. **Figure 2.4** shows the results of combining two variables, the 1981–82 data of Figure 2.3 and the same data for ten years earlier. The change from 1971–72 to

[5] Alphabetization is a boon when you are searching for one item among many. We decided that 12 was not many, leaving us free to choose some other order for this nominal variable.

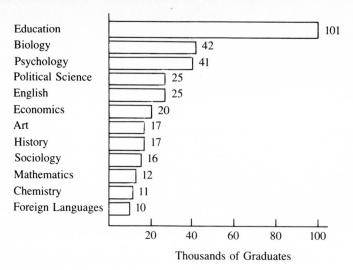

Figure 2.3   College graduates by major for the academic year 1981–82

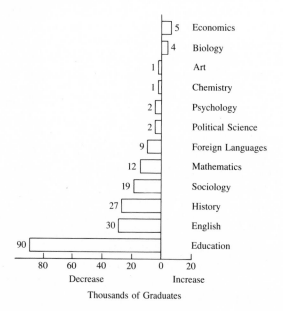

Figure 2.4   Change (in thousands of graduates) in college majors from 1971–72 to 1981–82

1981–82 in thousands is plotted on the *Y* axis. The result is a graphic that tells you about change, a variable that seems to fascinate humans.

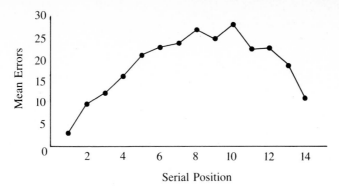

Figure 2.5   The serial position effect

The Line Graph

Perhaps the graph most frequently used by scientists is the line graph. A line graph is a picture of the relationship between two variables.[6] A single point on the graph may represent the two scores made by one subject on each of the two variables. Often the mean of a group of scores is used rather than the score of one subject, but the idea is the same: a group with a mean score of X on one variable had a mean score of Y on the other variable. Thus, a point on the graph lines up with the two scores made by the subject or group.

Figure 2.5 is an example of a line graph. It shows the phenomenon called the *serial position effect,* the fact that when you study a set of ordered material several times until you know it all, the most difficult part is just past the middle, with the first part of the material being the easiest. When you understand the serial position curve, you can figure out what material to study most.

A variation of the line graph places performance scores on the Y axis and some condition of training on the X axis. Examples of such training conditions are: number of trials, hours of food deprivation, year in school, and amount of reinforcement. The "score" on the training condition is assigned by the experimenter. **Figure 2.6** is a generalized learning curve with a performance measure (scores) on the Y axis and the number of reinforced trials on the X axis. Early in training (after only one or two trials), performance is poor. As trials continue, performance improves rapidly at first and then more and more slowly. Finally, at the extreme right-hand portion of the graph, performance has leveled off; continued trials do not produce further changes in the scores.

To summarize, a line graph presents a picture of the relationship between two variables. By looking at the line, you can tell what changes take place in the Y variable as the value of the X variable changes.

---

[6] A frequency distribution is a special case of a line graph in which one of the variables is frequency.

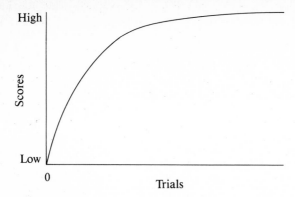

Figure 2.6 Generalized learning curve

DESCRIBING DISTRIBUTIONS

There are three ways to describe the form or shape of a distribution: verbally, pictorially, and mathematically. In this section we will use the first two of these ways. We will not cover mathematical methods that describe the form of a distribution except for one method that appears in Chapter 11, the chapter on chi square.

Bell-Shaped Distributions

Look back at Table 2.4, graphed as Figure 2.1. Notice that the largest frequencies are in the middle of the distribution. The same thing is true in Problem 1 of this chapter. These distributions are referred to as bell-shaped.

There is a special case of a bell-shaped distribution that you will soon come to know very well. It is called the **normal distribution** or **normal curve. Figure 2.7** is an example of this distribution.

Skewed Distributions

In some distributions the scores with the greatest frequency are not found in the middle but near one end. Such distributions are said to be **skewed.**

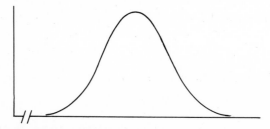

Figure 2.7 A normal distribution

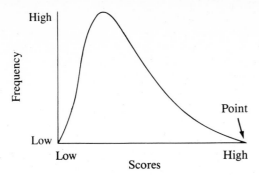

Figure 2.8   Example of a positively skewed distribution. Scores with the highest frequencies are concentrated among the low scores.

The word *skew* is similar to the word *skewer,* the name of the cooking implement used in making shish kebab. A skewer is thick at one end and pointed at the other (not symmetrical). Although skewed distributions do not function like skewers (you would have a terrible time poking one through a chunk of lamb), the name does help you remember that a skewed distribution has a thin point on one side.

Figures 2.8 and 2.9 are illustrations of skewed distributions. **Figure 2.8** is *positively skewed;* the thin point is toward the high scores, and the most frequent scores are the low ones. Note that the point is to the right—the positive direction. **Figure 2.9** is *negatively skewed;* the thin point or skinny end is toward the low scores, and the most frequent scores are the high ones. Note that the point is to the left—the negative direction.

Other Shapes

Curves of frequency distributions can, of course, take many different shapes. Some shapes are common enough to have names, and examples of these are presented in Figure 2.10.

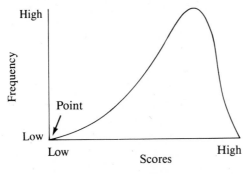

Figure 2.9   Example of a negatively skewed distribution. Scores with the highest frequencies are concentrated among the high scores.

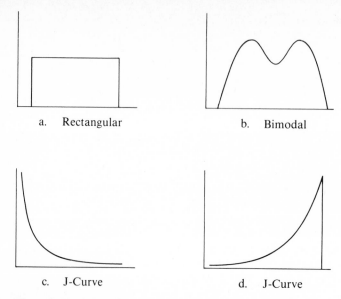

a.   Rectangular   b.   Bimodal

c.   J-Curve   d.   J-Curve

Figure 2.10   Examples of rectangular, bimodal, and J-curves

Rectangular Distribution.   Figure 2.10a, the **rectangular distribution,** occurs when the frequency of each value on the X axis is the same. You will see this distribution again in Chapter 5.

Bimodal Distribution.   To understand the name of **Figure 2.10b,** you need to know that the mode of a distribution is the score that occurs most frequently. If two scores share this distinction and there are scores with lower frequencies between them, the distribution is said to be **bimodal.** Figure 2.10b is a picture of such a distribution.

J-Curves.   Finally, the **c** and **d** portions of **Figure 2.10** are called **J-curves.** By looking at Figure 2.10d and using your imagination, the name will make sense to you. J-curves are severely skewed distributions.

PROBLEMS

6. Answer the following questions for Figure 2.1.
   a. What is the meaning of the number 55 on the X axis?
   b. What is the meaning of the number 8 on the Y axis?
   c. How many students scored in the class interval 48–50?
   d. What information can you obtain from the point directly above the number 25 on the X axis?
7. For Problem 1 you constructed a frequency distribution. Graph it, being careful to include the score with a frequency of zero.

8. Decide whether the following distributions should be graphed as frequency polygons or as bar graphs. Graph both distributions.

a.

Class interval	f
50–54	1
45–49	1
40–44	2
35–39	4
30–34	5
25–29	7
20–24	10
15–19	12
10–14	6
5–9	2

b.

Class interval	f
54–56	3
51–53	7
48–50	15
45–47	14
42–44	11
39–41	8
36–38	7
33–35	4
30–32	5
27–29	2
24–26	0
21–23	0
18–20	1

9. Determine the direction of the skew for the two curves in Problem 8 by examining the curves or the frequency distributions (or both).
10. Look at the simple frequency distribution that you constructed for Problem 2. Which kind of graph should be used to display these data? Graph the distribution.
11. Without looking at Figures 2.7 or 2.10 sketch the form of the normal distribution, a rectangular distribution, a bimodal distribution, and two J-curves.
12. Is the point of a positively skewed distribution directed to the right or left?
13. Distinguish between a line graph and a frequency polygon.
14. For each frequency distribution listed, tell whether it would be positively skewed, negatively skewed, or approximately symmetrical (bell-shaped or rectangular).
    a. age of all people alive today
    b. age in months of all first-graders
    c. number of children in families
    d. wages in a large manufacturing plant
    e. age at death of everyone who died in one year
    f. shoe size

MEASURES OF CENTRAL VALUE

So far in this chapter you have learned some ways of condensing, clarifying, and presenting data by using frequency distributions and graphs. These techniques will help you understand your data and make their meaning clear to others. In this section, you will learn other ways of summarizing data—the calculation of three measures of central value. Measures of **central value** (often called measures of central tendency) give you one score or measure that represents, or is typical of, an entire group of scores.

Recall from Chapter 1 that statistics are the numbers you get from samples and parameters are the numbers you get from populations. In the case of all three measures of central value, the formula for calculating the statistic is the same as that for calculating the parameter.[7] The interpretations, however, are different.

We will use the **mean,** which you are already somewhat familiar with, as an example. A sample mean, $\bar{X}$ (pronounced "mean" or "ex-bar"), is only one of many possible sample means from a population. Suppose you have a sample from a population and you calculate the sample mean. Since other samples from that same population would probably produce somewhat different $\bar{X}$'s, there is some uncertainty that goes with the one $\bar{X}$ you have. However, if you measured the entire population, the mean, μ (pronounced "mew"), would be the only one and would carry no uncertainty with it. Clearly, the parameter μ is more desirable than the statistic $\bar{X}$, but, unfortunately, it is often impossible to measure the entire population. Fortunately, $\bar{X}$ is the best estimator of μ.

The difference in interpretation, then, is that a statistic carries some uncertainty with it; a parameter does not.

The Mean

Let's begin with a very simple example. Suppose a college freshman arrives at school in the fall with a promise of a monthly allowance for spending money. Sure enough, on the first of each month the check comes in the mail. However, by Thanksgiving our student has discovered a recurring problem: there is too much month left at the end of each allowance.

In pondering his problem, it occurs to the student that lots of money is escaping from his pocket at the Student Center. So, for a two-week period, he keeps a careful accounting of every cent spent at the center on soft drinks, snacks, video games, coffee, and so forth. He then computes the mean amount spent per day. His data are presented in **Table 2.6.** You already know how to compute the mean of this set of scores. To find the mean, add the scores and divide that sum by the number of scores.

In terms of a formula:

$$\bar{X} = \frac{\Sigma X}{N}$$

where $\bar{X}$ = the mean
Σ = an instruction to add (Σ is an upper case Greek sigma)
X = the numbers symbolized by X; ΣX means to add all the X's
N = number; N is the number of X's
For the data in Table 2.6:

$$\bar{X} = \frac{\Sigma X}{N} = \frac{\$24.98}{14} = \$1.78$$

These are data for a two-week period, but our freshman is interested in his expenditures for at least one month and more likely for many months. Thus, this is a sample mean and the $\bar{X}$ symbol is appropriate. The amount $1.78 is an estimate of the mean amount per day that our friend spends in the Student Center.

---

[7] This is not true for the standard deviation (see Chapter 3).

Table 2.6 Amount of money spent per day at the Student Center over a two-week period

Day	Amount spent
1	$2.56
2	0.47
3	1.25
4	0.00
5	3.25
6	1.15
7	0.00
8	0.00
9	6.78
10	2.12
11	0.00
12	0.00
13	3.78
14	3.62
	$\Sigma = \$24.98$

Now we come to an important part of any statistical analysis, which is to answer the question, So what? Calculating numbers or drawing graphs is a part of almost every statistical problem, but unless you can tell the story of what the numbers or pictures mean, you won't find statistics worthwhile. We will deal with interpretation now and again later.

The first interpretation you can make from Table 2.6 is to estimate monthly Student Center expenses. This is easy to do. Thirty days times $1.78 is $53.40.

Now, let's suppose that our student decides that this $53.40 is an important part of the "out of money before the end of the month" phenomenon. It strikes us that the student has three options. The first is to get more money. A second is to spend less at the Student Center. The third is to justify leaving things as they are. For this third option, our student might perform an economic analysis to determine what he gets in return for his $50+ a month. His list might be pretty impressive: some excellent social times, several dates, occasional information about classes and courses and professors, a borrowed book that was just super, hundreds of calories, and more.

The point of all this is that part of the attack on his money problem involved calculating a mean. However, an answer of $1.78 doesn't have much meaning by itself. Interpretation and comparisons are called for.

Characteristics of the Mean. There are two characteristics of the mean that we need to introduce you to now. Both characteristics will come up again later.

First, if the mean of a distribution is subtracted from each score in that distribution and the differences are added algebraically, the sum will be zero: that is, $\Sigma(X - \bar{X}) = 0$. Each difference score is called a deviation and these will be explained more fully in Chapter 3. You might pick a few numbers to play with to *demonstrate* that $\Sigma(X - \bar{X}) = 0$. (1, 2, 3, 4, and 5 are easy to work with.) If you know about performing algebraic operations on summation (Σ) notation, you can *prove* the relationship that $\Sigma(X - \bar{X}) = 0$.

Second, the mean is defined as the point about which the sum of the squared deviations is minimized. (A deviation is the answer you get from $X - \bar{X}$.) If the mean is subtracted from each score and each deviation is squared and all squared deviations are added together, the resulting sum will be smaller than if any number other than the mean had been used; that is, $\Sigma(X - \bar{X})^2$ is a minimum. This "least squares" characteristic of the mean will come up in Chapters 3 and 4. Again, you can demonstrate this relationship by playing with some numbers or you can prove it to yourself (this time proving it will require calculus).

The Median

The **median** is the *point* that divides a distribution of scores exactly in half. To find the median of the Student Center expense data, arrange the daily expenditures from highest to lowest as we have done in **Table 2.7**. Since there are 14 scores, the halfway point, or median, will have seven scores above it and seven scores below it. The seventh score from the bottom is $1.15. The seventh score from the top is $1.25. The median, then, is halfway between these two, or $1.20. Remember, the median is a *point* in the distribution; it may or may not be an actual score.

What interpretation can we make of a median of $1.20? The simplest is that on half the days our student spends less than $1.20 in the Student Center, and on the other half he spends more. We will have more to say about the interpretation of the median later in the chapter.

What if there had been an odd number of days in our student's sample? Suppose he had chosen to sample half a month, or 15 days. Then the median would be the eighth score. This would leave seven scores above and seven below. For example, if an additional day had been included, during which $3.12 was spent, the median would be $1.25.

Table 2.7   Data of Table 2.6 arranged in descending order

X	
$6.78	
3.78	
3.62	
3.25	7 scores
2.56	
2.12	
1.25	
	Median = $1.20
1.15	
0.47	
0	
0	7 scores
0	
0	
0	

The Mode

The third central value statistic is the **mode.** As mentioned earlier, the mode is the most frequently occurring score—the score with the greatest frequency.

For the Student Center expense data, the mode is $0.00. This can be seen most easily in **Table 2.7.** The zero amount occurred five times and all other amounts occurred only once.

When a mode is given, it is often accompanied by the percentage of times it occurred. You would probably agree that "The mode was $0.00, which occurred on 36 percent of the days" is much more informative than "The mode was $0.00."

FINDING CENTRAL VALUES OF SIMPLE FREQUENCY DISTRIBUTIONS

Mean

Table 2.8 is an expanded version of Table 2.3, the distribution of self-esteem scores. We will use **Table 2.8** to illustrate the steps for calculating the mean from a simple frequency distribution.

First, multiply each score in the X column by its corresponding f value, so that all the people making a particular score are included. Sum the fX values and divide

Table 2.8   Simple frequency distribution of self-esteem scores; calculating the mean

Self-esteem scores (X)	f	fX
65	1	65
59	1	59
58	1	58
55	4	220
53	3	159
51	5	255
49	2	98
47	4	188
46	3	138
44	5	220
42	6	252
41	8	328
40	5	200
39	10	390
38	4	152
36	5	180
34	9	306
32	4	128
30	5	150
29	5	145
27	3	81
25	4	100
23	3	69
	$\Sigma = 100$	3941

$$\mu \text{ or } \bar{X} = \frac{\Sigma fX}{N} = \frac{3941}{100} = 39.41$$

by N. (N is the sum of the f values.) This will give you the mean of a simple frequency distribution. In terms of a formula,

$$\mu \text{ or } \bar{X} = \frac{\Sigma f X}{N}$$

For the data in Table 2.8,

$$\mu \text{ or } \bar{X} = \frac{3941}{100} = 39.41$$

To answer the question of whether this 39.41 is an $\bar{X}$ or a μ, we would need more information. If the 100 scores were a population, we would have a μ, but if they represent a sample of some larger population, $\bar{X}$ would be the appropriate symbol.

Median

Since there are 100 self-esteem scores, the median will be a point that has 50 scores above it and 50 below it. If you begin adding frequencies in Table 2.8 from the bottom $(3 + 4 + 3 + 5 + \cdots)$, you will find a total of 42 when you include a score of 38. To include 39 would make the total 52, more than you need. So the median is somewhere among those ten scores of 39.

Remember from Chapter 1 that any number actually stands for a range of numbers that has a lower and an upper limit. This number, 39, has a lower limit of 38.5 and an upper limit of 39.5. To find the exact median somewhere within the range of 38.5–39.5, use a procedure called **interpolation.** We will give you the procedure and the reasoning that goes with it at the same time. Study it until you understand it. It will come up again.

There are 42 scores below 39. To get to 50, where the median is, you need eight more scores $(50 - 42 = 8)$. Since there are ten scores of 39, you need $\frac{8}{10}$ of them to reach the median. Assume that those ten scores of 39 are distributed evenly throughout the interval of 38.5–39.5 and that, therefore, the median is $\frac{8}{10}$ of the way through the interval. Adding 0.8 to the lower limit of the interval, 38.5, gives 39.3, which is the median for these scores.

When the number of scores is odd, the same reasoning applies. However, the halfway point will be a number with .5 at the end of it. For example, if you drop the 59, the number of self-esteem scores would be 99 and the median would be the point with 49.5 scores above it. The median now is $\frac{7.5}{10}$ of the way through the interval. (Median = 39.25.) You will have a chance to practice this in the problems.

**ERROR
DETECTION**

> Calculating the median by starting from the top of the distribution will produce the same answer as calculating it by starting from the bottom.

Mode

It is easy to find the mode from a simple frequency distribution. In Table 2.8, more people had a score of 39 than any other score, so 39 is the mode.

PROBLEMS

15. Find the median for the following sets of scores.
  a. 2, 5, 15, 3, 9
  b. 9, 13, 16, 20, 12, 11
  c. 8, 11, 11, 8, 11, 8
16. Which of the following distributions would be described as bimodal?
  a. 10, 12, 9, 11, 14, 9, 16, 9, 13, 20
  b. 21, 17, 6, 19, 23, 19, 12, 19, 16, 7
  c. 14, 18, 16, 28, 14, 14, 17, 18, 18, 6
17. For Problem 4, find the mean, median, and mode.
18. For Problem 2:
  a. Determine which of the measures of central value is appropriate and find it.
  b. Is this central value measure a statistic or a parameter?
  c. Write a sentence of interpretation.
19. Examine Problems 3 and 5 to determine whether the numbers are samples or populations.
20. Find the median of each distribution below.

  a.

X	f
12	4
11	3
10	5
9	4
8	2
7	1

  b. 9, 4, 3, 6, 5, 3, 7, 5, 2, 2,
       3, 6, 5, 7, 4, 2, 5, 6, 4, 5

21. Write the two mathematical characteristics of the mean that were covered in this section.

FINDING CENTRAL VALUES OF GROUPED FREQUENCY DISTRIBUTIONS

As we mentioned earlier, the most common reason for constructing a grouped frequency distribution is to draw a graph or to present the data as a table. Sometimes, however, you need to find central values from such presentations. This involves only a step or two more than finding such values from a simple frequency distribution.

Mean

The procedure for finding the mean of a grouped frequency distribution is similar to that for determining the mean of a simple frequency distribution. In the grouped distribution, however, the midpoint of each interval represents all the scores in the interval. Look at **Table 2.9,** which is an expansion of Table 2.4. Assume that the scores in each interval are evenly distributed throughout the interval. Thus, X is the mean for all scores within the interval. Multiply each X by its f value in order to include all frequencies in that interval. Place the product in the fX column. Summing the fX column gives ΣfX, which, when divided by N, yields the mean.

Table 2.9 A grouped frequency distribution of self-esteem scores, expanded to include columns for calculating the mean

Class interval	Midpoint X	f	fX
63–65	64	1	64
60–62	61	0	0
57–59	58	2	116
54–56	55	4	220
51–53	52	8	416
48–50	49	2	98
45–47	46	7	322
42–44	43	11	473
39–41	40	23	920
36–38	37	9	333
33–35	34	9	306
30–32	31	9	279
27–29	28	8	224
24–26	25	4	100
21–23	22	3	66
		$\Sigma = 100$	3937

$$\mu \text{ or } \bar{X} = \frac{\Sigma fX}{N} = \frac{3937}{100} = 39.37$$

In terms of a formula,

$$\mu \text{ or } \bar{X} = \frac{\Sigma fX}{N}$$

For Table 2.9,

$$\mu \text{ or } \bar{X} = \frac{3937}{100} = 39.37$$

Note that grouping introduces minor inaccuracies into the results. The mean self-esteem score from the raw data or from the simple frequency distribution is 39.41. When the data are grouped, the mean is 39.37. (The mean for Table 2.5, which has coarser grouping, is off even more. That mean is 38.90.)

Median

Finding the median of a grouped distribution usually requires interpolation within the interval that contains the median. That is the case for Table 2.9. As before, you are looking for the point that has 50 frequencies above it and 50 frequencies below it. Adding frequencies from the bottom of the distribution, you find that there are 42 who scored below the interval 39–41. You need 8 more frequencies (50 − 42 = 8) to find the median. Since 23 people scored in the interval 39–41, you need 8 of these 23 frequencies or 8/23. *Again, assume that the 23 people in the interval are evenly distributed throughout the interval.* Thus, you need the same proportion of score points in the interval as you have frequencies—that is, 8/23 or 0.35 of the 3 score points in the interval. Since 0.35 × 3 = 1.05, you must go 1.05 score points into the interval to reach the median. Since the lower limit of the interval is 38.5, add 1.05 to find the median. The median is 39.55. **Figure 2.11,** which will require *careful study,* illustrates this procedure.

In summary, the steps for finding the median in a grouped frequency distribution are as follows:

1. Divide N by 2.
2. Starting at the bottom of the distribution, add the frequencies until you find the interval that contains the median.
3. Subtract from $N/2$ the total frequencies of all intervals below the interval that contains the median.
4. Divide the difference found in Step 3 by the number of frequencies in the interval that contains the median.
5. Multiply the proportion found in Step 4 by i.[8]
6. Add the product found in Step 5 to the lower limit of the interval that contains the median. That sum is the median.

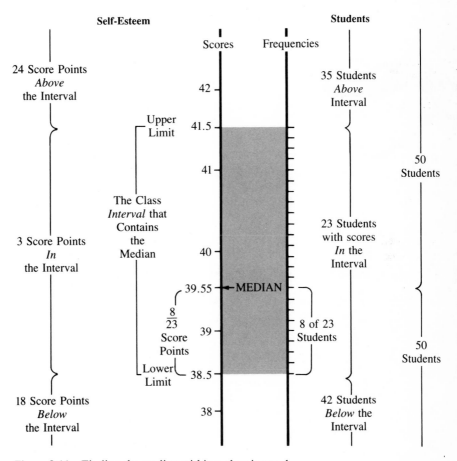

Figure 2.11   Finding the median within a class interval

---

[8] The most common error in calculating the median is forgetting to multiply by i.

If you like the idea of creating your own formula, you might reduce these six steps to a formula for the median.

Mode

The mode is the midpoint of the interval that has the highest frequency. In Table 2.9, the interval 39–41 has the highest frequency—23. The midpoint of that interval, 40, is the mode.

**ERROR
DETECTION**

> Eyeballing data is a valuable way to avoid gross errors. Begin a problem by making a quick estimate of the answers. If the answers you calculate differ from what you had estimated, wisdom dictates that you reconcile the difference. You have either overlooked something when eyeballing or made a mistake in your computations.

THE MEAN, MEDIAN, AND MODE COMPARED

A common question is, Which measure of central value should I use? The general answer is, given a choice, use the mean. Sometimes, however, the data give you no choice. Here are three considerations that limit your choice.

Scale of Measurement

A mean is appropriate for ratio or interval scale data, but not for ordinal or nominal distributions. A median is appropriate for ratio, interval, and ordinal data, but not for nominal. The mode is appropriate for any of the four scales of measurement.

You have already thought through part of this in working Problem 18. In it you found that the yard sign names (very literally, a nominal variable) could be characterized with a mode, but it would be impossible to try to add up the names and divide by N or to find the median of the names.

For an ordinal scale like class standing in college, either a median or a mode would make sense. The median would probably be part of the way through the classification sophomore, and the mode would be freshman.

Skewed Distributions

Even if you have interval or ratio data, a mean is not recommended if the distribution is severely skewed. The following story demonstrates why the mean gives an erroneous impression for severely skewed distributions. The story also presents an example of a population of data.

The developer of Swampy Acres Retirement Homesites is attempting, with a computer-selected mailing list, to sell the lots in a southern paradise to northern buyers. The "marks" express concern that flooding might occur. The developer reassures them by explaining that the average elevation of the lots is 78.5 feet and

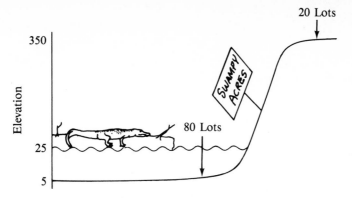

Figure 2.12   Elevation of Swampy Acres

that the water has never exceeded 25 feet in that area. On the average, the developer has told the truth; but this average truth is misleading. Look at the actual lay of the land in **Figure 2.12** and examine the frequency distribution in **Table 2.10.**

The mean elevation, as the developer said, is 78.5 feet; however, only 20 lots, all on a hill, are out of the flood zone. The other 80 lots are, on the average, under water. The mean for these data is misleading. The central value that best describes the typical case for these data is the median because it is unaffected by the size of the few extreme lots on the hill. The median elevation is 12.5 feet, well below the high-water mark.

A number of books with engaging titles tell how statistics can be chosen to convey one particular idea rather than another. The grandparent of such books is probably Huff's *How to Lie with Statistics* (1954). More recent versions are Campbell's *Flaws and Fallacies in Statistical Thinking* (1974) and Runyon's *How Numbers Lie* (1981). All three books are delightfully written.

Open-Ended Categories

There is another instance that requires a median, even though you have symmetrically distributed interval or ratio data. This is when the class interval with the

Table 2.10   Frequency distribution of lot elevations at Swampy Acres

Elevation in feet (X)	Number of lots (f)	fX
348–352	20	7000
13–17	30	450
8–12	30	300
3–7	20	100
	100	7850

$$\mu = \frac{\Sigma fX}{N} = \frac{7850}{100} = 78.50 \text{ feet}$$

highest (or lowest) scores is not limited. In such a case, you do not have a midpoint and therefore *cannot* compute a mean. For example, age data are sometimes reported with the highest category as "75 and over." The mean cannot be computed. Thus, when one or both of the extreme class intervals is not limited, the median is the appropriate measure of central value.

To reiterate: given a choice, use the mean.

DETERMINING SKEWNESS FROM THE MEAN AND MEDIAN

Examining the relationship of the mean to the median is a way of determining the direction of skew in a distribution without having to draw a graph. When the mean is smaller than the median, there is some amount of negative skew. When the mean is larger than the median, there is positive skew. The reason for this is that the mean is affected by the size of the numbers and is pulled in the direction of the extreme scores. The median is not influenced by the size of the scores. The relationship between the mean and the median is illustrated by **Figure 2.13.** The size of the difference between the mean and the median gives you an indication of how much the distribution is skewed. There is a mathematical way of measuring the *degree* of skewness that is more precise than eyeballing, but it is beyond the scope of this book.

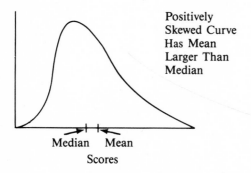

Positively Skewed Curve Has Mean Larger Than Median

Median   Mean

Scores

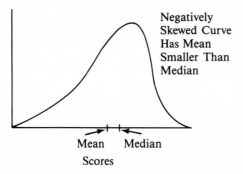

Negatively Skewed Curve Has Mean Smaller Than Median

Mean   Median

Scores

Figure 2.13   The effect of skewness on the relative position of the mean and median

THE MEAN OF A SET OF MEANS ($\bar{\bar{X}}$)

Occasions arise in which means are available from several samples taken from the same population. If these means are combined, the mean of the set of means ($\bar{\bar{X}}$) is the best estimator of the population parameter, μ. If every sample has the same N, you can compute $\bar{\bar{X}}$ by adding the means and dividing by the number of means. If the sample means are based on N's of different size, however, you cannot use this procedure. Here is a story that illustrates the right way and the wrong way to calculate the mean of a set of means. After his sophomore year one of your authors found that if a person had a cumulative grade point average of 3.25 in the middle of his junior year, he could begin a program to "graduate with honors." (In those days [1960] only about 10 percent of the class had GPA's above 3.25.) Calculating a cumulative GPA seemed easy enough to do: four semesters had produced GPA's of 3.41, 3.63, 3.37, and 2.16. Given another GPA of 3.80, the sum of the five semesters would be 16.37 and dividing by 5 gave an average of 3.27, well above the required 3.25.

Graduating with honors seemed like a great ending for college, so our hero embarked on a goal-oriented semester: a B in German and an A in everything else, a GPA of 3.80. And, at the end of the semester the goal had been accomplished. Unfortunately, "graduating with honors" was not to be.

There was a flaw in his method of calculating his cumulative GPA. The method assumed that all of the semesters were equal in weight, that they had all been based on the same number of semester hours. The calculations based on this assumption are shown on the left side of **Table 2.11.**

However, all five semesters were not the same. For example, the semester with the GPA of 2.16 was based on 19 semester hours, rather than the usual 16 or so.[9] Thus, this semester must be weighted more heavily than semesters in which only 16 hours were taken.

Table 2.11 Two methods of calculating a mean of a set of means from five semesters' GPA's; the method on the left is correct only if all semesters are based on the same number of credit hours

Flawed method	*Correct method*		
Semester GPA's	*Semester GPA's*	*Hours credit*	*GPA's × hours*
3.41	3.41	17	58
3.63	3.63	16	58
3.37	3.37	19	64
2.16	2.16	19	41
3.80	3.80	16	61
$\Sigma = 16.37$		$\Sigma = 87$	$\Sigma = 282$
$\bar{\bar{X}} = \dfrac{16.37}{5} = 3.27$		$\bar{\bar{X}} = \dfrac{282}{87} = 3.24$	

[9] The semester was more educational than a GPA of 2.16 would indicate. A great deal of American literature was consumed that spring, but unfortunately I was not registered for any courses in American literature.

The right side of Table 2.11 shows the correct way to calculate his cumulative GPA. Each semester's GPA is multiplied by its number of semester hours. These products are summed and that total is divided by the sum of the semester hours. As you can see from the numbers on the right, the actual cumulative GPA was 3.24, not high enough to qualify for the honors program.

More generally, to find the mean of a set of means, multiply each separate mean by its N, add these products together, and divide by the sum of the N's. Thus, means of 2.0, 3.0, and 4.0 based on N's of 6, 3, and 2 produce an overall mean (mean of a set of means) of 2.64 ($29 \div 11 = 2.64$).

CLUE TO THE FUTURE

The distributions that you have been working with in this chapter are **empirical distributions** based on scores actually gathered in experiments. This chapter and the next two are about these empirical frequency distributions. Starting with Chapter 5 and throughout the rest of the book, you will also make use of **theoretical distributions**— distributions based on mathematical formulas and logic rather than on actual observations.

PROBLEMS

22. Find the mean, median, and mode of the grouped frequency distribution you constructed for Problem 3.

23. Find the mean, median, and mode of the grouped frequency distribution you constructed for Problem 5

24. For the following situations, tell which central value is appropriate and why.
    a. As part of a study on prestige, an investigator sat on a corner in a high-income residential area and classified passing autombiles according to producers: General Motors, Ford, Chrysler, and foreign.
    b. In a study of helping behavior, an investigator pretended to have locked himself out of his car. Passersby who stopped were classified on a scale of 1 to 5 as (1) very helpful, (2) helpful, (3) slightly helpful, (4) neutral, and (5) discourteous.
    c. In a study of per capita income in a city, the following income categories were established: $0–$5000, $5001–$10,000, $10,001–$15,000, $15,001–$20,000, $20,001–$25,000, $25,001–$30,000, $30,001–$35,000, $35,001–$40,000, $40,001–$45,000, and $45,001 and up.
    d. In a study of per capita income in a city, the following income categories were established: $0–$5000, $5001–$10,000, $10,001–$15,000, $15,001–$20,000, $20,001–$25,000, $25,001–$30,000, $30,001–$35,000, $35,001–$40,000, $40,001–$45,000, and $45,001–$50,000.
    e. First admissions to a state mental hospital for the previous five years were classified as schizophrenic, paranoid, anxiety, dissociative, and other disorders.
    f. A teacher gave her class an arithmetic test, and most of the children scored in the range 70–79. A few were above this, and a few were below.
    g. The frequency distribution of pairs of shoes in Problem 5.

25. A senior psychology major performed the same experiment on three groups and obtained means of 74, 69, and 75. The groups consisted of 12, 31, and 17 subjects, respectively. What is the overall mean for all subjects?

26. In Problem 9, you eyeballed two distributions to determine the direction of the skew. Verify that judgment now with a comparison of the means and medians of the distributions.

27. A three-year veteran of the local baseball team was figuring out his lifetime batting average. During the first year he played for half the season and batted .350 (28 hits in 80 at-bats). The second year he had about twice the number of at-bats and his average was .325. The third year, although he played even more regularly, he was in a slump and batted only .275. By adding the three season averages and dividing by 3, he found his lifetime batting average to be .317. Is this correct? Justify your answer.

3 VARIABILITY

Objectives for Chapter 3: After studying the text and working the problems in this chapter, you should be able to:

1. Explain the concept of variability
2. Find the range of a distribution
3. Distinguish among the following: the standard deviation of a population, the standard deviation of a sample used to describe the sample, and the standard deviation of a sample used to estimate the population standard deviation
4. Calculate a standard deviation from grouped or ungrouped data and interpret its meaning
5. Add standard deviation measures to a line graph
6. Calculate the variance of a distribution
7. Convert raw scores to z scores
8. Use z scores to compare two scores in a distribution or a score in one distribution with a score in a second distribution

In Chapter 2 you studied several descriptive statistics. *Frequency distributions* and *graphs* presented all the data, and the *mean, median,* and *mode* each gave you one number that was typical of a distribution. In Chapter 3, we continue with more descriptive statistics. This time you get one number that expresses the **variability** of a distribution.

The value of knowing about variability is illustrated by the story of two brothers who went water skiing on Christmas Day (far north of Miami). On the *average,* each skier finished his turn 5 feet from the shoreline (where one may step off the ski into only 1 foot of very cold water). This bland average of the two actual stopping places does not convey the excitement of the day. In fact, the first brother, determined to avoid the cold water, overshot the beach so much that he finally stopped rolling, scraped and bruised, up on the rocky shore 35 feet from the water. The second brother, determined not to share that fate, stopped too soon. Although he swam the 45 feet to shore very quickly, his lips were very blue. No, the average stopping place of 5 feet from shore doesn't convey the excitement of the day. Knowing about variability is required.

Here are some other situations in which knowing the average leaves you without enough information.

1. You are an elementary school teacher who has been assigned a class of fifth-graders whose mean IQ is 115, well above the IQ of 100 that is the average for the general population. Because a child with an IQ of 115 can handle more complex, abstract material than the average child, you might be tempted to plan more sophisticated projects for the year. *If the variability of the IQs in the class is small, your projects will probably succeed. If the variability is large, however, the projects will be too complicated for some of the pupils, and for others, even these projects will not be challenging enough.*

2. You are a developer planning houses for the general public in a newly opened 160-acre tract of land. Knowing that the average size of a household is 2.69 persons, you order house plans designed for three people. *There is variability around that mean of 2.69 persons per household. To give you some idea of the variability, the general public includes one-person households (20 percent), three-person households (15 percent), five-person households (6 percent), and seven-or-more-person households (1 percent) (Statistical Abstract of the United States: 1986, 1985).*

3. Having graduated from college, you are considering offers from two companies, one in sales and the other in management. The pay is about the same in both. After checking out the statistics for salespersons and managers at the library, you find that those who have been working for 5 years also have similar averages. You might conclude that the pay for the two occupations is equal. *Pay is more variable for those in sales than for those in management. Some in sales make much more than the average (and some make much less), whereas the pay of those in management is clustered together. Your reaction to this difference in variability might help you choose.*

Clearly, variability is important. This chapter is about statistics and parameters that measure the variability of a distribution.

The range is the first we will describe. The second, the standard deviation, is the most important. Most of this chapter will be about the standard deviation. A third way to measure variability is with the variance. Finally, at the end of this chapter, z scores will be described. The standard deviation is necessary in order to calculate z scores, which are used to compare the relative standing of individual scores. A z score, however, is not a measure of variability.

THE RANGE

The **range** is the upper limit of the highest score minus the lower limit of the lowest score; that is,

$$\text{range} = X_H - X_L$$

where X_H = upper limit of the highest score in the distribution
X_L = lower limit of the lowest score in the distribution

Thus, the range from 5 to 10 is 6. You can check this for yourself by counting the spaces in the following illustration. In terms of the formula, $10.5 - 4.5 = 6$.

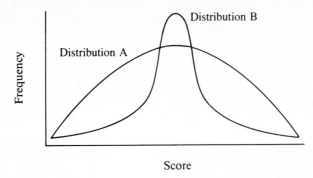

Figure 3.1   Two frequency distributions with the same mean and range but with different variability

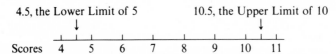

Applying this same logic to a distribution with a high score of 3.7 and a low score of 2.0, the range is 1.8 (3.75 – 1.95). Similarly, the range is 0.26 when the high score is 0.70 and the low score is 0.45. Often the range is reported with a statement like, "The scores ranged from 55 to 89."

As a matter of fact, you have already been exposed to the range as a measure of variability. You used it in Chapter 2 when you learned to set up frequency distributions. You will recall that this same procedure was used to find out how many class intervals to establish.

The range is also used in quality control procedures in manufacturing. A small sample is taken and the range is quickly calculated and plotted. A range larger than expected means that there is too much variability in the process. Adjustments are called for (see Chapter 6 or Berger and Hart, 1986).

The range is very easy to compute; it depends on only the two *extreme* scores. Because of this, it is easy to picture two very *different* distributions with the *same* mean and the *same* range. **Figure 3.1** illustrates such a situation.

THE STANDARD DEVIATION

A second measure of variability, the standard deviation, distinguishes between the two distributions in Figure 3.1. The standard deviation is larger for the more variable Distribution A than it is for Distribution B. Unlike the range, every score is used in the calculation of the standard deviation.

Standard deviations are very useful and informative measures of variability. They can be used to distinguish between distributions (as in Figure 3.1), and they are used to measure distances along the baseline of an important theoretical distribution, the normal curve. You will use the normal curve throughout your statistical

Table 3.1   Symbols, purposes, and descriptions of three standard deviations

Symbol	Purpose	Description
σ	Describe a population's variability	This lowercase Greek *sigma* is the symbol for the standard deviation of a population. σ is a parameter and is used to *describe* variability when the population is available.
s	Estimate a population's variability	This lowercase s is an *estimate* of σ, in the same way that $\bar{X}$ is an estimate of μ. As you know, parameters are rarely known. Often the best you can do is draw a sample from the population and use statistics such as $\bar{X}$ and s as estimates of parameters such as μ and σ. s, the estimate of σ, is the variability statistic you will use most often in this book.
S	Describe a sample's variability	There are occasions when you want to describe the variability of a sample but have no interest in estimating σ. In such cases, S is the statistic to use. In this book you will encounter S in this chapter and in places where correlation is stressed (principally in Chapter 4).

life. We think it would be difficult for us to overemphasize the importance of standard deviations, "the yardsticks of variability."[1] Your principal task in this section is to learn the distinctions among three different standard deviations. For a particular situation or problem, the standard deviation you use will be determined by your purpose. Table 3.1 lists purposes, symbols, and descriptions. It will be worth your time to study **Table 3.1** thoroughly.

Distinguishing among these three standard deviations and clearly understanding them are sometimes a problem for beginning students. Our advice is to study Table 3.1 and then become alert at the situations in the text where a standard deviation is used. With each situation you will acquire more understanding. We will first discuss the calculation of σ and S (which are computed in the same way except for μ and $\bar{X}$ in the formulas) and then deal with s.

THE STANDARD DEVIATION AS A DESCRIPTIVE INDEX OF VARIABILITY

Both σ and S are used to *describe* the variability of some data on hand. σ is a parameter of a population; S is a statistic of a sample. The two are calculated with similar formulas. We will show you two ways to arrange the arithmetic of this formula—the raw-score method and the deviation-score method. The quick and accurate raw-score method is the one used by experienced researchers and statisticians, and it is the method you will use after you complete this chapter. The deviation-score method, however, will give you a much better understanding of what a standard deviation is measuring. Algebraically, the two methods are identical.

---

[1] "Deviation" is part of the name because calculation requires you to compute the deviation of each score from the mean.

So we will begin your introduction to the standard deviation with the deviation-score method. But before you can use this method, you need to be introduced to a common statistic, the deviation score, which you will also use in situations other than the calculation of standard deviations.

Deviation Scores

A **deviation score** is a raw score minus the mean, either $X - \bar{X}$ or $X - \mu$. It is simply the difference between a score in the distribution and the mean of that distribution. Deviation scores are encountered so often that they have a special symbol, lowercase x. Note that a capital X is used for a score and a lowercase x is used for a deviation score. Not only must you know this, you must also be sure that you (and your instructor) can tell the difference between your written versions of X and x. Write them. Can your roommate tell the difference?

Since $x = X - \bar{X}$, raw scores that are larger than the mean will have positive deviation scores, raw scores that are smaller than the mean will have negative deviation scores, and raw scores that are equal to the mean will have a deviation score of zero.

Table 3.2 gives you a brief demonstration of computing deviation scores for a small population of data. In Table 3.2, we first computed the mean (which we arranged to be a whole number). We then subtracted the mean from each score to obtain the deviation scores, which appear in the right-hand column.

A deviation score tells you the number of points that a particular score deviates from, or differs from, the mean. In Table 3.2, the x value for John tells you that he scored six points above the mean, Joshua scored at the mean, and Jeri was five points below the mean.

ERROR DETECTION	Notice that the sum of the deviation scores is zero. This will always be the case, so summing the deviation scores serves as a useful check on computations. If the sum is not zero, you have made an error.

Table 3.2   The computation of deviation scores from raw scores

Name	Score	$X - \mu$	x
John	14	$14 - 8$	6
Shelly	10	$10 - 8$	2
Joshua	8	$8 - 8$	0
Melissa	5	$5 - 8$	-3
Jeri	3	$3 - 8$	-5
	$\Sigma X = 40$		$\Sigma x = 0$

$$\mu = \frac{\Sigma X}{N} = \frac{40}{5} = 8$$

PROBLEMS   For each of the four distributions below, find the range and the deviation scores. Check to see that $\Sigma x = 0$.

1. 15, 13, 12, 10, 8, 7, 5
2. 17, 5, 1, 1
3. 3.4, 3.1, 2.7, 2.7, 2.6
4. 0.45, 0.30, 0.30
5. Give the symbol and purpose of the three standard deviations.

Computing σ and S Using Deviation Scores

The deviation-score formula for computation of the standard deviation as a descriptive index is

$$\sigma = \sqrt{\frac{\Sigma(X - \mu)^2}{N}} = \sqrt{\frac{\Sigma x^2}{N}} \quad \text{or} \quad S = \sqrt{\frac{\Sigma(X - \bar{X})^2}{N}} = \sqrt{\frac{\Sigma x^2}{N}}$$

where σ = standard deviation of a population
  S = standard deviation of a sample
  N = number of deviations (same as the number of scores)
  Σx^2 = sum of the squared deviations[2]

Look at $\Sigma x^2/N$. It is not too different from $\Sigma X/N$. Indeed, $\Sigma X/N$ is an average of the scores (X), and $\Sigma x^2/N$ is an average of the squared deviations (x^2).

Table 3.3 shows the results when a word processing task was given to ten secretaries who work for a law firm. The scores represent the time in minutes necessary to complete the task and print the results. We will use these scores to illustrate the computation of σ. If the secretaries had been a sample and your only interest was in that sample, the standard deviation computed would be identified by S instead of σ. The computations, however, would be the same.

To compute σ for the data in Table 3.3, first compute the mean (which we again arranged to be an integer). Subtract the mean from each of the scores to obtain x for each score. Add these deviation scores to be certain that their sum is zero. Next, square each x and sum the x^2 values to obtain Σx^2. Finally, divide Σx^2 by N and extract the square root; $\sigma = 2.68$ minutes.

Now what does this mean? Knowing that $\sigma = 2.68$ minutes, what can be said? The 2.68 minutes is a measure of the variability of the time it takes the ten secretaries to perform the word processing task. If σ had been 0, you would know that each person completed the task in exactly the same amount of time. (There would have been no variability.) The closer σ is to 0, the more confidence you would have in predicting that any one secretary's time was the mean for the group. Conversely, the farther σ is from 0, the less confidence you would have that the mean was descriptive for a particular secretary. Starting in Chapter 5 you will learn new concepts that will

---

[2] Σx^2 is shorthand notation for $\Sigma(X - \bar{X})^2$ [or $\Sigma(X - \mu)^2$]. $\Sigma(X - \bar{X})^2$ is shorthand notation that tells you to subtract the group mean from each score in the group, producing N deviation scores, and then square each deviation score and add all the squares together. One rule for working with summation notation is to perform the operations in the parentheses first.

Table 3.3 The deviation-score method of computing σ for the time scores of a population of ten secretaries

X	x	x^2
16	5	25
14	3	9
13	2	4
12	1	1
11	0	0
11	0	0
9	-2	4
9	-2	4
8	-3	9
7	-4	16
$\Sigma = 110$	0	72

$$\mu = \frac{\Sigma X}{N} = \frac{110}{10} = 11$$

$$\sigma = \sqrt{\frac{\Sigma(X - \mu)^2}{N}} = \sqrt{\frac{\Sigma x^2}{N}} = \sqrt{\frac{72}{10}} = \sqrt{7.20} = 2.68$$

allow you to tell a more complete story. For now, though, our interpretation has been an almost definitional one.

Now look back at Table 3.3 and the formula for σ. Notice what is happening. The mean is being subtracted from each score. This difference, whether positive or negative, is squared and these squared differences are added together. This sum is divided by N and the square root is found. Every score in the distribution contributes to the final answer.

Notice the contribution made by a score like 16 that is far from the mean: it is large. This makes sense because the standard deviation is a measure of variability and if there are scores far from the mean, they cause the standard deviation to be larger. Take a moment to think through the contribution to the standard deviation made by a score near the mean.[3]

ERROR DETECTION

All standard deviations are positive numbers. If you find yourself trying to take the square root of a negative number, you've made an error.

PROBLEMS   Compute σ or S for each of the three distributions, using the deviation-score method.

6. 7, 6, 5, 2
7. 14, 11, 10, 8, 8

---

[3] If you play with a formula you will become more comfortable with it and understand it better. Make up a small set of numbers and calculate a standard deviation. See what happens when you change one of the numbers, add a number, or leave out a number.

8. 107, 106, 105, 102
9. Compare the standard deviation of Problem 6 with that of Problem 8. What conclusion can you draw about the effect of the size of the numbers on the standard deviation?
10. Does the size of the numbers in a distribution have any effect on the mean?
11. The temperatures in this problem are averages for the months of March, June, September, and December. Calculate the mean and standard deviation for each city. Summarize your results as a sentence.

| San Francisco | 54° | 59° | 62° | 52° |
| Albuquerque | 46° | 75° | 70° | 36° |

12. No computation is needed for this one. Eyeball the following pairs of distributions to determine which of the two has the greater variability if they are not equal.
   a. 1, 2, 4, 1, 3 and 9, 7, 3, 1, 0
   b. 9, 10, 12, 11 and 4, 5, 7, 6
   c. 1, 3, 9, 6, 7 and 14, 15, 14, 13, 14
   d. 114, 113, 114, 112, 113 and 14, 13, 14, 12, 13
   e. 8, 4, 6, 3, 5 and 4, 5, 7, 6, 15

Computing σ and S by the Raw-Score Method

The deviation-score method of computing the standard deviation requires that you "cook" the raw scores by converting them to deviation scores. This "cooking" usually requires that every deviation score be rounded off, which introduces some error into the final answer. With the raw-score method, there are only three calculations that might require rounding. The raw-score method, then, is more exact than the deviation-score method.

The formula for the raw-score method is

$$\sigma \text{ or } S = \sqrt{\frac{\Sigma X^2 - \frac{(\Sigma X)^2}{N}}{N}}$$

Table 3.4 The raw-score method of computing σ for time scores of a population of ten secretaries

X	X^2
16	256
14	196
13	169
12	144
11	121
11	121
9	81
9	81
8	64
7	49
$\Sigma = 110$	1282

$\Sigma X = 110 \qquad \Sigma X^2 = 1282 \qquad (\Sigma X)^2 = (110)^2 = 12{,}100$

$$\sigma = \sqrt{\frac{1282 - \frac{(110)^2}{10}}{10}} = \sqrt{\frac{1282 - 1210}{10}} = \sqrt{\frac{72}{10}} = \sqrt{7.2} = 2.68$$

where ΣX^2 = sum of the squared scores

$(\Sigma X)^2$ = square of the sum of the raw scores

N = number of scores

Although the raw-score formula may at first glance appear forbidding, it is actually easier to use than the deviation-score formula because you don't have to compute deviation scores. The numbers you work with will be larger, but your calculator probably won't mind (unless you exceed its capacity).

We will use the data of Table 3.4 to illustrate the calculation of σ and S by the raw-score method. These data are the same as those of Table 3.3—the word processing time scores of the ten secretaries.

Follow the steps of the formula in **Table 3.4** on page 57. Square the sum of the X column and divide by N. Subtract this quotient from the sum of the X^2 column. Divide this difference by N and extract the square root. The result is σ or S.[4] Notice that the answers for σ in Tables 3.4 and 3.3 are the same. In this case, the mean is an integer so the deviation scores did not have to be rounded. No rounding errors were introduced.

ΣX^2 and $(\Sigma X)^2$. Did you notice the difference in these two terms when you were working with the data in Table 3.4? We hope you did. They are different and you cannot calculate a standard deviation correctly unless you understand the difference. Reexamine Table 3.4 if you aren't sure of the difference between ΣX^2 and $(\Sigma X)^2$. Be alert for these two sums in the problems that are coming up.

ERROR DETECTION

The range is usually two to five times larger than the standard deviation when $N = 100$ or less. The range (which can be calculated quickly) serves as a useful check on any large errors you may have made in calculating the standard deviation.

CLUE TO THE FUTURE

You will be glad to know that in your efforts to calculate a standard deviation, you have produced two other statistics along the way. Each of these has a name and they will turn up again, either in this chapter or in a future one. The number that you took the square root of is called the **variance** (also called a **mean square**), symbolized σ^2. The expression in the numerator, Σx^2, is called the **sum of squares**. You will see both of these terms in Chapters 9 and 10, which are about the analysis of variance.

[4] Some textbooks and statisticians prefer the algebraically equivalent formula

$$\sigma \text{ or } S = \sqrt{\frac{N\Sigma X^2 - (\Sigma X)^2}{N^2}}$$

We are using the formula in the text because the same form, or parts of it, will be used in other procedures.

Yet another formula is often used in the field of testing. This formula, which requires you to begin by calculating the mean, is

$$\sigma = \sqrt{\frac{\Sigma X^2}{N} - \mu^2} \quad \text{and} \quad S = \sqrt{\frac{\Sigma X^2}{N} - \bar{X}^2}$$

All three of these arrangements of the arithmetic are algebraically equivalent.

PROBLEMS

13. Look at the two sample distributions below. Without any calculation (just eyeball the data), decide which one has the larger standard deviation. Guess the size of *S* for each distribution. (You may wish to calculate the range before you choose a number.) Compute *S* for each distribution. Compare your computation with your guess.
    a. 6, 5, 4, 3, 2   **b.** 6, 6, 6, 2, 2

14. For each of the distributions in Problem 13, divide the range by the standard deviation. Is the result between 2 and 5?

15. By now you can look at the two distributions below and see that (a) is more variable than (b). The difference in the two distributions is in the lowest scores (2 and 6). Calculate σ for each distribution, using the raw-score formula. Notice the difference the one score makes in a distribution of five scores.
    a. 9, 8, 8, 7, 2   **b.** 9, 8, 8, 7, 6

---

s AS AN ESTIMATE OF σ

We want to emphasize again that *s* is the principal statistic you will learn in this chapter. It will be used again and again throughout the rest of this text. (Write *S* and *s*. Are they so different that you need not check with your roommate again?)

As we explained in Chapters 1 and 2, the purpose of a sample is usually to find out something about a population; that is, a statistic from a sample is used to estimate a parameter of the population. The most obvious pitfall in this reasoning is that two samples from the same population are often slightly different, yielding two different statistics. Which one of the statistics is closest in value to the parameter? There is no way to know other than to measure the entire population, which is usually impossible. The best you can do is to calculate the statistic in such a way that, *on the average,* its value is equal to the parameter. In the language of the mathematical statistician, you want a statistic that is an "unbiased estimator" of the corresponding population parameter.

It turns out that if you have sample data and you want to calculate an estimate of σ, you should use the statistic

$$\sqrt{\frac{\Sigma(X - \bar{X})^2}{N - 1}}$$

This statistic is symbolized with a lowercase *s*. Note that the difference between *s* and σ is that *s* has $N - 1$ in the denominator, whereas σ has *N*.

This issue of dividing by *N* or by $N - 1$ can leave students shrugging their shoulders and muttering, "OK, I'll memorize it and do it however you want." We would like to explain, however, why you use $N - 1$ for *s*.

As you already know, the formula for σ is

$$\sigma = \sqrt{\frac{\Sigma(X - \mu)^2}{N}}$$

To find the standard deviation of a sample, it would seem logical to just substitute $\bar{X}$ in the numerator of the formula. This seems appealing, especially if you also know that $\bar{X}$ is an *unbiased estimator* of μ. However, if you use $\bar{X}$, the value of the numerator will be consistently too small.

To understand this surprising state of affairs, recall from Chapter 2 that one of the characteristics of the mean is that for any set of scores, the numerical value of $\Sigma(X - \text{mean})^2$ is minimized. Think about having a population of scores, which we will symbolize by X_p. The value of $\Sigma(X_p - \mu)^2$ is minimized by using μ compared to any number other than μ. Now, from that population, consider a sample of scores, symbolized by X_s. For this sample, the value of $\Sigma(X_s - \bar{X})^2$ will be a minimum. The question is, How does the value of $\Sigma(X_s - \bar{X})^2$ compare to that of $\Sigma(X_p - \mu)^2$, the value you are trying to find? Naturally, different samples (each with its own $\bar{X}$) will produce different standard deviations, but on the average, these standard deviations will be less than that produced by $\Sigma(X_p - \mu)^2$, which is the value you are after.[5] Because $\Sigma(X_s - \bar{X})^2$ systematically underestimates $\Sigma(X_p - \mu)^2$, a correction must be made. The best correction is to divide by $N - 1$ instead of by N. Thus, the generally underestimated numerator is divided by a smaller denominator, resulting in a much better estimator of σ.[6]

(We know that you just finished a dense paragraph—lots of ideas per square inch. You may understand it already, but if you don't, take 10 or 15 minutes to reread, do the exercise, and think.)

Note also that as N gets larger, the subtraction of 1 from N has less and less effect on the size of the estimate of variability. This makes sense because the larger the sample size is, the closer $\bar{X}$ will be to μ.

There is one new task that comes with the introduction of s. It is the decision whether to calculate σ, S, or s on a given set of data. Your decision will be based on your purpose for the standard deviation. If your purpose is to describe, use σ or S, depending on whether you have population data. If your purpose is to estimate a population σ from sample data (a common requirement in inferential statistics), use s.

Calculating s

A raw-score formula is recommended for calculating s:

$$s = \sqrt{\frac{\Sigma X^2 - \dfrac{(\Sigma X)^2}{N}}{N - 1}}$$

This is practically the same as the formula for σ, so you will have no trouble calculating s (assuming you mastered the calculation of σ).[7]

This raw-score formula is the one you will probably use for your own data. Sometimes, though, you may be confronted with someone else's frequency distri-

---

[5] You can demonstrate this for yourself by taking a small population of scores (say, 1, 2, 3) and calculating the value of $\Sigma(X_p - \mu)^2$. Then for the three $N = 2$ samples in the population, calculate $\Sigma(X_s - \bar{X})^2$. Find the average of these three samples and compare that to the value of $\Sigma(X_p - \mu)^2$.

[6] Here are two technical points that you should know if you plan to learn more about statistics after this course. First, even with $N - 1$ in the denominator, s is not an unbiased estimator of σ. Since the bias is not very serious, s is used as the best estimator of σ. Second, s^2 (see the next section) is an unbiased estimator of σ^2.

[7] Calculators that have a standard deviation function differ. Some use N in the denominator and some use $N - 1$. You will have to check yours to see how it is programmed.

bution for which you want to calculate a standard deviation. For a simple frequency distribution or a grouped frequency distribution, the formula is

$$s = \sqrt{\frac{\Sigma fX^2 - \dfrac{(\Sigma fX)^2}{N}}{N - 1}}$$

where f is the number of frequencies in an interval.

Here is a problem that illustrates the calculation of s both for ungrouped raw scores and for a frequency distribution. Consider puberty. As you know, females reach puberty earlier than males (about two years earlier on the average). Is there any difference between the sexes in the *variability* of reaching this developmental milestone? Comparing standard deviations will give you an answer.

We have only a sample of ages for both sexes, and since the interest is in all females and males, s is the appropriate standard deviation. **Table 3.5** shows the calculation of s for the females. Follow the calculations to satisfy yourself that the standard deviation is 2.19 years.

We will illustrate the calculation of s for a simple frequency distribution using the data for males in **Table 3.6.** Note as you work through these calculations that grouping causes two extra columns in the calculations. The standard deviation for males for these data is 1.44.

Based on the sample data, we concluded that there is more variability among females in reaching puberty than there is among males. You might stop for a moment and think about your observations of teens and preteens and whether these numbers are consistent with your observations.

Here are three final points about working with simple and grouped frequency distributions. Recognize that ΣfX^2 is found by squaring X, multiplying by f, and summing. And $(\Sigma fX)^2$ is found by multiplying f by X, summing, and then squaring.

For grouped frequency distributions, use the midpoints of the class intervals as

Table 3.5   Calculation of s for ages at which females reach puberty (raw-score method)

Age (X)	X^2
17	289
15	225
13	169
12	144
12	144
11	121
11	121
11	121
$\Sigma X = 102$	$\Sigma X^2 = 1334$

$(\Sigma X)^2 = 10{,}404$

$$s = \sqrt{\frac{\Sigma X^2 - \dfrac{(\Sigma X)^2}{N}}{N - 1}} = \sqrt{\frac{1334 - \dfrac{(102)^2}{8}}{7}} = \sqrt{\frac{1334 - 1300.50}{7}} = 2.19$$

Table 3.6   Calculation of s for ages at which males reach puberty (simple frequency distribution)

Age (X)	f	fX	fX^2
18	1	18	324
17	1	17	289
16	2	32	512
15	4	60	900
14	5	70	980
13	3	39	507
	$N = 16$	$\Sigma = 236$	$\Sigma = 3512$

$$s = \sqrt{\dfrac{\Sigma fX^2 - \dfrac{(\Sigma fX)^2}{N}}{N-1}} = \sqrt{\dfrac{3512 - \dfrac{(236)^2}{16}}{15}} = \sqrt{\dfrac{3512 - 3481}{15}} = 1.44$$

the X values, as you did in Chapter 2 when you worked with grouped frequency distributions. (Refer to Table 2.9 for a review.)

For both grouped and simple frequency distributions, most calculators with memory will give you ΣfX and ΣfX^2 if you properly key in X and f for each line of the distribution. Procedures differ depending on the brand, so you will have to figure out how to do it with your calculator, but the time invested will be repaid several times in future chapters (not to mention the satisfying feeling you will get).

GRAPHING STANDARD DEVIATIONS

We have already suggested that graphing is a powerful technique for helping you and your reader better understand a set of data. The information conveyed by standard deviations can often be added to line graphs; you then get a picture of variability as well as central value.

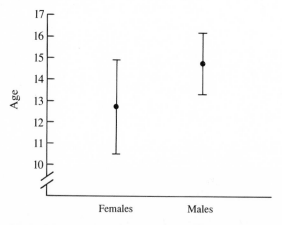

Figure 3.2   Puberty in females and males—means and standard deviations

Figure 3.2 is a graph of the puberty data, showing the standard deviations as well as the means. The dot shows the mean age (read from the Y axis), and the lines extend one standard deviation in either direction. From this graph you can see at a glance that the mean age of puberty is younger for females than for males and that for females there is more variability about the mean than there is for males.

PROBLEMS

16. A researcher had a sample of scores from the freshman class on a test that measured attitudes toward authority. She wished to estimate the standard deviation of the entire freshman class, since she believed that the current group of students was more homogeneous than students in the past. Given the summary statistics below, calculate the standard deviation.

$$N = 21 \qquad \Sigma X = 304 \qquad \Sigma X^2 = 5064$$

17. Here are those data on the heights of 18- to 24-year-old Americans that you graphed and found a mean for in the last chapter. Choose the more interesting group and find s.

Females		*Males*	
Height (in.)	*f*	*Height (in.)*	*f*
72	1	77	1
70	1	76	1
69	2	75	1
68	1	74	1
67	3	73	4
66	5	72	5
65	7	71	7
64	10	70	6
63	9	69	8
62	6	68	7
61	3	67	2
60	1	66	2
59	1	65	3
		64	1
		62	1

18. In manufacturing, engineers strive for consistency. The data below are the errors in millimeters of giant doodads manufactured by two different processes. Choose S or s and determine which process produces the more consistent doodads.

Process A	0	1	-2	0	-2	3
Process B	1	-2	-1	1	-1	2

19. A high-school English teacher measured the attitudes of 11th-grade students toward poetry. After a nine-week unit on poetry, she measured the students' attitudes again. She was disappointed to find that the mean change was 0. Below are some representative scores. (High scores indicate favorable attitudes.) Calculate s for both before and after, and write a conclusion based on the standard deviations.

Before	7	5	3	5	5	4	5	6
After	9	8	2	1	8	9	1	2

THE VARIANCE

In a "Clue to the Future" you were told that you produce the **variance** as part of your calculation of the standard deviation. The variance is the number you take the square root of to get the standard deviation. The symbols for the variance are σ^2 (population variance) and s^2 (sample variance used to estimate the population variance). By formula,

$$\sigma^2 = \frac{\Sigma(X - \mu)^2}{N} \quad \text{and} \quad s^2 = \frac{\Sigma(X - \bar{X})^2}{N - 1}$$

The difference between σ^2 and s^2 is the term in the denominator. The population variance uses N and the sample variance has $N - 1$.[8]

The variance is not very useful as a *descriptive* statistic. It is, however, of enormous importance in inferential statistics. You will see more of the variance in Chapters 9 and 10, which cover the **analysis of variance.**

z SCORES

You have used measures of central value and measures of variability to describe a *distribution* of scores. The next statistic, z, is used to describe *a single score.*

Suppose one of your friends tells you he made a 95 on a math exam. What does that tell you about his mathematical ability? Due to your previous experience with tests, 95 may seem like a pretty good score, but unless you make a couple of assumptions, a score of 95 is *meaningless.* Read on.

After you say, "95! Congratulations," suppose he tells you that 200 points were possible. Now a score of 95 seems like something to hide. "My condolences," you say. But then he tells you that the highest score on that difficult exam was 101. Now 95 has regained respectability and you chortle, "Well, all right!" In response, he shakes his head and tells you that the mean score was 98. 95 takes a nose dive. As a final blow, you find out that 95 was the lowest score, that nobody was worse than your friend. With your hand on his shoulder, your final remark is, "Come on, I'll buy you an ice cream cone."

This example illustrates that 95 acquires meaning only when it is compared with the rest of the test scores. In particular, a score gets its meaning from its relation to the mean and the variability of its fellow scores. A **z score** is a mathematical way to change a raw score so that it reflects its relationship to the mean and standard deviation of its fellow scores. The formula is

$$z = \frac{X - \bar{X}}{S} = \frac{x}{S}$$

Remember that x is an acquaintance of yours, the deviation score.

---

[8] Many calculators have a variance key. One that we know of uses $N - 1$ to calculate the standard deviation and N to calculate the variance. Given a square key and a square root key, this arrangement covers σ, σ^2, s, and s^2.

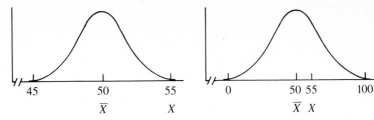

Figure 3.3 A comparison of $X - \bar{X} = 5$ in two distributions with different standard deviations

A z score describes the relation of an X to $\bar{X}$ with respect to the general width (variability) of the distribution. For example, if you know that a score (X) is five units from the mean ($X - \bar{X} = 5$), you know only that the score is better than average, but you have no idea how far above average it is. The distance 5 by itself isn't very meaningful. If the distribution is *ten* units wide and $\bar{X} = 50$, then an X of 55 is a very high score. On the other hand, if the distribution is *100* units wide and $\bar{X} = 50$, an X of 55 is barely above average. Thus, the variability of the distribution must be taken into account in order to find a particular score's position in the distribution. The way to do this is to divide $X - \bar{X}$ by a unit that measures variability, the standard deviation. This results in a deviation per unit of standard deviation.[9] When an X is converted to a z score, the z represents the number of standard deviations the score is above or below the mean. **Figure 3.3** is a picture of the ideas in this paragraph.

A z score is sometimes referred to as a **standard score** because it is a deviation score expressed in standard deviation units. Any distribution of raw scores can be converted to a distribution of z scores; for each raw score, there is a z score. Raw scores above the mean will have positive z scores; those below the mean will have negative z scores.

When used as a descriptive measure of one distribution of 100 or so scores, z scores range from approximately -3 to $+3$. Thus, a z score of -2.5 is at or near the bottom of the distribution. However, for some distributions, the highest z score may be as low as 1, and the lowest z score may be -1.

In Chapters 6 and 7, z scores are used for inferential purposes. Much larger z scores may occur then.

As we have indicated, a z score gives the relative position of a raw score in a distribution. It is a short step to the realization that if two raw scores are each converted to z scores, you can know the positions of the scores relative to *each other* as well as to the distribution.

In addition, z scores are used to compare two scores from different distributions, even when the scores are measuring different things. (If this sounds like trying to compare apples and oranges, see Problem 24.) We will illustrate these last two paragraphs with the description that follows and the data in Table 3.7.

[9] This is the same concept as a percentage (per centum, or per hundred). Thus, 24 out of 50 and 12 out of 25 may appear to be different, but you recognize them as the same when you convert both figures to per hundred (by multiplying numerators and denominators by the number necessary to make the denominator 100).

Table 3.7 The z scores of selected students on two 100-point tests in General Psychology

Student	Test 1 Learning and memory		Test 2 Psychopathology	
	Raw score	z Score	Raw score	z Score
—	—	—	—	—
—	—	—	—	—
Kris	76	+2.20	76	−1.67
Robin	54	.00	86	.00
—	—	—	—	—
Marty	58	+.40	82	−.67
Terry	58	+.40	90	+.67
—	—	—	—	—
	Test 1: $\bar{X} = 54$		Test 2: $\bar{X} = 86$	
	Test 1: $S = 10$		Test 2: $S = 6$	

In the General Psychology course taken years ago by one of your authors, the professor returned tests with a z score rather than a percentage score. This z score was *the* key to figuring your grade. A z score of $+1.50$ or higher was an A and -1.50 or lower was an F. (z scores between $+.50$ and $+1.50$ received B's.)

Table 3.7 shows the raw scores (percentage correct) and z scores for four of the many students who took two of the tests in that class. Consider the first student, Kris, who scored 76 on both tests. The two 76's appear to be the same, but the z scores show that they are not. The first 76 was a high A and the second 76 was a high F. The second student, Robin, appears to have improved on the second test, going from 54 to 86, but in fact, the scores were the test averages both times. Robin stayed the same. Marty also appears to have improved if you examine only the raw scores. However, the z scores reveal that Marty did *worse* on the second test. Finally, comparing Terry's and Robin's raw scores, it appears that Terry was 4 points higher than Robin on each test, but the z scores show that Terry's edge was greater on the second test.

The reason for the surprising comparisons in the previous paragraph is that the means and standard deviations were so different for the two tests. Reasons for this might be that the second test was easier, the material was more motivating to students, the students studied more, or even that the teacher was better. Perhaps all of these reasons may have been true.

Regardless of the reason for the differences in means and standard deviations, z scores give you a way to compare raw scores. The basis of the comparison is the distribution itself rather than some external standard (such as a grading scale of 90–80–70–60 percent for A's, B's, and so on). Here are some problems to finish the chapter.

PROBLEMS

20. The following scores are the number of eggs found by each of five children on an Easter egg hunt. No attempt is being made to generalize from this sample to a population.

   1   3   4   7   10

   a. Convert each score to a z score.
   b. Determine the variance of the egg scores.

21. What conclusion can you come to about Σz?

22. For any distribution, what is the z score of a raw score that is equal to the mean?

23. Hattie and Missy, twin sisters, were intense competitors, but they never competed against each other. Hattie specialized in long-distance running and Missy was an excellent sprint swimmer. As you can see from the distributions in the table, each was the best in her event. Take the analysis one step farther and use z scores to determine who is the more outstanding twin. You might start by looking at the data and making an estimate.

10K runners	Time (min)	50M swimmers	Time (sec)
Hattie	37	Missy	24
Julie	39	Jo	26
Liz	40	Sue	27
Ann	42	Judi	28

24. Tobe grows apples and Zeke grows oranges. In the local orchards, the mean weight of apples is 5 ounces, with $S = 1.0$ ounces. For oranges the mean weight is 6 ounces, with $S = 1.2$ ounces. At harvest time, each enters his largest specimen in the Warwick County Fair. Tobe's apple was 9 ounces and Zeke's orange was 10 ounces. This particular year Tobe fell ill on the day of judgment, so he sent his friend Hamlet to inquire who had won. Adopt the role of judge and use z scores to determine the winner. Hamlet's query to you as judge must be, "Tobe, or not Tobe; that is the question."

25. Milquetoast's anthropology professor announced that he would drop the poorest exam grade for each student. Milquetoast scored 79 on the first anthropology exam. The mean was 67 and the standard deviation 4. On the second exam, he made 125. The class mean was 105 and the standard deviation 15. On the third exam, the mean was 45 and the standard deviation 3. Milquetoast made 51. On which test was Milquetoast's performance poorest?

Transition Page

Thus far in our exposition of the wonders of statistical methods, we have presented data on only one variable at a time. On this *univariate* data you now know how to perform basic statistical manipulations. You can construct an appropriate graph and compute some essential statistics such as means and standard deviations.

As we explained in Chapter 1 most research involving statistical analysis attempts to answer questions about how variables are related or how groups differ on certain variables. In the next chapter, you will take up the first of these questions—how variables are related. To do this, you will have to deal with two variables at the same time. Here are some pairs of variables that might be related. By the time you finish the following chapter you will know whether or not they are related and *to what degree.*

Verbal ability and mathematical ability
Height of daughters and height of their fathers
Amount of motivation and performance
Church membership and homicide
Inches of rainfall and bushels of wheat per acre
Level of income and probability of being diagnosed
 as psychotic

The next chapter, "Correlation and Regression," explains a method that is used to determine the degree of a relationship (correlation) and a method that is used to make predictions about one variable when you have measurements on another variable (regression).

4 CORRELATION AND REGRESSION

Objectives for Chapter 4: After studying the text and working the problems in this chapter, you should be able to:

1. Explain the difference between univariate and bivariate distributions
2. Explain the concept of correlation
3. Draw scatterplots
4. Explain the difference between positive and negative correlation
5. Compute a Pearson product-moment correlation coefficient (r)
6. Interpret correlation coefficients using the terms *reliability*, *causation*, and *variance in common*
7. Identify situations in which the Pearson r will not accurately reflect the degree of relationship
8. Name and explain the elements of the regression equation
9. Compute regression coefficients and fit a regression line to a set of data
10. Interpret the appearance of a regression line
11. Make predictions about one variable from measurements of another variable

Correlation and regression: our guess is that you have some understanding of the concept of correlation and that you are less comfortable with the word regression. Speculation aside, correlation is simpler. Correlation is a statistical technique for measuring the *degree* of relationship between two variables.

Regression is a more complex set of ideas. In this chapter we will illustrate just two uses of regression—*drawing* the line that best fits the data and *predicting* a person's score on one variable when you know that person's score on a second, correlated variable. Regression has other, more sophisticated uses, but you will have to put those off until you study more advanced statistics.

Sir Francis Galton of England gets the credit for developing the ideas of correlation and regression about 100 years ago. Galton was a genius (he could read at age 3) with an amazing variety of interests, many of which he actively pursued during his 89 years. He once listed his occupation as "private gentleman," which meant that he had inherited money and did not have to work at a job. Lazy, however, he was not. Galton wrote 16 books and more than 200 articles.

From an early age, Galton was enchanted with counting and quantification. Among the many things he tried to quantify were weather, individuals, beauty, boringness of lectures, and efficacy of prayer. He was successful in many of his attempts. For example, it was Galton who discovered that atmospheric pressure highs produce clockwise winds around a calm center, and one of his efforts at quantifying individuals resulted in a book on fingerprints. Because of his many successes, Galton actively promoted the philosophy of **quantification,** the idea that you can understand a phenomenon much better if you can translate its essential parts into numbers.

Now, as you well know, simply measuring some interesting phenomenon like a breeze and getting a number like 12 doesn't leave you sighing the sigh of the newly enlightened. According to researchers, such satisfying sighs come when the data you get confirm the existence of a *relationship* that you suspected all along, and Galton was a master at suspecting relationships.

Many of the relationships that interested Galton were connected with heredity. Although it was common in the 19th century to comment on physical similarities within a family (height and facial characteristics, for example), Galton thought that psychological characteristics, too, tended to run in families. Specifically, he thought that characteristics like genius, musical talent, sensory acuity, and quickness had a hereditary basis. (A clue to the position that Galton took on this matter is that he and his illustrious cousin, Charles Darwin, shared an illustrious grandfather, Erasmus Darwin [although they had no grandmother in common].) Galton's 1869 book, *Hereditary Genius,* listed many families and their famous members.

Galton wasn't satisfied with the lists in that early book; he wanted to express the relationships in quantitative terms. So he established an anthropometric (people-measuring) laboratory for a year at a health exposition (a fair) and later at the South Kensington Museum in London. Approximately 17,000 people who stopped at a booth paid three pence to be measured. They left with self-knowledge; Galton left with data and a pocketful of coins. In today's research the investigator usually pays the subjects. For us researchers, the old days were better in some ways. (For one summary of Galton's results, see Johnson et al., 1985.)

The most important result of Galton's effort to quantify, however, was the invention of the concepts of correlation and regression. With correlation, he could express the degree of relationship between *any* two paired variables. (The relationship between the height of fathers and the height of their adult sons is the classic example.)

Galton was not enough of a mathematician to work out the theory and formulas for his concepts; this task fell to Galton's young friend and protégé, Karl Pearson, a professor of applied mathematics and mechanics at University College in London.[1] Pearson's 1896 *product-moment correlation coefficient* and other **correlation coefficients** he and his students developed were quickly adopted by researchers in many fields and are widely used today in psychology, sociology, education, political science, the biological sciences, and other areas.

Finally, although Galton and Pearson became famous for their data gathering

---

[1] We have some biographical information on Pearson in Chapter 11, the chapter on chi square. Chi square is another statistical invention of Professor Pearson.

and statistics, this was not what they set out to do. For both, the principal goal was to find ways to improve the human condition. Recommendations required a better understanding of heredity and evolution, and statistics was simply the best way to arrive at better understanding. As Galton put it in 1889:

> Some people hate the very name of statistics, but I find them full of beauty and interest. . . . Their power of dealing with complicated phenomena is extraordinary. They are the only tools by which an opening can be cut through the formidable thicket of difficulties that bars the path of those who pursue the Science of [Human Beings].[2]

Our plan in this chapter is for you to read about bivariate distributions (necessary for both correlation and regression), to learn to compute and interpret Pearson product-moment correlation coefficients, and to use the regression technique to draw the best-fitting straight line and predict outcomes.

BIVARIATE DISTRIBUTIONS

In the chapters on central value and variability, you worked with one variable at a time **(univariate distributions).** Height, time, test scores, and errors all received your attention. If you look back at those problems, you'll find a string of numbers under one heading (see, for example, page 38). Compare those distributions with the one in **Table 4.1.** In Table 4.1 there is a set of test scores under "Humor Test *X*." In addition, there is another set of scores under the variable "Intelligence Test *Y*." You could find the mean and standard deviation of either of these variables. The additional characteristic of Table 4.1, the characteristic that makes it a **bivariate distribution,** is that the scores on the two variables are *paired*. The 50 and the 8 go together, and the 20 and 4 go together. The reason they are paired is, of course, that the same person made the two scores. As you will see, there are also other reasons for pairing scores.

The essential idea of a bivariate distribution (which is required for correlation and regression techniques) is that there are two variables with values that are paired for some logical reason. Many situations produce bivariate distributions.

A bivariate distribution may show positive correlation, negative correlation, or zero correlation. We will discuss each of these possibilities in turn.

Table 4.1 A bivariate distribution of scores on two tests taken by the same individuals

Person	Humor test *X*	Intelligence test *Y*
Elgin	50	8
Greg	40	9
Debbie	30	5
Ursula	20	4

[2] For a short biography of Galton, we recommend David (1968).

POSITIVE CORRELATION

In the case of a *positive correlation* between two variables, high measurements on one variable tend to be associated with high measurements on the other variable and low measurements on one variable with low measurements on the other. In other words, the two variables vary together in the same direction. This is the case for the manufactured data in **Table 4.2**. Tall fathers tend to have sons who grow up to be tall men. Short fathers tend to have sons who grow up to be short men. If the true relationship was so undeviating that every son grew to be exactly his father's height (as in our manufactured data), the correlation would be perfect, and the correlation coefficient would be 1.00. A graph plotting the relationship would look like Figure 4.1. If such was the case (which, of course, is ridiculous; mothers and environments have their say, too), then it would be possible to predict perfectly the adult height of an unborn son simply by measuring his father.

Table 4.2   Manufactured data on two variables: heights of fathers and their sons*

Father	Height (in.) X	Son	Height (in.) Y
Michael Smith	64	Mike, Jr.	64
Matthew Johnson	66	Matt, Jr.	66
Christopher Williams	68	Chris, Jr.	68
Brian Brown	70	Brian, Jr.	70
David Jones	72	Dave, Jr.	72
Adam Miller	74	Adam, Jr.	74

*The first names are, in order, the six most common in the United States (Dunkling and Gosling, 1983). Rounding out the top ten are Andrew, Daniel, Jason, and Joshua. The surnames are also the six most common (Smith, 1969). Completing this top ten are Davis, Wilson, Anderson, and Taylor.

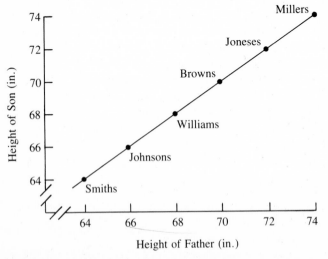

Figure 4.1   A scatterplot with a perfect positive correlation ($r = 1.00$)

Examine **Figure 4.1** carefully. Each point represents a pair of scores—the height of a father and the height of his son. Such an array of points is called a **scatterplot** or scattergram. These scatterplots will be used throughout this chapter. Incidentally, it was when Galton cast his data as a scatterplot graph that the idea of a co-relationship began to become clear to him.

The line that runs through the points in Figure 4.1 (and in Figures 4.2, 4.3, 4.4, and 4.5) is called a **regression line** or "line of best fit." When there is perfect correlation ($r = 1.00$), all points fall exactly on the line. When the points are scattered away from the line as in Figure 4.3, correlation is less than perfect and the correlation coefficient falls between .00 and 1.00. It is from Galton's use of the term *regression* that we get the symbol r for correlation.[3]

In the past generation or so, with changes in nongenetic factors such as nutrition and medical care, sons tend to be somewhat taller than their fathers, except for extremely tall fathers. If every son grew to be exactly 2 inches taller than his father (or 1 inch or 6 inches, or 5 inches shorter), the correlation would still be perfect, and the coefficient would again be 1.00. **Figure 4.2** demonstrates this point. It is not necessary that the numbers on the two variables be exactly the same in order to have perfect correlation. The only requirement is that the differences between pairs of scores all be the same. Another way of stating this requirement is to say that the

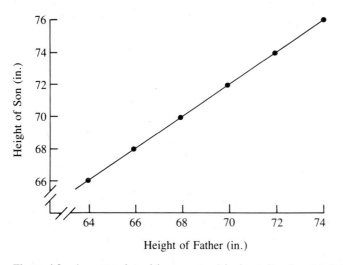

Figure 4.2   A scatterplot with every son 2 inches taller than his father ($r = 1.00$)

---

[3] The term *regression* can lead to confusion because it has two separate meanings. As you already know, it is a statistical method that allows you to draw the line of best fit and to make predictions when you are working with bivariate data. Regression also refers to a phenomenon that occurs when a group is tested a second time. Suppose that, from a large sample of scores, you select a subsample of extreme scores for retesting (those who did either very well or very poorly the first time). On the retest the mean of the subsample tends to be closer to the larger group's mean than it was on the first test. Galton was the first to notice this effect. He found that the mean height of sons of extremely tall men was shorter than the mean height of their fathers and also that the mean height of the sons of extremely short men was greater than the mean height of their fathers. In both cases the sons' mean is closer to the population mean. Since the mean of an extreme group, when measured a second time, tended to regress toward the population mean, Galton named this phenomenon *regression*.

relationship must be such that all points in a scatterplot will lie on the regression line. If this requirement is met, correlation will be perfect and an exact prediction can be made.

Of course, people cannot predict their sons' heights precisely. The points do not all fall on the regression line; some miss it by far. However, as Galton found so long ago, there is some positive relationship; the correlation coefficient is about .50. The points do tend to cluster around the regression line.

By this time in your academic career you have taken an untold number of aptitude and achievement tests. For several of these tests, separate scores have been computed for verbal aptitude and mathematics aptitude. You probably have an answer for the question, What is the relationship between verbal aptitude and math aptitude? That is, if people are good in one, are they good in the other, are they poor in the other, or is there no relationship? Of course, if there is a relationship (either positive or negative), you can predict a math score if you know a person's verbal score. (This example will be used in several places in this chapter.)

As you may have suspected, the next graph shows a scatterplot of data that will begin to answer the questions we have posed. **Figure 4.3** shows the scores of eight high-school seniors who took the Scholastic Aptitude Test (SAT). The SAT-verbal scores are on the X axis, and the SAT-math scores are on the Y axis. As you can see in Figure 4.3, there is a positive relationship, though not a perfect one. Verbal and mathematics aptitude scores tend to vary together; if the score on one is high, the other tends to be high, and if one is low, the other tends to be low. Since there is a relationship, you can predict students' math scores if you know their verbal scores. Later in this chapter you'll learn to calculate the precise *degree* of relationship and to use a formula for making precise *predictions*.[4]

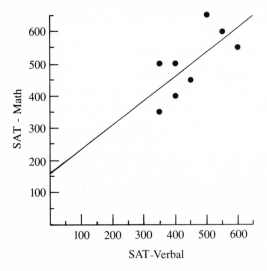

Figure 4.3   Scatterplot and regression line for SAT-verbal and SAT-math scores for eight high-school students ($r = .69$)

[4] Examining Figure 4.3, you might complain that the graph is oddly shaped; it violates the 60 to 75 percent rule for the Y axis and all the data points are stuck up in one corner. We acknowledge your complaints, but we had a dilemma, and later in the chapter we'll expain why we drew such an ungainly graph.

Here is another bivariate distribution for you to think about. For a particular wheat field over a period of several years, what is the relationship between yield and rainfall? Do you suppose that it is positive like that of verbal and math aptitude scores, that there is no relationship, or that the relationship is negative?

NEGATIVE CORRELATION

As you have probably figured out from the heading, the answer to the preceding question is, Negative. **Figure 4.4** is a scatterplot of yield and rainfall for a wheat field over a 35-year period. These data came from the first edition (1925) of a book by R. A. Fisher, *Statistical Methods for Research Workers*. This book became a bible for research workers in many fields, going into a 14th edition in 1973.

As you can see in Figure 4.4, the relationship between rainfall and wheat yield is that as rainfall goes up, yield goes down. Any relationship in which one variable goes down as the other goes up will produce a negative correlation coefficient. When the correlation is negative, the regression line goes from the upper left corner of the graph to the lower right corner. As you may recall from algebra, such lines have a negative slope.

Some other examples of negative correlation are

1. driving speed and miles per gallon,
2. daily rain and daily sunshine, and
3. grouchiness and friendships.

As was the case with perfect positive correlation, there is such a thing as perfect negative correlation. In cases of perfect negative correlation, all the data points of the scatterplot also fall on the regression line.

You will recall that the purpose of a correlation coefficient is to measure the strength of the relationship between two variables. The two coefficients .69 and

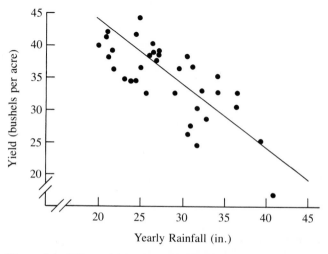

Figure 4.4 Wheat yield and rainfall for 35 years

−.69 indicate the same degree of relationship. Thus, with correlation coefficients, positive is *not* better than negative. The larger the *absolute size* of the number, the stronger the relationship. The algebraic sign (+ or −) indicates simply the direction of the relationship.

ZERO CORRELATION

A *zero correlation* means that there is no relationship between the two variables. High and low scores on the two variables are not associated in any predictable manner.

There is almost no relationship between church membership and homicide in the 50 American states. (The correlation coefficient is −.04.) **Figure 4.5** shows a scatterplot that produces a zero correlation coefficient.

Although a regression line can be drawn when *r* = 0, it is not useful for making predictions. In the case of zero correlation, the best prediction from any *X* score is the mean of the *Y* scores. The regression line, then, runs parallel to the *X* axis at a height of $\bar{Y}$ on the *Y* axis.

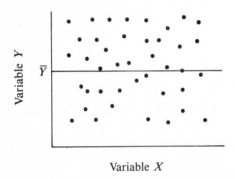

Figure 4.5   Scatterplot demonstrating zero correlation

**CLUE TO
THE FUTURE**

Correlation will come up again in future chapters. If there is correlation between two sets of measurements, and you want to compare the *mean* of one set with the mean of the other set, you can make a more accurate comparison if you take the correlation into account (see Chapters 8 and 12).

PROBLEMS

1. What are the characteristics of a bivariate distribution?
2. What is meant by the statement "Variable *X* and variable *Y* are correlated"?
3. Describe the difference between positive and negative correlation.
4. Can the following variables be correlated and, if so, would you expect the correlation to be positive or negative?
   a. Heights and weights of a group of adults

b. Height of oak trees and height of pine trees

c. Average daily temperature and the cost of heating a home

d. IQ and reading comprehension

e. Scores made by two sections of a math course on the same quiz

THE CORRELATION COEFFICIENT

A correlation coefficient provides a quantitative way to express the *degree* of relationship that exists between two variables. In terms of a definitional formula:

$$r = \frac{\Sigma(z_X z_Y)}{N}$$

where r = Pearson product-moment correlation coefficient

z_X = a z score for variable X

z_Y = a corresponding z score for variable Y

N = number of pairs of X and Y values

You can think through the formula and discover what happens when high scores on one variable are paired with high scores on the other variable (positive correlation). The large positive z scores are multiplied together, and the large negative z scores are multiplied together. Both of these products are positive and, when added together, make a large positive numerator. The result is a large positive value of r. Think through for yourself what happens in the formula when there is a negative correlation or a zero correlation.

This definitional formula using z scores is cumbersome if you want to calculate a value for r. Other formulas have been developed that require less computation (a boon for all) so our examples will use those.

Computational Formulas

We will illustrate two formulas for r, a blanched (partially cooked) formula and a raw-score formula. The blanched formula requires you to calculate means and standard deviations from the summary data before you calculate r. It is useful if you are going to do a regression analysis (draw a regression line and predict Y scores) because one formula for the regression line requires means and standard deviations. The blanched formula also gives you a better feel for the data (because you know means and standard deviations) and is less likely to overflow a calculator that has limited capacity. The raw-score formula, on the other hand, does not require any intermediate steps between the summary data and r. If all you want is r and you have a calculator with large capacity, the raw-score formula is the quicker method.

Blanched Formula.   This procedure requires you to "cook out" the means and standard deviations of both X and Y before computing r. Researchers often use means and standard deviations when telling the story of the data, so this formula is used by many:

$$r = \frac{\dfrac{\Sigma XY}{N} - (\bar{X})(\bar{Y})}{(S_X)(S_Y)}$$

where X and Y are paired observations

   XY = product of each X value multiplied by its paired Y value

   $\bar{X}$ = mean of variable X

   $\bar{Y}$ = mean of variable Y

   S_X = standard deviation of variable X

   S_Y = standard deviation of variable Y

   N = number of pairs of observations

The only really new term in this formula is ΣXY. To obtain this value, it is necessary to multiply each of the X values by its paired Y value and then sum those products. Do *not* try to obtain ΣXY by multiplying $(\Sigma X)(\Sigma Y)$. *It won't work*. The following table demonstrates this fact. Work through it.

X	Y	XY
2	5	10
3	4	12
$\Sigma = 5$	$\Sigma = 9$	$\Sigma = 22$

$\Sigma XY = 22$

$(\Sigma X)(\Sigma Y) = (5)(9) = 45$

The term ΣXY is called "the sum of the cross-products." Some form of the sum of the cross-products is necessary for computing the Pearson r. One other caution about formulas for a correlation coefficient: remember that N is the number of *pairs* of observations.

Table 4.3 demonstrates the steps you use to compute r by the blanched procedure. The data are the ones we used to draw Figure 4.3, the scatterplot of the SAT-verbal and SAT-math scores for the eight students. We made the numbers up so that the correlation coefficient that results will be equal to the "real" coefficient (Wallace, 1972). Work through the numbers in **Table 4.3,** paying careful attention to the calculation of ΣXY.

Raw-Score Formula.   With the raw-score formula, you start with the raw scores and calculate r directly without computing means and standard deviations. The formula is

$$r = \frac{N\Sigma XY - (\Sigma X)(\Sigma Y)}{\sqrt{[N\Sigma X^2 - (\Sigma X)^2][N\Sigma Y^2 - (\Sigma Y)^2]}}$$

You have already learned what all the terms of this formula mean. Remember that N is the number of *pairs* of values.

All calculators with two or more memory storage registers—and some with one—permit the simultaneous accumulation of values for ΣX and ΣX^2. Next, ΣY and ΣY^2 may be computed simultaneously. This leaves only ΣXY to be computed.

With the raw-score formula you may run into overflow problems if your calculator has a limited capacity. One solution is to change to the blanched formula. A second solution is to reduce the size of the X and Y values by dividing each by a constant (or subtracting a constant from each). Reducing the size of the numbers

Table 4.3 Calculation of *r* for SAT-verbal and SAT-math aptitude scores by the blanched formula

Student	SAT-verbal X	SAT-math Y	X^2	Y^2	XY
1	350	500	122,500	250,000	175,000
2	550	600	302,500	360,000	330,000
3	400	400	160,000	160,000	160,000
4	350	350	122,500	122,500	122,500
5	500	650	250,000	422,500	325,000
6	600	550	360,000	302,500	330,000
7	450	450	202,500	202,500	202,500
8	400	500	160,000	250,000	200,000
	3600	4000	1,680,000	2,070,000	1,845,000

$$\bar{X} = \frac{\Sigma X}{N} = \frac{3600}{8} = 450 \qquad \bar{Y} = \frac{\Sigma Y}{N} = \frac{4000}{8} = 500$$

$$S_X = \sqrt{\frac{\Sigma X^2 - \frac{(\Sigma X)^2}{N}}{N}} = \sqrt{\frac{1,680,000 - \frac{(3600)^2}{8}}{8}} = 86.60$$

$$S_Y = \sqrt{\frac{\Sigma Y^2 - \frac{(\Sigma Y)^2}{N}}{N}} = \sqrt{\frac{2,070,000 - \frac{(4000)^2}{8}}{8}} = 93.54$$

$$r = \frac{\frac{\Sigma XY}{N} - (\bar{X})(\bar{Y})}{S_X S_Y} = \frac{\frac{1,845,000}{8} - (450)(500)}{(86.60)(93.54)} = .69$$

does not affect the value you get for *r*, but it does affect the mean and standard deviation.

Some calculators have a function for *r* built in; by entering *X* and *Y* values and pressing the *r* key, the coefficient is displayed. If you have such a calculator, we recommend that you use this labor-saving device after you have used the computation formulas a number of times. Working directly with terms like ΣXY leads to an understanding of what goes into *r*.

Table 4.4 illustrates the use of the raw-score procedure for computing *r*. The data are the same as those used in Table 4.3, demonstrating that the value of *r* is the same for both methods.

The final step in any statistics problem is interpretation. What is the story that goes with a correlation coefficient of .69 between SAT-verbal scores and SAT-math scores? The story is that there is a substantial correlation between verbal aptitude and math aptitude. Students with high aptitude in one *tend* to have a high aptitude in the other, although there are exceptions. If the correlation had been near zero, we could say that the two abilities are independent or unrelated. If the coefficient had been strong and negative, we could say, Good in one, poor in the other.

You may have noticed that we are describing *r* as a sample statistic; there are no σ_X, σ_Y, μ_X, or μ_Y symbols in this chapter. Of course, if a population of data is available that can be correlated, the parameter may be computed. The procedure is the same as for sample data. Most recent statistical texts use ρ (the Greek letter rho) as the symbol for this parameter, although you must be cautious in your outside

Table 4.4   Calculation of r for SAT-verbal and SAT-math aptitude scores by the raw-score formula using data from Table 4.3

$$\Sigma X = 3600 \qquad \Sigma Y = 4000 \qquad \Sigma X^2 = 1,680,000 \qquad \Sigma Y^2 = 2,070,000 \qquad \Sigma XY = 1,845,000$$

$$r = \frac{N\Sigma XY - (\Sigma X)(\Sigma Y)}{\sqrt{[N\Sigma X^2 - (\Sigma X)^2][N\Sigma Y^2 - (\Sigma Y)^2]}}$$

$$= \frac{(8)(1,845,000) - (3600)(4000)}{\sqrt{[(8)(1,680,000) - (3600)^2][(8)(2,070,000) - (4000)^2]}}$$

$$= \frac{360,000}{518,459} = .69$$

reading; ρ has also been used to symbolize Spearman's correlation coefficient, which you will learn about in Chapter 12.

Correlation coefficients should be based on an "adequate" number of pairs of observations. As a general rule of thumb, "adequate" means 30 or more. We have used an example with fewer than 30 and we will ask you to work problems with fewer than 30 pairs. This is a textbook device that allows you to spend your time on interpretation and understanding rather than "number crunching." In Chapter 8 you will learn the reasoning behind our admonition that N be adequate.

ERROR DETECTION

> The Pearson correlation coefficient ranges between -1.00 and 1.00. Values smaller than -1.00 or larger than 1.00 are not possible.

Now it is time for you to try your hand at computing r. We suggest that you compute it using both the raw-score formula and the blanched formula. You should get the same answer both ways. We will use small numbers so that you can work with the raw-score formula even if your calculator has limited capacity.

PROBLEMS

5. This problem is based on data published in 1903 by Karl Pearson and Alice Lee. In the original article, 1376 pairs of father–daughter heights were analyzed. The scores here produce the same means and the same correlation coefficient that Pearson and Lee obtained. For these data, draw a scatterplot and calculate r.

Father's Height, X (in.)	69	68	67	65	63	73
Daughter's Height, Y (in.)	62	65	64	63	58	63

6. This problem is based on actual data also. By now you know how to obtain the summary values for ΣX, ΣX^2, ΣY, ΣY^2, and ΣXY, so we are providing them for you. The raw data are scores on two different tests that measure self-esteem. Subjects were seventh-grade students. Since the two tests were designed to measure the same trait, they should be positively correlated. Summary values are: $N = 38$, $\Sigma X = 1755$, $\Sigma Y = 1140$, $\Sigma X^2 = 87,373$, $\Sigma Y^2 = 37,592$, $\Sigma XY = 55,300$. Compute r.

7. The X variable is the population of each of the states in the United States in millions. The Y variable is expenditure per pupil in public schools for each of the states. Calculate r

using either method. $N = 50$, $\Sigma X = 202$, $\Sigma Y = 41{,}048$, $\Sigma X^2 = 1740$, $\Sigma Y^2 = 35{,}451{,}830$, $\Sigma XY = 175{,}711$.

SCATTERPLOTS

You already have an introductory knowledge of scatterplots—what their elements are and what they look like when $r = 1.00$, $r = .00$, and $r = -1.00$. In this section, we will illustrate some intermediate cases and reiterate our philosophy about the value of pictures.

Figure 4.6 shows scatterplots of data that have correlation coefficients of .20, .40, .60, and .80. If you draw an envelope around the points in a scatterplot, the picture that the data present becomes clearer. Obviously, the thinner the envelope, the higher the correlation. To say this in more mathematical language, the closer the points are clustered around the regression line, the higher the correlation.

Pictures help you understand. Scatterplots are easy to construct. Although they require some time, the benefits are worth it. If we had given you the raw data on the populations of states and their school expenditures (see Problem 7) and you had drawn a scatterplot, you would have plotted one point in the upper left-hand corner of the graph (a place where data points contribute to a *negative r*). Very likely you would have raised your eyebrows at the unusual point and wondered what state it was.[5] So this is a paragraph that suggests: draw pictures—they're worth it.

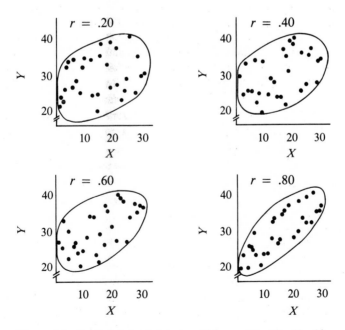

Figure 4.6 Scatterplots of data in which $r = .20$, .40, .60, .80

[5] Alaska

THE USE AND INTERPRETATION OF *r*

The basic and simple interpretation of *r* is probably familiar to you at this point. A correlation coefficient is a measure of the relationship between the two variables of a bivariate distribution. It describes how the two variables vary together (covary); that is, it describes the tendency of high or low values of one variable to be associated regularly with either high or low values of the other variable. The absolute size of the coefficient (from .00 to 1.00) indicates the strength of the tendency to covary.

Fortunately for those eager to understand, correlation coefficients can produce more information than that indicated by the preceding paragraph. However, the interpretation of *r* can be a tricky business; there are several common pitfalls. The next few pages are planned so that you can learn the basics of further interpretation and avoid some of the those pitfalls.

Reliability

One important use of correlation coefficients is to assess the **reliability** of devices that measure things. A device (a test is an example) is reliable if the score you get is not subject to chance variation.

One way to assess chance variation is to measure a number of individuals and then measure them a second time. If the measurement you get the first time has not been influenced by chance, then you would expect to get the same measurement the second time. If the second measurement is *exactly* the same as the first for each individual, it is easy to conclude that the method of measurement is completely reliable, that chance has nothing to do with the score you get. However, if the measurements are not exactly the same, you experience some uncertainty. Fortunately, a correlation coefficient will tell you the degree of agreement between the test and the retest scores. (Thus, you can measure your uncertainty.) High correlations mean lots of agreement and therefore high reliability, and low correlations mean lots of disagreement and therefore low reliability.

For example, in Galton's data, for 435 adults whose height was measured twice, the correlation was .98. Not surprisingly, Galton's method of measuring height was very reliable. The correlation, however, for "highest audible tone," a measure of pitch perception, was only .28 for the 349 who were tested a second time within a year (Johnson et al., 1985). One of two interpretations is possible. Either people's ability to hear high sounds changes up and down during a year, or the test was not reliable. In this case, the test lacked reliability. Two possible explanations are that the test environment was not as quiet from one time to the next and that the instruments they used were not calibrated exactly the same on each test.

Every investigator who measures must be concerned about reliability. If a scale is not reliable, then the next step in the project is to find better ways to measure. And for those of you who are asking, What size coefficient indicates reliability? a rule of thumb is .80 or better for social science measurements.

Correlation Versus Causation

A high correlation coefficient does not tell you that one of the variables is *causing* the variation in the other. Quite possibly some third variable is responsible for the variation in both. For example, in most parts of the United States, there is a positive correlation between the number of birds in deciduous trees and the number of leaves on those trees; however, a change in one of these variables does not *cause* a change in the other variable. It is foolish to think that falling leaves frighten away the birds or that the birds knock off leaves as they fly away. Variation in the number of birds does not cause variation in number of leaves, nor does variation in leaves cause variation in birds. A third variable, seasonal changes, causes variation in both. Thus, *r* can be interpreted in terms of only relationship and prediction, not in terms of cause and effect.

Jumping to a cause-and-effect conclusion is easy to do; it usually seems so reasonable. And, of course, a cause-and-effect relationship may in fact exist. However, a correlation coefficient *alone* cannot establish a causal relationship. For example, the early statements about cigarette smoking causing lung cancer were based on correlational data. Persons with cancer were often heavy smokers and national comparisons indicated a relationship (see Problem 10). However, as careful thinkers (and the cigarette companies) pointed out, both cancer and smoking might have been caused by a third variable; stress and anxiety were often suggested as possible causes. What was required to establish the cause-and-effect relationship was data from controlled experiments, not correlational data. The experimental data, complete with control groups, established the cause-and-effect relationship between smoking and lung cancer. (Chapter 7, "Differences Between Means," includes a discussion of experimental data.)

Coefficient of Determination

This entire chapter is about ways to express the relationship between two variables. Except in this section, we express these relationships in terms of scores. The concept that if one score is high the other is low is fairly simple. In a similar way, predicting sales that would result if an advertising budget is increased to $10,000 and predicting the year that there will be 1 million baccalaureate degrees granted in the United States are both rather easy to understand. In this section, you will learn of a more complicated concept.

The concept is **common variance.** You can think of a test score as being the result of many separate elements. Some of these elements are characteristics of the person taking the test, some are characteristics of the situation in which the test is taken, and some are characteristics of the test itself. Now, if a person takes two tests, some elements will have their effect on both scores—that is, cause each score to vary in the same way. This variance in common can be measured, and the name given to this measurement is the **coefficient of determination.** The coefficient of determination is r^2, a very easy statistic to calculate; just square *r*.

A coefficient of determination interpretation of the correlation of .69 between SAT-verbal scores and SAT-math scores is that 48 percent ($.69^2 = .48$) of the variance of each test is related to the same factors.[6] Of the several factors common to the two tests, the principal one was probably the general intellectual ability of the test takers. Other factors such as distractions in the room, pleasantness of the

---

[6]If you have some skill at algebra and some curiosity about this, you can find proofs in Minium (1978, p. 209) or Edwards (1984, p. 32).

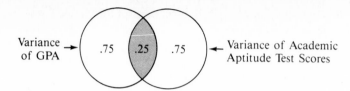

Variance of GPA → .75 | .25 | .75 ← Variance of Academic Aptitude Test Scores

Figure 4.7 Illustration of the common variance in academic aptitude test scores and first-semester freshman grade point averages

teacher, and health of the students may have affected both tests and thus be part of the common variance.

On the other hand, 52 percent of the variance was not held in common. Fifty-two percent of the variance of each test was specific to that test, not predictable from the other test. This *independent* variance presumably was due to skills that are specific to verbal tasks and to math tasks.

Note what happens to a fairly strong correlation of .69 when it is interpreted in terms of variance. To the novice, such a correlation may seem to mean that the two tests are measuring pretty much the same things, but according to a variance intepretation, they have only about half of their variance in common. To predict 69 percent of the variance requires a correlation of .83, a rather high correlation in much research.

Here is another example to help you conceptualize the notion of variance held in common by two variables. First-semester college grade point averages usually correlate with academic aptitude test scores, with an *r* equal to about .50. Squaring .50 gives a coefficient of determination of .25. **Figure 4.7** presents a picture of this concept. The shaded area represents the variance held in common by both variables and is related to the same factors. Intelligence and previous learning influence both variables and thus enter into the common variance. The rest of the variance (75 percent of both variables) is independent. As examples, part of the variance of GPA may be due to such factors as illness, falling in love (or is that just another illness?), developing a taste for beer, or financial problems. Academic aptitude tests are never likely to predict such factors.

The coefficient of determination is also useful in comparing correlation coefficients. When one compares an *r* of .80 with an *r* of .40, the tendency is to think of the .80 as being twice as high as .40, but that is not the case. Correlation coefficients are compared in terms of the amount of common variance, as shown:

$$.80^2 = .64$$
$$.40^2 = .16$$
$$\frac{.64}{.16} = 4$$

Thus, two variables that are correlated with $r = .80$ have four times as much common variance as two variables correlated with $r = .40$.

PROBLEMS   **8.** You found a correlation of .57 between two measures of self-esteem. What is the coefficient of determination, and what does it mean?

9. What percent of variance in common do two variables have if their correlation is (a) .40? (b) .10?

10. Eleven countries are listed in the table. To the right of each is the cigarette consumption per capita in 1930 and the male death rate from lung cancer 20 years later in 1950 (Doll, 1955; reprinted in Tufte, 1983). Calculate a Pearson *r* and write a statement telling what the data show.

Country	Per capita cigarette consumption	Male death rate (per million)
Iceland	217	59
Norway	250	91
Sweden	308	113
Denmark	370	167
Australia	455	172
Holland	458	243
Canada	505	150
Switzerland	542	250
Finland	1112	352
Great Britain	1147	467
United States	1283	191

11. Interpret the following statements:
  a. *r* = .72 was found between scores on a test of intolerance of ambiguity and scores on a test of authoritarianism.
  b. The correlation between vocational-interest scores at age 20 and at age 40 for the same subjects was found to be .70.
  c. The correlation between intelligence test scores of identical twins is in the high .90s.
  d. The correlation between IQ and family size is about −.30.
  e. *r* = .22 between height and IQ for 20-year-old men.
  f. *r* = −.83 between income level and probability of psychosis.

STRONG RELATIONSHIPS BUT LOW CORRELATIONS

One good thing about understanding something is that you come to know what's going on beneath the surface. Knowing the inner workings, you can judge whether the surface appearance is to be trusted or not. You are about to learn of two "inner workings" of correlation. These will help you evaluate the meaning of low correlations. Low correlations do not always mean that there is no relationship between two variables.[7]

Nonlinearity

For *r* to be a meaningful statistic, the best-fitting line through the scatterplot of points must be a *straight line*. If a curved regression line fits the data better than a straight line, *r* will be low, not reflecting the true degree of relationship between the two variables.

[7] Correlations that do not reflect the true degree of relationship are said to be *spuriously* low or high.

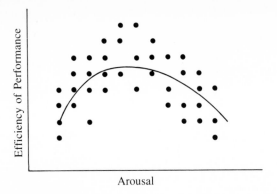

Figure 4.8   Generalized relationship between arousal and efficiency of performance

Figure 4.8 is an example of a situation in which *r* is inappropriate because the best-fitting line is curved. The *X* variable is arousal and the *Y* variable is efficiency of performance. At low levels of arousal (sleepy, for example), performance is not very efficient. Likewise at very high levels of arousal (agitation, for example), people don't perform efficiently. However, in the middle range there is a degree of arousal that is optimum; performance is most efficient at moderate levels of arousal.

There is obviously a strong relationship between arousal and efficiency of performance, but *r* for these two variables would be low, not reflecting the strong relationship. The product-moment correlation coefficient is just not useful as a measure of the strength of curved relationships. Special nonlinear correlation techniques for such relationships do exist and are described in texts such as Howell (1987) and Guilford and Fruchter (1978).

Truncated Range

Besides nonlinearity, there is a second situation that can give you a low Pearson coefficient even though there is a strong relationship between the two variables. Spuriously low *r* values can occur when the sample range is much smaller than the population range (a **truncated range**).

Suppose you wanted to know the relationship between IQ and self-esteem. To get data you used students from an Introduction to Psychology class and got the picture shown in **Figure 4.9.** These data look like a snowstorm; the conclusion would be that there is no relationship between IQ and self-esteem.

However, college students have IQ scores higher than those of the general population, so the study does not include those with lower IQs. What effect does this restriction of the range have? You can get an answer to this question by looking at **Figure 4.10,** which shows a hypothetical scatterplot of scores for the *population,* with the scores of the psychology students again shown as open circles. This scatterplot shows a moderate relationship in the population. So, unless you recognized that your sample of college students truncated the population range, you would not find the relationship that exists.

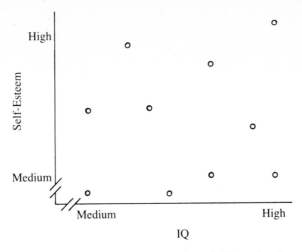

Figure 4.9 Hypothetical scatterplot of IQ and self-esteem scores for Introduction to Psychology students

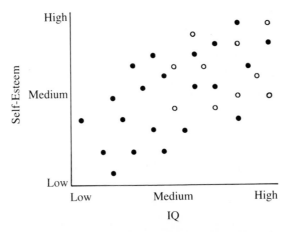

Figure 4.10   Hypothetical scatterplot of IQ and self-esteem scores for the population

OTHER KINDS OF CORRELATION COEFFICIENTS

Correlation is a widely used statistical technique. You have learned about one of the several kinds of linear correlation—the Pearson product-moment coefficient. Here is a list of some other kinds of data for which a descriptive index number can be calculated. These numbers are also correlation coefficients, but they are not Pearson product-moment coefficients.

1. Correlations may be computed on data for which one or both of the variables are **dichotomous** (having only two possible values). An example is the correlation of the dichotomous variable gender and the quantitative variable grade point average.

2. Several variables can be combined, and the resulting combination can be correlated with one variable. With this technique, called **multiple correlation,** a more precise prediction can be made. Performance in school usually can be predicted better by using several measures of a person rather than just one.

3. A technique called **partial correlation** allows you to separate or partial out the effects of one variable from the correlation of two variables. For example, if you want to know the true correlation between achievement test scores in two school subjects, it will probably be necessary to partial out the effects of intelligence, since IQ and achievement are correlated.

4. Chapter 12 will teach you how to compute a statistic called Spearman's r_s, which is used when the data are ranks rather than raw scores.

5. If the relationship between two variables is curved rather than linear, the correlation ratio *eta* (η) gives the degree of association.

These and other correlational techniques are covered in intermediate-level textbooks. (See Christensen and Stoup, 1986; Downie and Heath, 1983; and Edwards, 1984.)

PROBLEMS

12. The correlation between number of older brothers and sisters and degree of acceptance of personal responsibility for one's own successes and failures is $-.37$.
   a. How would you interpret this correlation?
   b. What can you say about the cause of this correlation?
   c. What practical significance might it have?

13. A correlation of .97 has been found between anxiety and neuroticism.
   a. Interpret the meaning of this *r*.
   b. Are you justified in assuming that being neurotic is responsible in large part for manifestations of anxiety?

14. Examine the following data, make a scatterplot, and compute *r* if appropriate.

Serial Position	1	2	3	4	5	6	7	8
Errors	2	5	6	9	13	10	6	4

CORRELATION AND REGRESSION

You are now prepared to learn how to make predictions. The technique you will use is called *linear regression*. As you will see, what you have learned about correlation (the degree of relationship between variables) will be most helpful when you interpret the results of a regression analysis (a formula that allows predictions).

A few paragraphs ago we said that the correlation between college entrance examination scores and first-semester grade point averages is about .50. Thus, you can predict that those who score high on the entrance examination are more likely to succeed as freshmen than those who score low. All this is rather general, though. Usually you want to predict a *specific* grade point for a *specific* applicant. For example, if you were in charge of admissions at Collegiate U, you would want to know the entrance examination score that predicts a GPA of 2.00, the minimum required for graduation. To make specific predictions, you must do a regression analysis.[8]

THE REGRESSION EQUATION—A LINE OF BEST FIT

You are used to making predictions; some you make with a great deal of confidence. "If I get my average up to 80 percent, I'll get a B in this course." "If I spend $15 plus tax on this compact disc, I won't have enough left from my $20 to go to a movie."

Often predictions are based on an assumption that the relationship between two variables is linear, that a straight line will tell the story exactly. Frequently, this assumption is quite justified. For the short-term economics problem above, imagine a set of axes with "amount spent" on the X axis ($0 to $20) and "amount left" on the Y axis ($0 to $20). A straight line that connects the two $20 marks on the axes tells the whole story. (This is a line with a negative slope that is inclined 45 degrees from horizontal.) Draw this picture.

Part of your education in algebra was about straight lines. You may recall that the slope-intercept formula for a straight line is

$$Y = mX + b$$

where Y and X are variables representing scores on the Y and X axes
   m = slope of the line (a constant)
   b = intercept of the line with the Y axis (a constant)

For a formula like $Y = 3X + 6$, you can easily find the value of Y if you are given a value of X. If X is 4, Y is 18. Thus, if you have the specific formula for a line, you can make predictions about any Y for any given value of X.

A more difficult question is how to go from the general formula $Y = mX + b$ to a specific formula like $Y = 3X + 6$—that is, how to find the values for m and b. A common solution involves a rule and a little algebra. The rule is that if a point lies on the line, the point satisfies the equation of the line. Thus, the point where $X = 5$ and $Y = 8$, represented as $(5, 8)$, produces $8 = 5m + b$ when substituted into the general equation. If you are given a second point that lies on the line, you will get another equation with m and b as unknowns. Now you have two equations with two unknowns and you can solve for each in turn, giving you values for m and b. If you would like to check your understanding of this on a simple problem, you might figure out the formula for the line that tells the story of the $20 problem. For simplicity, use the two points where the line crosses the axes, $(0, 20)$ and $(20, 0)$.

There are many situations in which a practical prediction about Y is needed. For example, suppose a truck's springs are half compressed by its load (say, 10,000 pounds of pulp wood that is to be used to make paper for statistics books). How much more can the truck hold?

If you added 2000 pounds and then 2000 pounds more and found compression percentages of 60 percent and 70 percent, you would have enough information to

---

[8] For an entire chapter on prediction using regression, see Ferguson (1981, chap. 9).

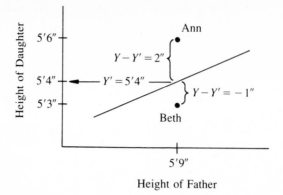

Figure 4.11 Hypothetical case of two daughters with a father 5′9″ tall

leap to the conclusion that there is a linear relationship between the compression of a spring and its load.[9] And sure enough, if you predict that the truck can carry 20,000 pounds (enough for a first year's supply—6000 copies—of this book), you will be correct.

Once you have assumed or have satisfied yourself that the relationship is linear, *any* two points will allow you to draw the line that tells the story of the relationship between spring compression and load.

With this background in place, you are in a position to appreciate the problem Karl Pearson faced at the end of the 19th century when he looked at father–daughter height data like those in Problem 5. Look at your scatterplot of those data (or at Figure 4.3 or 4.4). The assumption that the relationship is linear is reasonable, but you quickly realize that the two-point solution will not work. The equation would depend on which two particular points you chose to plug into the formula. The two-point solution produces dozens of different lines. The question is how to find the best line. Do you have a suggestion?

One idea is to find the mean Y value for each X value, connect those means with straight lines, and then choose two points that make the line fall within this more narrow range of the means. Such a line could be thought of as a "moving average" and might be thought of as a best line. For every X value there would be a predicted Y value (symbolized by Y' and pronounced "Y prime"). The Y values (your observations) would vary around the Y' values (your predictions).

Pearson's solution, which statistics has embraced, is to use the **least squares** method, which is a mathematically sophisticated version of the thinking in the previous paragraph. Look at **Figure 4.11,** which shows two of the daughters from Pearson and Lee's 1903 data, both of whom had fathers who were 5′9″ tall. The regression line calculated by the least squares method has been drawn, and as you can see, the height predicted for a daughter of a father 5′9″ is 5′4″. Although both fathers were 5′9″ tall, Ann was 5′6″ and Beth was 5′3″. Thus, the prediction was in error 2 inches for Ann and 1 inch for Beth. You can imagine that if the other

---

[9] This relationship is known as Hooke's law of elasticity, which applies to all solid bodies like metal and wood. Robert Hooke articulated this law in 1678, some years after he observed and named as "cells" the fundamental stuff studied by biologists.

daughters' heights were on the graph, most of them would not fall on the regression line. With all this error, why use the least squares method for drawing the line?

The answer is that a regression line drawn by the method of least squares *minimizes* error in a particular way. To explain, you can calculate an error for each person using the formula

$$\text{error} = Y - Y'$$

Using inches for the two daughters in our example, you get

Ann: error $= Y - Y' = 66 - 64 = 2$
Beth: error $= Y - Y' = 63 - 64 = -1$

In a similar way there is an error for each person on the scatterplot (though for some the error would be zero). The least squares method places a straight line such that the *sum of the squares* of the errors is a minimum. In symbol form, $\Sigma(Y - Y')^2$ is a minimum for a straight line calculated by the least squares method.

As a result of applying the least squares method to a set of bivariate data, you get *the slope and the intercept* of a particular straight line. With a slope and an intercept, you can write the equation for a line and this will be a line that *best* fits the data.

To summarize, the problem was to find a way to express mathematically what is clear in the scatterplot of the Pearson and Lee data (and in Figure 4.3), which is that the scores vary together. By assuming that a straight line will be adequate, a slope-intercept formula results. To actually write a formula for a specific bivariate distribution, use the least squares method to find the slope and the intercept (because a two-point solution will not work).

One more transitional point is necessary. In the language of algebra, the idea of a straight line is expressed as $Y = mX + b$. In the language of statistics, exactly the same idea is expressed as $Y = a + bX$. Y and X are used the same way in the two formulas, but different letters are used for the slope and the intercept. Unfortunately, the terminology is well established in both fields. Fortunately, the translation is easy and doesn't cause many problems. Thus, in statistics the letter b stands for the slope of the line and a is the intercept of the line with the Y axis.

Now the **regression equation** is

$$Y' = a + bX$$

where $Y' = Y$ value predicted from a particular X value
   $a =$ point at which the regression line intersects the Y axis
   $b =$ slope of the regression line
   $X = X$ value for which you wish to predict a Y value

In a correlation problem, the symbols X and Y can be assigned arbitrarily, but in a regression equation, Y is assigned to the variable you wish to predict.

The Regression Coefficients

To make predictions of Y using the regression equation, you need to calculate the values of the constants a and b, which are called **regression coefficients.**

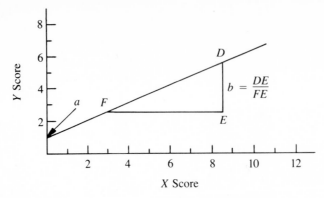

Figure 4.12   The regression coefficients, a and b

If you have already computed r and the standard deviations for both X and Y,[10] b can be obtained very simply by the formula

$$b = r\frac{S_Y}{S_X}$$

where r = correlation coefficient for X and Y
S_Y = standard deviation of the Y variable
S_X = standard deviation of the X variable
Notice that for positive correlation, b will be a positive number. For negative correlation, b will be negative.

You can compute a from the formula

$$a = \bar{Y} - b\bar{X}$$

where $\bar{Y}$ = mean of the Y scores
b = regression coefficient computed previously
$\bar{X}$ = mean of the X scores

Figure 4.12 illustrates the meaning of the regression coefficients. The regression line in Figure 4.12 is the line that goes up at an angle. The point marked a in Figure 4.12 is the point at which the regression line crosses the Y axis. It is called the "Y intercept." For these data, $a = 1.00$. The b stands for the slope of the regression line. Regression slopes for positive correlations range from 0 to $+\infty$; for negative correlations, they range from 0 to $-\infty$. (A line horizontal to the X axis has a slope of 0; a vertical line has a slope that approaches $+\infty$.) The slope of a line is measured by the vertical distance the line rises divided by the horizontal distance the line covers. If you measure the line DE (vertical rise of the line FD) in Figure 4.12, you will find it is half the length of FE (the horizontal distance of line FD). Thus, the slope of the regression line (DE/FE) is 0.50 ($b = 0.50$). Put another way, the value of Y increases

[10] If you have not computed r, S_X, and S_Y, b can be obtained by the formula
$$b = \frac{N\Sigma XY - (\Sigma X)(\Sigma Y)}{N\Sigma X^2 - (\Sigma X)^2}$$

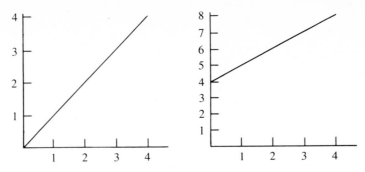

Figure 4.13 Two regression lines with the same slope ($b = 1.00$) but with different appearances. The difference is caused by the units chosen for the Y axis.

one-half point for every one-point increase in X. Expressed in a third way, the regression line is "going out" horizontally twice as fast as it is "going up" vertically.

The Appearance of the Regression Line

A word of caution is in order about appearances of graphs. The *appearance* of the slope of a line drawn on a scatterplot or other graph depends heavily on the units chosen for the X and Y axes. Look at **Figure 4.13.** Although the two lines appear different, $b = 1.00$ for both. They appear different because the space allotted to each Y unit on the right is half that allotted to each Y unit on the left.

We can now return to the dilemma we acknowledged earlier in this chapter (see footnote 4 of the section "Positive Correlation"), where we described Figure 4.3 as ungainly. It was ungainly because we drew it with equal units on the X and Y axes, and we did that so the regression line would cross the Y axis at a and so the *appearance* of the slope of the line would be the same as b (in case you used our illustration method in Figure 4.12 to check out our earlier Figure 4.3).

For a sophisticated, educated audience (the group you now belong to), we can draw an attractive, well-proportioned graph like **Figure 4.14,** knowing that you won't be confused because the regression line crosses the Y axis at 400 (although $a = 162.5$) and the slope appears to be 0.24 (when $b = 0.75$).

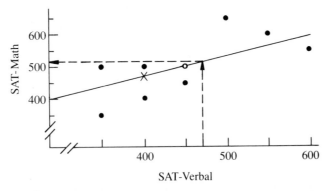

Figure 4.14 Scatterplot and regression line for the verbal and math aptitude data (Table 4.3)

Writing a Regression Equation and Drawing a Regression Line

In this section, you will learn the specifics of writing the regression equation and drawing the regression line for a set of data. For the SAT data on verbal aptitude (X) and math aptitude (Y), the summary data were

$$r = .69 \qquad S_X = 86.60 \qquad S_Y = 93.54 \qquad \bar{X} = 450 \qquad \bar{Y} = 500$$

Using the formula

$$b = r\frac{S_Y}{S_X}$$

you get

$$b = (.69)\frac{93.54}{86.60} = (.69)(1.08) = .75$$

Using the formula

$$a = \bar{Y} - b\bar{X}$$

you get

$$a = 500 - (.75)(450) = 500 - 337.50 = 162.50$$

The b coefficient tells you that when verbal aptitude increases by one point, math aptitude increases by three-fourths of a point. The a coefficient tells you that the regression line will intersect the Y axis at a value of 162.50. Thus, the formula for the regression line that relates math achievement and verbal achievement is

$$Y' = 162.50 + .75X$$

Now all you need to draw the regression line is a straightedge and two points that fall on the line. Any two points will do. One point that is always on the regression line is $\bar{X}, \bar{Y}$. Thus, the point $(450, 500)$ serves as one point and is marked on Figure 4.14 with a circle.

A second point that is always on the regression line is $X = 0$, $Y = a$. Unfortunately for us, our well-proportioned graph does not include this point (though if you were to draw a regression line on Figure 4.3, you could easily find it).

A second point can always be found by choosing an X value and substituting it into the previous formula. By choosing X as 400 the arithmetic can be done at a glance.

$$Y' = 162.50 + .75X = 162.50 + 300 = 462.50$$

Our second point is $(400, 462.50)$, which is marked on Figure 4.14 with an X.

Predicting a Y Score—Finding Y

If you are given a particular X score, what Y score should be predicted? If you have a regression line drawn accurately on a scatterplot, you may use it for quick predictions. Start with the given X score and draw a vertical line up to the regression line. Then draw a horizontal line to the vertical axis. The Y score there is the predicted Y for the X score you began with. This is demonstrated by the broken lines in Figure 4.14 for a verbal aptitude score of 470. The projection to the Y axis shows a predicted math aptitude score of just over 500.

A more precise way to find Y' is to use the regression equation. To predict a math aptitude score for a verbal aptitude score of 470, the regression equation is

$$Y' = a + bX = 162.50 + (.75)(470)$$
$$= 162.50 + 352.50$$
$$= 515$$

Thus, given an X score of 470, the predicted Y score is 515.

To predict Y' values from your summary data without the intermediate step of calculating a and b, a little algebraic manipulation produces the following formula:

$$Y' = r\frac{S_Y}{S_X}(X - \bar{X}) + \bar{Y}$$

To demonstrate how this formula works, we'll again use the data from Table 4.3 to predict a math aptitude score for a verbal aptitude score of 470:

$$Y' = (.69)(1.08)(20) + 500$$
$$= 14.90 + 500$$
$$= 515$$

This is the same value we got before.

Now you know how to make predictions. However, predictions are cheap; anyone can make them. Respect accrues only when the predictions come true. So far, we have dealt with accuracy by simply pointing out that when r is high, accuracy is high, and when r is low, you cannot put much faith in your predicted values of Y'. The other variable that influences the accuracy of Y predictions is the standard deviation of Y. The smaller the standard deviation of Y, the greater the accuracy of prediction.

We recognize that listing the factors that influence the accuracy of regression analysis predictions does not get you very far toward "true understanding." However, that list is as far as we are going in this introductory book. To go further, to *measure* the accuracy of predictions made from a regression analysis, you need to know about the **standard error of estimate.** This statistic is discussed in most intermediate-level textbooks and in textbooks on testing. After you have covered the material in Chapters 5 and 6 of this book, you will have the background necessary to understand the standard error of estimate.

PROBLEMS

15. In Problem 5, the father–daughter data, you computed r using six pairs of numbers.
   a. Compute the regression coefficients a and b.
   b. Use your scatterplot from Problem 5 and draw the regression line.
16. In Problem 6, you also computed an r.
   a. Compute a and b.
   b. What Y score would you predict for a child who scores 42 on X?

17. Regression is a technique that economists and businesspeople rely on heavily. To illustrate, think about the relationship between advertising expenditures and sales. It would be helpful to make predictions about the effect of increasing (or decreasing) advertising expenditures. For the data in the table, (a) write the regression equation, (b) plot the regression line on the scatterplot, and (c) predict sales that would result if $10,000 were spent on advertising. Should any confidence at all be put in this prediction? (d) Write a sentence that answers this question.

Advertising, X ($ thousands)	Sales, Y ($ thousands)
3	70
4	120
3	110
5	100
6	140
5	120
4	100

18. The correlation between Stanford-Binet IQ scores and Wechsler Adult Intelligence Scale (WAIS) IQs is about .80. Both tests have a mean of 100. The standard deviation of the Stanford-Binet is 16. For the WAIS, $S = 15$. What WAIS IQ would you predict for a person scoring 65 on the Stanford-Binet? (An IQ score of 70 has been used by some schools as a cutoff point between regular classes and special education classes.)

19. From 1977 through 1982, more than 900,000 people a year graduated from college with a baccalaureate degree. Using the following data from the *Statistical Abstract of the United States*, predict the year when 1 million will graduate with a bachelor's degree. Carefully choose which variable to call X and which to call Y.

Year	'77	'78	'79	'80	'81	'82
Graduates (x 100,000)	9.2	9.2	9.2	9.3	9.4	9.5

20. Now it is time for an *integrative* question. We suggest that you plan to devote at least 30 minutes to this one (either now or sometime soon). Write an essay on descriptive statistics. We suggest that you start by jotting down from memory things you'd like to include. Then look back over these four chapters, noting additional facts, considerations, or organizational ideas. Write the essay and revise it. Rest. Revise and write a final answer.

Transition Page

You are now through with the part of the book devoted to descriptive statistics. You should be able to describe a set of data by using a graph and a few descriptive numbers, such as a mean, a standard deviation, and (if appropriate) a correlation coefficient. The remainder of the book is devoted to inferential statistics.

Inferential statistics help you come to a decision about populations. As an example, it is common to want to know whether the means of two populations are different, whether one is "better" than the other. Using the techniques of inferential statistics, samples from each population are drawn, sample means found, an inferential statistic calculated, and a decision reached about the populations (on the basis of the sample data).

To expand on the decision-making part of this process, the decision is often whether the difference between the two sample means is probably due to chance or to some other factor. Inferential statistics help you in this decision by giving you the probability of such a difference if only chance is at work. If the probability is very high, a decision that the difference is due to chance is supported. If the probability is very low, a decision that the difference is due to some other factor is supported.

Chapter 5 shows you how statisticians find probabilities. Chapter 6 deals with samples and a simple case of using samples to decide about populations. Chapter 7 puts the whole story together with special attention to the logic that all this decision-making is based on.

As you will see rather quickly, most of the descriptive statistics that you've learned will be used in this decision-making process.

5

THEORETICAL DISTRIBUTIONS INCLUDING THE NORMAL DISTRIBUTION

Objectives for Chapter 5: After studying the text and working the problems in this chapter, you should be able to:

1. Distinguish between a theoretical and an empirical distribution
2. Distinguish between theoretical and empirical probability
3. Predict the probability of certain events from your knowledge of the theoretical distribution of those events
4. List the characteristics of the normal distribution
5. Find the proportion of a normal distribution that lies between two scores
6. Find the scores between which a certain proportion of a normal distribution falls
7. Find the number of scores associated with a particular proportion of a normal distribution

In this chapter (and the following two chapters), your most important task is to understand the concepts. There are calculations for you to do, but the main purpose of the calculations is to help you understand the ideas that are at the heart of inferential statistics. Once you grasp these ideas and their logical progression, the rest of this book (as well as more advanced books) will be much easier for you to understand.

We will begin by distinguishing between empirical distributions and theoretical distributions. In Chapter 2, you learned to arrange scores into frequency distributions. The scores you worked with were selected because they were representative of scores from actual research. Distributions of such observed scores are **empirical distributions.**

This chapter is about theoretical distributions. Like the empirical distributions in Chapter 2, a theoretical distribution is a presentation of all the scores, arranged from the highest to the lowest. **Theoretical distributions,** however, are based on mathematical formulas and logic rather than on empirical observations. From the theoretical distributions we will be discussing in this chapter, you will be able to determine probabilities.[1] Probabilities are most valuable in decision making. As

---

[1] Mathematical statisticians use the term *probability density functions* when they refer to these theoretical distributions.

one example, if there is a correspondence between an empirical distribution and a theoretical distribution, you can use the theoretical distribution to arrive at probabilities about future empirical events. On the basis of the probabilities, a decision can be made.

In this chapter, we will introduce the topic *probability* and then cover three theoretical distributions. The first two of these distributions will be used to illustrate probability more fully and to establish some points that are true for all theoretical distributions. The third distribution, the normal distribution, will occupy the bulk of your time and attention in this chapter.

PROBABILITY

The concept of probability is already somewhat familiar to you. You know, for example, that probability values range from .00 (there is no possibility that an event will occur) to 1.00 (the event is certain to happen). Events are referred to as "successes" or "failures," and calculating the actual probability of a success takes one of two courses.

The *theoretical* course is to enumerate all the ways a success *can* occur. Then enumerate all the events that can occur (whether successes or failures). Finally, form a ratio with successes on top (the numerator) and total events on the bottom (the denominator). This fraction, changed to a decimal, is the theoretical probability of the event.

In coin-flipping the theoretical probability of "head" is .50. A head is a success and it can occur in only one way. The total number of possible outcomes is two (head and tail), and the ratio $\frac{1}{2}$ is .50. In a similar way the probability of a six on a die is $\frac{1}{6} = .167$. For playing cards, the probability of a jack is $\frac{4}{52} = .077$.

The *empirical* course for finding probability is to *observe* actual events, some of which are successes and some of which are failures. The end result is similar, a decimal number between .00 and 1.00. Again, the decimal is based on the ratio of two numbers, but to find an empirical probability, you use observations rather than logic to get the numbers.

Remember the data in Figure 2.3, the majors chosen by college graduates? A probability question like, What is the probability of choosing a graduate at random and getting a certain major? can be answered by processing numbers from that figure. Here's how. Choose the major you are interested in and call the frequency of that major the number of successes. Divide that number by 953,000, the total number of baccalaureate degrees granted in 1981–82. The figure you get will answer the probability question. If the major was sociology, then 16,000/953,000 = .02 is the answer; if the major was psychology, then 41,000/953,000 = .04 is the answer.[2]

This chapter is about theoretical probability. You will work with coins and cards early in the chapter, but before you are finished, we promise you a much wider variety of applications.

_____

[2] You can understand why the empirical probability approach is referred to as the relative frequency approach.

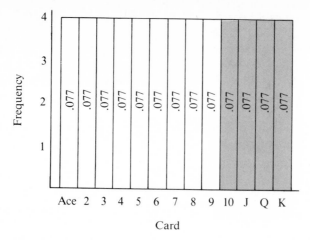

Figure 5.1   Theoretical distribution of 52 draws from a deck of playing cards

A RECTANGULAR DISTRIBUTION

We'll start with an example of a theoretical distribution that you are already familiar with to show you the relationship between theoretical distributions and probabilities. **Figure 5.1** is a histogram that shows the distribution of types of cards in an ordinary deck of playing cards. There are 13 kinds of cards, and the frequency of each card is 4. This theoretical curve is rectangular in shape. (The line that encloses a frequency polygon is called a curve, even if it is straight.) The number in the area above each card is the probability of obtaining that card in a chance draw from the deck. That theoretical probability (.077) was obtained by dividing the number of cards that represent the event (4) by the total number of cards (52).

Probabilities are often stated as "chances in a hundred." The expression $p = .077$ means that there are 7.7 chances in 100 of the event in question occurring. Thus, from Figure 5.1 you can tell at a glance that there are 7.7 chances in 100 of drawing an ace from a deck of cards. This knowledge might be helpful in a poker game.

With this theoretical distribution, you can determine other probabilities. Suppose you wanted to know your chances of drawing a face card or a 10. These are the shaded events in Figure 5.1. Simply add the probabilities associated with a 10, jack, queen, and king. Thus, $.077 + .077 + .077 + .077 = .308$. This knowledge might be helpful in a game of blackjack, in which a face card or a 10 is an important event (and may even signal "success").

One property of the distribution in Figure 5.1 that is *true for all theoretical distributions is that the total area under the curve is 1.00.* In Figure 5.1, there are 13 kinds of events, each with a probability of .077. Thus, $(13)(.077) = 1.00$. With this arrangement, any statement about area is also a statement about probability. Of the total area under the curve, the proportion that signifies ace is .077, and that is also the probability of drawing an ace from the deck.[3]

**CLUE TO
THE FUTURE**

The probability of an event or a group of events corresponds to the *area* of the theoretical distribution associated with the event or group of events. This idea will be used throughout this book.

PROBLEMS

1. What is the probability of drawing a card that falls between 3 and jack, excluding both?
2. If you drew a card at random, recorded the result, and replaced it, how many 7's would you expect in 52 draws?
3. What is the probability of drawing a card higher than a jack *or* lower than a 3?
4. If you made 78 draws from a deck, replacing each card, how many 5's and 6's would you expect?

A BINOMIAL DISTRIBUTION

The **binomial distribution** (two names) is another example of a theoretical distribution. Suppose you took three new quarters and tossed them in the air. What is the probability that all three will come up heads? As you may know, the answer is found by multiplying together the probabilities of each of the independent events. For each coin, the probability of a head is $\frac{1}{2}$, so the probability that all three will be heads is $(\frac{1}{2})(\frac{1}{2})(\frac{1}{2}) = \frac{1}{8} = .1250$.

Here are two other questions about tossing those three coins. What is the probability of two heads? What is the probability of one head or zero heads? Since you could answer these questions easily if you had a theoretical distribution of the probabilities, we will construct one for you. We will start by listing in **Table 5.1** the

Table 5.1   All possible outcomes when three coins are tossed

Outcomes	Number of heads	Probability of outcome
Heads, heads, heads	3	.1250
Heads, heads, tails	2	.1250
Heads, tails, heads	2	.1250
Tails, heads, heads	2	.1250
Heads, tails, tails	1	.1250
Tails, heads, tails	1	.1250
Tails, tails, heads	1	.1250
Tails, tails, tails	0	.1250

---

[3] In gambling situations, uncertainty is commonly expressed in odds. The expression 5:1 means that there are five ways to fail and one way to succeed; 3:2 means three ways to fail and two to succeed. To convert odds to a probability of success, add the two numbers together and divide into the second number.

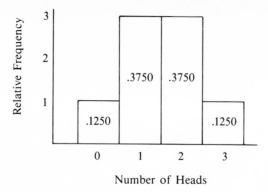

Figure 5.2   A binomial distribution showing the number of heads when three coins are tossed

eight possible outcomes of tossing the three quarters in the air. Each of these eight outcomes is equally likely, so the probability for any one of them is $\frac{1}{8} = .1250$. There are three outcomes in which two heads appear, so the probability of two heads is $.1250 + .1250 + .1250 = .3750$. This is the answer to the first question. Based on Table 5.1, we constructed **Figure 5.2,** which is the theoretical distribution of probabilities we need. You can use it to answer the following questions.[4]

PROBLEMS

5. If you toss three coins in the air, what is the probability of a success if success is (a) either one head or two heads? (b) all heads or all tails?

6. If you threw the three coins in the air 16 times, how many times would you expect to find zero heads?

COMPARISON OF THEORETICAL AND EMPIRICAL DISTRIBUTIONS

We have carefully called Figures 5.1 and 5.2 theoretical distributions. They may not, in fact, reflect exactly what would happen if you drew cards from an actual deck of playing cards or tossed quarters in the air. Actual results could be influenced by lost or sticky cards, sleight of hand, unbalanced coins, or chance deviations. Now we'll turn to the empirical question of what a frequency distribution of actual draws from a deck of playing cards looks like. **Figure 5.3** is a histogram based on 52 draws from a used deck shuffled once before each draw.

As you can see, Figure 5.3 is not exactly like Figure 5.1. In this case, the differences between the two distributions are due to chance or worn cards and not to lost cards or sleight of hand (at least not conscious sleight of hand). Of course, if we made 52 more draws from the deck and constructed a new histogram, the picture would probably be different from both Figures 5.3 and 5.1. However, if we con-

---

[4] The binomial distribution is discussed by Howell (1987), Loftus and Loftus (1988), and Downie and Heath (1983).

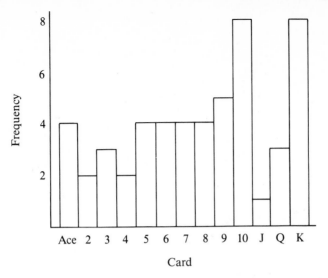

Figure 5.3   Empirical frequency distribution of 52 draws from a deck of playing cards

tinued, drawing 520 or 5200 or 52,000 times,[5] and only chance was at work, we would get a curve that was practically flat on the top; that is, the empirical curve would look like the theoretical curve.

The major point here is that a theoretical curve represents the "best estimate" of how the events would actually occur. As with all estimates, the theoretical curve is somewhat inaccurate, but in the world of real events it is better than any other estimate.

In summary, then, a theoretical distribution is one based on logic and mathematics rather than on observations. It shows you the probability of each event that is part of the distribution. When it is similar to an empirical distribution, the probability figures obtained from the theoretical distribution are accurate predictors of actual events.

There are a number of theoretical distributions that applied statisticians have found useful. In the rest of this chapter and the next two, you will work with a theoretical distribution called the normal distribution. In later chapters you will deal with others. At this point, we will simply name some of these other theoretical distributions: *t* distribution, *F* distribution, chi square distribution, and *U* distribution.

THE NORMAL DISTRIBUTION

One theoretical distribution that has proved to be extremely valuable is called the **normal distribution.** In this case, *normal* has few of its usual connotations. The name was established by early statisticians, who found that frequency distributions of data

[5] Statisticians describe this extensive sampling as "the long run."

gathered from a wide variety of fields were similar. Thus, *normal* simply means that this distribution is found frequently.

The normal distribution is sometimes called the Gaussian distribution after Carl Friedrich Gauss (1777–1855), who developed the curve in about 1800 as a way to represent the random error in astronomy observations. (See Stewart, 1977.) Because this curve was such an accurate picture of the effects of random variation, early writers referred to the curve as the *law* of error.[6]

Besides astronomy observations, many other measurements produced frequency distributions that corresponded to the normal curve. There was a time when some scientists jumped to the unwarranted conclusion that data "should" be distributed normally.

Description of the Normal Distribution

Figure 5.4 is a normal distribution. It is a bell-shaped, symmetrical distribution, a theoretical distribution based on a mathematical formula rather than on any empirical observations. (Even so, if you peek ahead to Figures 5.7, 5.8, and 5.9, you will see that empirical curves often look similar to this theoretical distribution.) When the theoretical curve is drawn, the Y axis is sometimes omitted. On the X axis, z scores are used as the unit of measurement for the standardized normal curve, where

$$z = \frac{X - \mu}{\sigma}$$

We will discuss the standardized curve first and later add the raw scores to the X axis.

There are several other things to note about the normal distribution. The mean, the median, and the mode are the same score—the score on the X axis at which the curve is at its peak. If a line was drawn from the peak to the mean score on the X axis, the area under the curve to the left of the line would be half the total

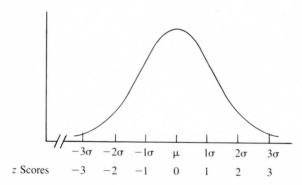

| | -3σ | -2σ | -1σ | μ | 1σ | 2σ | 3σ |

z Scores   −3   −2   −1   0   1   2   3

Figure 5.4   The normal distribution

---

[6] This word, *error*, meaning random variations, remains in statistical terminology. In the next chapter you will learn about a statistic called the *standard error*.

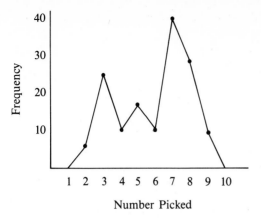

Figure 5.5   Frequency distribution of choices of numbers between 1 and 10

area—50 percent—leaving half the area to the right of the line. The tails of the curve are **asymptotic** to the X axis; that is, they never actually cross the axis but continue in both directions indefinitely with the distance between the curve and the X axis becoming less and less. Although theoretically the curve never ends, it is convenient to think of (and to draw) the curve as extending from -3σ to $+3\sigma$. (The *table* for the normal curve, however, covers the area from -4σ to $+4\sigma$.)

Another point about the normal distribution is that the two inflection points in the curve are at exactly -1σ and $+1\sigma$. The inflection points are where the curve is the steepest. (See the points above -1σ and $+1\sigma$ on Figure 5.4.)

Unfortunately, the antonym for *normal* is *abnormal*. The word *abnormal* has connotations that are not appropriate in statistics. *Curves that are not normal distributions are definitely not abnormal.* There is nothing abnormal about the distribution of playing cards in Figure 5.1, and it is not a normal curve. There is nothing abnormal about the distribution in **Figure 5.5**. It simply shows what numbers were picked when an instructor asked introductory psychology students to pick a number between 1 and 10. (This is an example of a bimodal distribution with modes at 3 and 7.)

Use of the Normal Distribution

The theoretical normal distribution is used to determine the probability of an event just as Figure 5.1 was. **Figure 5.6** is a picture of the normal curve, showing the probabilities associated with certain areas. These probability figures were obtained from Table C in Appendix C. Look at **Table C** now. It is arranged so that you can begin with a z score (column A) and find the following:

1. The area between the mean and the z score (column B)
2. The area beyond the z score (column C)

In column A find the z score of 1.00. The proportion of the curve between the mean and a z score of 1.00 is .3413. The proportion beyond the z score of 1.00 is .1587. Since the normal curve is symmetrical and since the area under the entire

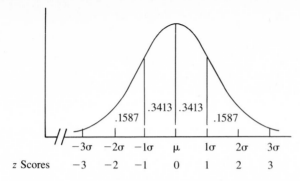

Figure 5.6   The normal distribution showing the probabilities of certain *z* scores

curve is 1.00, you will probably find it satisfying to add .3413 and .1587 together. Also, since the curve is symmetrical, these same proportions hold for $z = -1.00$. Thus, all the proportions in Figure 5.6 were derived by looking up the proportions associated with a *z* value of 1.00 in Table C. Don't just read this paragraph; do it. Understanding the normal curve *now* will pay you dividends throughout the book.

Notice that the proportions in Table C are carried to four decimal places and that we used all of them. This is customary practice in dealing with the normal curve.

PROBLEMS

7. In Chapter 3 you read of a professor who gave A's to those with *z* scores of $+1.50$ or higher.
   a. What proportion of a class would be expected to make A's?
   b. What assumption must you make in order to arrive at the proportion you found in part a?
8. What proportion of the normal distribution is found in the following areas:
   a. Between the mean and $z = .21$
   b. Between the mean and $z = -2.01$
   c. Beyond $z = .55$

Many empirical distributions are approximately normally distributed. **Figure 5.7** shows a set of 261 IQ scores, **Figure 5.8** shows the diameter of 199 ponderosa pine trees, and **Figure 5.9** shows the hourly wage rates of 185,822 union truck drivers in 1944. As you can see, these distributions from diverse fields are similar to Figure 5.4, the theoretical normal distribution. Please note that all of these empirical distributions are based on a "large" number of observations. Usually more than 100 observations are required for the curve to fill out nicely.

So far in this section, we have made two points: first, that Table C can be used to determine areas (proportions) of a normal distribution, and, second, that many empirical distributions are approximately normally distributed. Our final point is that any normally distributed empirical distribution can be made to correspond to the standardized normal distribution (a theoretical distribution) by using *z* scores. Converting the raw scores of *any* empirical normal distribution to *z* scores will give the distribution *a mean equal to zero* and *a standard deviation equal to 1.00,* and

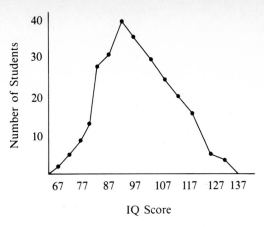

Figure 5.7 Frequency distributions of IQ scores of 261 fifth-grade students (unpublished data of J. O. Johnston)

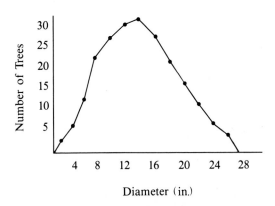

Figure 5.8 Frequency distribution of diameters of 100-year-old ponderosa pine trees on one acre, $N = 199$ (Forbs and Meyer, 1955)

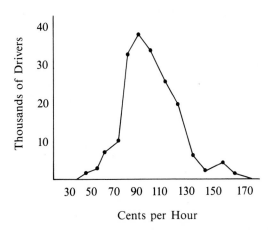

Figure 5.9 Frequency distribution of hourly wage rates of union truck drivers on July 1, 1944, $N = 185,822$ (Bureau of Labor Statistics, December 1944)

that is exactly the scale used in the theoretical normal distribution. With this correspondence established, the theoretical normal distribution can be used to determine the probabilities of empirical events, whether they are IQ scores, tree diameters, or hourly wages.

In the following examples, the normal curve will be used with IQ scores. IQ scores are distributed normally (Figure 5.7 is evidence for that) with a mean of 100 and a standard deviation of 15.[7]

Finding the Proportion of a Population That Has Scores of a Particular Size or Greater

Suppose you were interested in knowing the proportion of all people who have IQs of 120 or higher. The normal distribution shown in **Figure 5.10** has IQs on the X axis. You need to know the proportion under the curve to the right of IQ = 120 (the shaded area in Figure 5.10). For an IQ score of 120, the z score would be $(120 - 100)/15 = 20/15 = 1.33$. Table C shows that the proportion beyond $z = 1.33$ is .0918. Thus you would expect a proportion of .0918 or 9.18 percent of the population to have an IQ of 120 or higher. *Since the size of an area under the curve is also a probability statement about the events in that area, we can say that there are 9.18 chances in 100 that any randomly selected person will have an IQ of 120 or above.* **Figure 5.11** shows the proportions just determined.

Table C gives the proportions of the normal curve for positive z scores only. However, since the distribution is symmetrical, knowing that .0918 of the population has an IQ of 120 or higher tells you that .0918 has an IQ of 80 or lower. An IQ of 80 has a z score of -1.33.

You can answer questions of "how many" as well as questions of proportions. Suppose there were 500 first-graders entering school. How many would be expected to have IQs of 120 or higher? You just found that 9.18 percent of the population would have IQs of 120 or higher. If the population is 500, then calculating 9.18

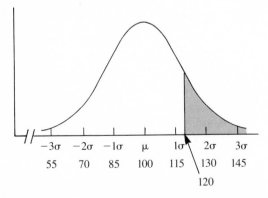

Figure 5.10   Theoretical distribution of IQ scores

[7] This is true for the Wechsler intelligence scales (WAIS, WISC, and WPPSI). The Stanford-Binet has a mean of 100 also, but the standard deviation is 16. Recent evidence (Flynn, 1987) suggests that the population mean IQ is well above 100 in many countries.

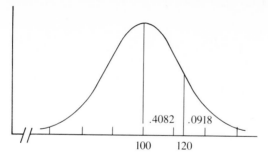

Figure 5.11 Proportion of the population with an IQ of 120 or higher

percent of 500 would give you the number of children. Thus, $(.0918)(500) = 45.9$. So 46 of the 500 first-graders would be expected to have an IQ of 120 or higher.

There are 19 problems left in this chapter for you to do. You can do every one of them correctly by using just the z-score formula and Table C. However, if you will first sketch a normal curve for each problem, write in the givens and the unknown, and finally apply the z-score formula, you are more likely to get the correct answers and, as a bonus, you will be able to conceptualize the normal curve more quickly. Drawing pictures of the normal curve is very helpful.

PROBLEMS

   9. Is the distribution in Figure 5.5 theoretical or empirical?

  10. Calculate the z scores for IQ scores of 55, 110, 103, and 100.

  11. Many school districts place children with IQs of 70 or lower in special education classes. What proportion of the general population would be expected in these classes?

  12. In a school district of 4000 students, how many would be expected to be in special education?

  13. What proportion of the population would be expected to have IQs of 110 or higher?

  14. Answer the following questions for 250 first-grade students.

    a. How many would you expect to have IQs of 110 or higher?

    b. How many would you expect to have IQs lower than 110?

    c. How many would you expect to have IQs lower than 100?

Finding the Score That Separates the Population into Two Proportions

Instead of starting with an IQ score and calculating proportions, you can also work backward and answer questions about scores if you are given proportions. For example, what IQ score is required to be in the top 10 percent of the population? The problem is pictured in **Figure 5.12.**

ERROR DETECTION

> Drawing pictures of normal distributions is the best way to understand these problems and avoid errors.

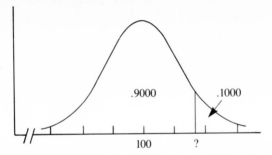

Figure 5.12   Theoretical distribution of IQ scores divided into an upper 10 percent and a lower 90 percent

Look at Figure 5.12. You can see that what you need to know is the IQ score associated with the point that divides the distribution into an upper 10 percent and a lower 90 percent. Begin by finding the z score that separates the upper 10 percent of the distribution from the rest. Look in column C of Table C for .1000. It is not there. You have a choice between .0985 and .1003. Since .1003 is closer to the desired .1000, use it.[8] The corresponding z score is 1.28.

Since

$$z = \frac{X - \mu}{\sigma}$$

multiplying both sides by σ produces

$$(z)(\sigma) = X - \mu$$

Adding μ to both sides isolates X:

$$X = \mu + (z)(\sigma)$$

Substituting numbers for the mean, the z score, and the standard deviation,

$$X = 100 + (1.28)(15)$$
$$= 100 + 19.20$$
$$= 119.2$$
$$= 119 \quad \text{(IQs are usually expressed as whole numbers.)}$$

Therefore, the IQ score required to be in the top 10 percent of the population is 119.

Here is a similar problem. Suppose a mathematics department wants to restrict the remedial math course to those who really need it. The department has the scores on the math achievement exam taken by entering freshmen for the past 10 years. The scores on this exam are distributed in an approximately normal fashion with $\mu = 58$ and $\sigma = 12$. The department wants to make the remedial course available to those students whose mathematical achievement places them in the bottom third of the

---

[8] You might use interpolation to determine the *exact z* score associated with a proportion of .1000. This extra precision (and labor) is unnecessary in this problem because the final result is rounded to the nearest whole number. For IQ scores, the extra precision does not make any difference in the final answer.

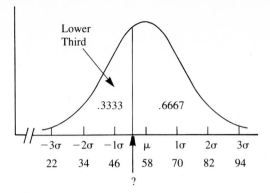

Figure 5.13 Distribution of scores on a math achievement exam

freshman class. The question is, What score will divide the lower third from the upper two-thirds? **Figure 5.13** should help to clarify the question.

The first step is to look in column C to find .3333. Again, such a proportion is not listed. The nearest proportion is .3336, which has a z value of $-.43$. (This time you are dealing with a z score below the mean, where all z scores are negative.) Applying $z = -.43$, we find that

$$X = \mu + (z)(\sigma)$$
$$= 58 + (-.43)(12)$$
$$= 58 - 5.16 = 52.84$$

If the scores on the math achievement exam are always given in whole numbers, a decision must be made either to let persons with scores of 53 or lower enroll (which would make the course available to slightly more than one-third of the freshmen) or to require a score of 52 or lower (which would make the course available to less than one-third).

Using the theoretical normal curve to establish a cutoff score is efficient. All you need is the mean, the standard deviation, and confidence in your assumption that the scores are distributed normally. The empirical alternative for the mathematics department is to sort physically through all scores for the past 10 years, arrange them in a frequency distribution, and calculate the score that separates the bottom one-third.

PROBLEMS

15. Mensa is an organization of people who have high IQs. To be eligible for membership, a person must have an IQ "higher than 98 percent of the population." What IQ is required to qualify?

16. The mean height of American women aged 18–24 is 64.3 inches, with a standard deviation of 2.5 inches (*Statistical Abstract of the United States: 1986*, 1985).
   a. What height divides the tallest 5 percent of the population from the rest?
   b. The minimum height required for women to join the U. S. Army is 58 inches. What proportion of the population will be excluded?

17. The mean height of American men aged 18–24 is 69.7 inches, with a standard deviation of 3.0 inches (*Statistical Abstract of the United States: 1986*, 1985).
    a. The minimum height required for men to join the U.S. Army is 60 inches. What proportion of the population will be excluded?
    b. What proportion of the population is taller than Napoleon Bonaparte, who was 5′2″?
18. The weight of new U.S. pennies is approximately normally distributed with a mean of 3.11 grams and a standard deviation of .05 gram (Youden, 1962).
    a. What proportion of all new pennies would you expect to weigh more than 3.20 grams?
    b. What weights separate the middle 80 percent of the pennies from the lightest 10 percent and the heaviest 10 percent?

Finding the Proportion of the Population Between Two Scores

Table C in Appendix C can also be used to determine the proportion of the population between two scores. For example, scores that fall in the range of IQ scores from 90 to 110 are often called "average." What proportion of the population falls in this range? **Figure 5.14** is a picture of the problem.

In this problem, you must add an area on the left of the mean to an area on the right of the mean. First, you need z scores that correspond to the IQ scores of 90 and 110:

$$z = \frac{90 - 100}{15} = \frac{-10}{15} = -.67$$

$$z = \frac{110 - 100}{15} = \frac{10}{15} = .67$$

The proportion of the distribution between the mean and $z = .67$ is .2486, and the same proportion is found between the mean and $z = -.67$. Therefore, $(2)(.2486) = .4972$ or 49.72 percent. So approximately 50 percent of the population is classified as "average," using the "IQ = 90 to 110" definition.

What percent of the population would be expected to have IQs between 90 and 125? **Figure 5.15** illustrates this question. Again, you must add an area that is on the left of the mean to an area on the right of the mean, but this time the areas are unequal. The corresponding z scores are

$$z = \frac{90 - 100}{15} = -.67 \quad \text{and} \quad z = \frac{125 - 100}{15} = 1.67$$

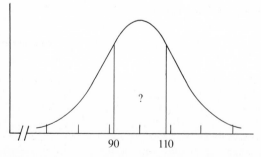

Figure 5.14 The normal distribution showing the IQ scores that define the "average" range

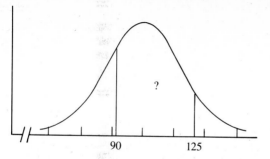

Figure 5.15   The normal distribution showing the area bounded by IQ scores of 90 and 125

Using Table C, you can find that the proportions between the mean and z scores of $-.67$ and 1.67 are $.2486$ and $.4525$, respectively. When these two proportions are added, the result, $.7011$, is the proportion of the population with IQs between 90 and 125.

PROBLEMS

19. The distribution of 800 test scores in an Introduction to Psychology course was approximately normal, with $\mu = 35$ and $\sigma = 6$.
   a. What proportion of the students had scores between 30 and 40?
   b. What is the probability that a randomly selected student would score between 30 and 40?

20. Now that you know the proportion of students with scores between 30 and 40, would you expect to find the same proportion between scores of 20 and 30? If so, why? If not, why not?

21. Calculate the proportion of scores between 20 and 30. Be careful with this one; drawing a picture may prevent an error.

22. How many of the 800 would be expected to have scores between 20 and 30?

Finding the Extreme Scores in a Population

What IQ scores are so extreme that only 1 percent of the population could be expected to have them? This is a little different from the earlier questions about the "lower third" and the "upper 2 percent" because this question does not specify which end of the curve is of interest. To answer this type of question, divide the proportion in half, placing each part at one extreme end of the curve. For the current question, the problem is to find the scores that leave one-half of 1 percent (.0050) in each tail of the curve, as seen in **Figure 5.16**. This 1 percent problem turns up in other contexts in statistics.

Since you are starting with a proportion and looking for a score, look in column C for .0050. By convention, the z score to use is 2.58. You can again use $X = \mu + (z)(\sigma)$ to determine the raw score (IQ). Thus,

$$X = 100 + (2.58)(15) \quad \text{and} \quad X = 100 + (-2.58)(15)$$
$$= 100 + 38.7 \qquad\qquad\qquad = 100 - 38.7$$
$$= 138.7 \text{ or } 139 \qquad\qquad\quad = 61.3 = 61$$

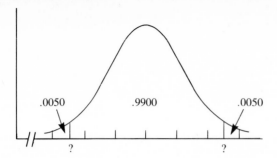

Figure 5.16   Distribution of IQs showing 1 percent of the scores divided between the two extremes

Thus, IQs of 139 or higher and 61 or lower are so extreme that they occur in only 1 percent of the population. Ninety-nine percent of the population, of course, has scores in between.

**CLUE TO
THE FUTURE**

> The idea of finding scores and proportions that are extreme in either direction will come up again after Chapter 6. In particular, the extreme 5 percent and the extreme 1 percent will be calculated.

PROBLEMS

23. What IQ scores are so extreme that they are achieved by only 5 percent of the population? Set this problem up using the "extreme 1 percent" example as a model.

24. What is the probability that a randomly selected person has an IQ higher than 129 *or* lower than 71?

25. Back in Chapter 3 you calculated a standard deviation for the time required to complete a word processing task. That standard deviation was 2.68 minutes (the mean was 11.00 minutes; see Table 3.3). At that time we promised you that $\sigma = 2.68$ would become more meaningful after you had studied Chapter 5. Suppose the firm manager decided that any secretary who could not do the problem in 13 minutes or less would be transferred to a lower position. What proportion of the secretaries would be affected?

26. Look at Figure 5.9 and suppose that the union leadership decided to ask for $0.85 per hour as a minimum wage. For those 185,822 workers, the mean was $0.99 with a standard deviation of $0.17. If $0.85 per hour was established as a minimum, how many workers would be affected?

27. Look at Figure 5.8 and suppose that a timber company decided to harvest all trees 8 inches DBH (diameter breast height) or larger from a 100-acre tract. On a 1-acre tract there were 199 trees with $\mu = 13.68$ and $\sigma = 4.83$. How many trees would be expected to be harvested from 100 acres?

COMPARISON OF THEORETICAL AND EMPIRICAL ANSWERS

You have been using the theoretical normal distribution to find probabilities and to calculate scores and proportions of IQs, diameters, wages, and other measures.

Table 5.2 Comparison of predicted and actual proportions

IQs	Predicted from normal curve	Calculated from actual data	Difference
Higher than 120	.0918	.0919	.0001
Lower than 90	.2514	.2069	.0445
Between 90 and 110	.4972	.5249	.0277

Earlier in this chapter, the claim was made that *if* the empirical observations are distributed like a normal curve, accurate predictions can be made. A reasonable question is, How accurate are all these predictions I've just made? A reasonable answer can be fashioned from a comparison of the predicted proportions (from the theoretical curve) to the actual proportions (computed from empirical data). Figure 5.7 is based on 261 IQ scores of fifth-grade public-school students. You can easily calculate the proportion who had IQs higher than 120 or lower than 90 or between 90 and 110. These actual proportions can then be compared with those predicted from the normal distribution. **Table 5.2** shows these comparisons.

As you can see by examining the "difference" column of Table 5.2, the accuracy of the predictions ranges from excellent to not so good. Some of this variation can be explained by the fact that the mean IQ of the fifth-grade students was 101 and the standard deviation 13.4. Both the higher mean (101, compared with 100 for the normal curve) and the lower standard deviation (13.4, compared with 15) are due to the systematic exclusion of very-low-IQ children from regular public schools. Thus, the actual proportion of students with IQs lower than 90 is less than predicted, which is the result of the fact that our school sample is not representative of all 10- to 11-year-old children. This problem of sampling is the major topic of the next chapter.

PROBLEMS

28. For human infants born weighing 5.5 pounds or more, the mean gestation period is 268 days, which is just less than 9 months. The standard deviation is 14 days (McKeown and Gibson, 1951). What proportion of the gestation periods would be expected to last 10 months or longer (300 days)?

29. An imaginative anthropologist measured the stature of 100 hobbits (using the proper English measure of inches) and found the following:

$$\Sigma X = 3600 \qquad \Sigma X^2 = 130,000$$

Assume that the height of hobbits is normally distributed. Find μ and σ and answer the questions.

a. The Bilbo Baggins Award for Adventure is 32 inches tall. What proportion of the hobbit population is taller than the award?

b. Three hundred hobbits entered a cave that had an entrance 39 inches high. The orcs chased them out. How many could exit without ducking?

c. Gandalf is 44 inches tall. What is the probability that he is a hobbit?

6 SAMPLES AND SAMPLING DISTRIBUTIONS

Objectives for Chapter 6: After studying the text and working the problems in this chapter, you should be able to:

1. Define a random sample
2. Obtain a random sample if you are given a population
3. Define and identify biased sampling methods
4. Define a sampling distribution and a sampling distribution of the mean
5. Calculate a standard error of the mean if you are given σ and N or if you are given s and N
6. Use the z-score formula to determine the probability of a certain sample mean being drawn from a population with a certain (known) mean
7. Verbally define the standard error of any statistic
8. Calculate 95 and 99 percent confidence intervals about a sample mean and explain what they mean

We want to begin by asking you to be especially attentive to the concept of a sampling distribution. In the topic of inferential statistics, the idea of a **sampling distribution** is clearly the *central concept*. This book on "Tales of Distributions" is a book of tales of sampling distributions. Every chapter after this one is about some kind of sampling distribution; thus, you might consider this paragraph to be a superclue to the future.

An understanding of sampling distributions requires an understanding of samples. A sample, of course, is some part of the whole thing; in statistics the "whole thing" is a population. The population is always the thing of interest; a sample is used only to estimate what the population is like. One obvious problem is to get samples that are representative of the population.

Here is a story illustrating the problems of getting a representative sample and then knowing how accurately it represents the population.

Late one afternoon, two economics students were discussing the average family income of students on their campus.

"Well, the average family income for college students is $38,500, nationwide" (Astin, 1987), asserted the junior, "and I'm sure the mean for this campus is at least that high."

"I don't think so," the sophomore replied. "I know lots of students who have only their own resources or come from pretty poor families. I'll bet you a dollar the mean for students here at State U. is below the national average."

"You're on," grinned the junior.

Together the two went out with a pencil and pad and asked 40 students how much their family income was. Thirty-seven students made estimations, and the mean of these was $33,300.

Now the sophomore grinned, "I told you so; the mean here is $5200 less than the national average of $38,500."

"Curses," cursed the junior. While thinking about those 37 interviews, however, he began to smile. "Actually," he said, "this mean of $33,300 is meaningless and here's why. Those 37 students are not representative of the entire student body. They are late-afternoon students, and many of them support themselves with temporary jobs while they go to school. Most of the student body is supported from home by parents who have permanent and better-paying jobs. That sample is no good, and I won't pay until we get results for the whole campus or at least from a representative sample."

To get the results for the whole student body, the two went the next day to the director of the financial aid office, who told them that family incomes for the student body are not public information.

Sampling became a necessity.

With the help of a friendly dean, a random sample of 40 students was selected. (A random sample has the best chance of being representative.) The two investigators then got replies from 38 and found the mean family income to be $36,900.

"Pay," demanded the sophomore.

"OK, OK, ... here!"

Later, the junior began again to think about that random sample. He wondered what the mean would be if he polled another random sample. It could be higher. Maybe the mean of $36,900 was just a fluke—a chance event.

Shortly afterward, the junior confronted his co-investigator with his thoughts about repeated sampling. "How do we know if that random sample we got told us the truth about the population? For example, suppose the mean for the entire student body *is* $38,500. Maybe a sample with a mean of $36,900 would occur just by chance fairly often. I wonder what the chances are of getting a sample mean of $36,900 from a population with a mean of $38,500?"

"To answer that, you would have to know about the sampling distribution of the mean," wisely intoned the sophomore, hurrying from the room.

The two basic points of the story are that samples that are random have the best chance of being representative (the junior paid off only after the results were based on a random sample) and that something called a sampling distribution can tell you how much faith (probability-wise) you can put in results based on a random sample.

CLUE TO THE FUTURE

In the story, the junior introduced a **hypothesis** about the State U. population—namely, that the mean family income for the student body was $38,500. The basic idea throughout the rest of this book will be to introduce a hypothesis about a population and then, based on the results of a sample, decide either that the hypothesis is reasonable or that it should be rejected.

Before we begin to explain these ideas, we'll expand a little on some terms we introduced in Chapter 1. *Population,* as you know, means all the members of a specified group. Examples of specifications are self-esteem scores of freshmen who were enrolled at State University on October 1, total time of hospitalization of deceased schizophrenics in the state of New York, and weights of the brains of all well-fed albino rats. It is the investigator who defines the population of interest. Sometimes the population can actually be measured, given plenty of time and money, as in the first and second examples. Sometimes, however, measuring the population is logically impossible, as in the third example. Many of those rats are already dead, and besides, rats reproduce faster than they can be measured. In the story, the population was the family income of all currently enrolled students at State University and it wasn't practical to get data on all of them. *Inferential statistics* are used when it is not possible or practical to measure an entire population.

So decisions about unmeasurable populations can be made by using samples and the methods of inferential statistics. Unfortunately, there is some peril in this. Samples are variable, changeable things. Each one produces a different statistic. How can you be sure that the sample you draw will produce a statistic that will lead to a correct decision about the population? Unfortunately, you *cannot* be absolutely sure. **To draw a sample is to agree to accept some uncertainty about the results.** In this chapter, you will learn how to *measure* this uncertainty. If a great deal of uncertainty exists, the sensible thing to do is suspend judgment. On the other hand, if there is very little uncertainty, the sensible thing to do is to reach a conclusion, even though there is a small risk of being wrong. *Reread this paragraph. It is important.*

The following problems give you an opportunity to review some concepts.

PROBLEMS

1. Define statistic.
2. Define parameter.
3. What is the difference between μ and $\bar{X}$?
4. A salesman presented a quality control engineer at a light-bulb factory with some new packaging material. To determine whether the old material was doing its job of protecting bulbs from damage in shipping, the engineer sampled from the previous month's production by going to 25 stores in the area and exchanging the bulbs on the shelves for new bulbs. The "recalled" bulbs were tested. A proportion of them, .005 (one-half of 1 percent), would not work. Name the population, the parameter of interest, the sample, and the statistic. What is the numerical value of the statistic?

REPRESENTATIVE AND NONREPRESENTATIVE SAMPLES

Let's begin by assuming that you want to know about an unmeasurable population, a desire common among researchers both young and old. Unmeasurable populations require you to draw a sample and, naturally, you would like a representative sample. The key to having a representative sample is to use a *method of obtaining samples* that is more likely to produce a representative sample than any other method.[1]

We will name two methods of sampling that are most likely to produce a representative sample, discuss one in detail, and then discuss some ways in which nonrepresentative samples are obtained when the sampling method is biased.

Random Samples

Random sampling is a method that produces a sample that is most likely to be representative of the population. *Random* has a technical meaning in statistics. It does not mean haphazard or unplanned. A **random sample** in most research situations is one in which every potential sample of size *N* has an equal probability of being selected. To obtain a random sample, you must do the following:

1. Define the population of scores.
2. Identify every member of the population.
3. Select scores in such a way that every sample has an equal probability of being chosen.

We will go through these steps with a set of actual data—the self-esteem scores of 24 fifth-grade children.[2] We define these 24 scores as our population. From these we will pick a random sample of seven scores.

39	43	40	25	42	42	32	36
36	24	30	31	35	31	45	36
22	31	35	37	46	42	33	36

One method of picking a random sample is to write each self-esteem score on a slip of paper, put the 24 slips in a box, jumble them around, and draw out seven. The scores on the chosen slips are a random sample. This method works fine if the slips are all the same size, they are jumbled thoroughly, and there are only a few members of the population. If there are many members, this method is tedious.

Another (easier) method of getting a random sample is to use a table of random numbers, such as Table B in Appendix C. To use the table, you must first assign an identifying number to each of the 24 self-esteem scores, as in **Table 6.1.**

Each score has been identified by a two-digit number. Now turn to Table B and pick a row and a column in which to start. Any haphazard method will work; close your eyes and stab a place with your finger. Suppose you started at row 35, columns 70–74. Reading horizontally, you'll find the digits 21105. Since you need only two digits to identify any member of our population, use the first two digits, 21. That identifies one score for the sample—a score of 46. From this point, you can read

---

[1] How well a particular method works can be assessed either mathematically or empirically. For an empirical assessment, start with a population of numbers, the parameter of which can be easily calculated. The particular method of sampling is used repeatedly, and the corresponding statistic is calculated for each sample. The mean of these sample statistics can then be compared with the parameter.

[2] The self-esteem scores are based on the Coopersmith Self-Esteem Inventory and are from Spatz and Johnston (1973). The Coopersmith inventory consists of 50 items that the student designates as "like me" or "unlike me." Items are similar to "I am often ashamed of myself" and "I am a fairly happy person."

Table 6.1   Assignment of identifying numbers
to population scores

ID number	Score	ID number	Score
01	39	13	35
02	43	14	31
03	40	15	45
04	25	16	36
05	42	17	22
06	42	18	31
07	32	19	35
08	36	20	37
09	36	21	46
10	24	22	42
11	30	23	33
12	31	24	36

two-digit numbers in any direction—up, down, or sideways—but the decision should have been made before you looked at the numbers. If you had decided to go down, the next number is 33. No self-esteem score has an identifying number of 33, so skip it and go to 59, which gives you the same problem as 33. In fact, the next five numbers are too large. The sixth number is 07, which identifies the score of 32 for the random sample. The next usable number is 13, which identifies a score of 35. Continue in this way until you arrive at the bottom. At this point, you can go in any direction. We will skip over two columns to columns 72 and 73 (you were in columns 70 and 71) and start up. The first number is 12, which identifies a score of 31. The next usable numbers are 19, 05, and 10, which give scores of 35, 42, and 24. Thus, the random sample of seven consists of the following scores: 46, 32, 35, 31, 35, 42, and 24. If Table B had produced the same identifying number twice, you would have ignored it the second time.

PROBLEM

5. A random sample is supposed to produce a statistic similar to the parameter. The population μ for the 24 self-esteem scores is 35.375. What is the mean of the random sample of seven?

What is this table of random numbers? In Table B (and in any table of random numbers), the probability of occurrence of any digit from 0 to 9 at any place in the table is the same—.10. Thus, you are just as likely to find 000 as 123 or 381. Incidentally, you cannot generate random numbers out of your head. Certain sequences begin to recur, and (unless warned) you will not include enough repetitions like 666 and 000. If warned, you will produce too many.

Here are some hints for using a table of random numbers.

1. Make a check beside the identifying number of a score when it is chosen for the sample. This will help prevent duplications.
2. If the population is large (more than 100), it is more efficient to get all the identifying numbers from the table first. As you select them, put them in some

rough order. This will help prevent duplications. After you have all the identifying numbers, go to the population to select the sample.

3. If the population has exactly 100 members, let 00 be the identifying number for 100. In this way, you can use two-digit identifying numbers, each one of which matches a population score. This same technique can be applied to populations of 10 or 1000 members.

Finally, we want to distinguish between random samples and random assignment. Random samples, as you know, are samples chosen randomly from the population. Random assignment is a method of partitioning into subsamples a sample that may or may not be a random sample. If this partitioning is based on chance (a table of random numbers), then the characteristics of the sample that are variable (things like gender, height, IQ, and experience) will fall more or less equally in each subsample.

PROBLEMS

6. Draw a random sample with $N = 5$ from the population of self-esteem scores in Table 6.1.
7. Draw a random sample of ten from the 24 self-esteem scores. This time, begin at a different place in the table of random numbers. Calculate the mean of the sample.
8. Why should you begin at a different place in the table for the sample in Problem 7?
9. Draw a random sample with $N = 12$ from the following scores:

76	47	81	70	67	80	64	57	76	81
68	76	79	50	89	42	67	77	80	71
91	72	64	59	76	83	72	63	69	
78	90	46	61	74	74	74	69	83	

Stratified Samples

A method called *stratified sampling* is another way to produce a sample that is very likely to mirror the population. It can be used when an investigator knows the numerical value of some important characteristic of the population. A **stratified sample** is controlled so that it exactly reflects some known characteristic of the population. Thus, in a stratified sample, not everything is left to chance.

American public opinion polls on political party identification show that approximately 40 percent are Democrats, 30 percent are Republicans, and 30 percent are Independents. For some political issues (like voting for President), party identification may be important, so the investigator draws a sample that will reflect the proportions found in the population. The same may be done for variables such as gender, age, and socioeconomic status. After the stratification of the samples has been determined, sampling within each stratum is usually random.[3]

To justify a stratified sample, the investigator must know what variables will affect the results and what the population characteristics are for those variables. Sometimes the investigator has this information (as from census data), but many times such information is just not available (as in most research situations).

---

[3] See Jaeger (1984) or Mendenhall, Ott, and Scheaffer (1971) for chapters on stratified sampling.

Biased Samples

A **biased sample** is one that is drawn using a method that systematically underselects or overselects from certain groups within the population. Thus, in a biased sampling technique, every sample of a given size does *not* have an equal opportunity of being selected. With biased sampling techniques, you are much more likely to get a nonrepresentative sample than you are with random or stratified sampling techniques.

For example, it is reasonable to conclude that some results based on mailed questionnaires are not valid because the samples are biased. Usually an investigator defines the population, identifies each member, and mails the questionnaire to a randomly selected sample. Suppose that 70 percent of the recipients respond. Can valid results for the population be based on the questionnaires returned? Probably not. There is good reason to suspect that the 70 percent who responded are different from the 30 percent who did not. Thus, although the population is made up of both kinds of people, the sample reflects only one kind. Therefore, the sample is biased. The probability of bias is particularly high if the questionnaire elicits feelings of pride or despair or disgust or apathy in *some* of the recipients.

PROBLEMS

10. Suppose a large number of questionnaires about educational accomplishments were mailed out. Do you think that some recipients would be more likely to return the questionnaire than others? Which ones? If the sample is biased, will it overestimate or underestimate the educational accomplishments of the population?

11. Sometimes newspapers sample opinions of the local population by printing a "ballot" and asking readers to mark it and mail it in. Evaluate such a sampling technique.

A very famous case of a biased sample occurred in a poll that was to predict the results of the 1936 election for President of the United States. The *Literary Digest* (a popular magazine) mailed 10 million "ballots" to those on their master mailing list, a list of more than 10 million people compiled from "all telephone books in the US, rosters of clubs, lists of registered voters," and other sources. More than 2 million "ballots" were returned and the prediction was clear: Alf Landon by a landslide over Franklin Roosevelt. As you may have learned, the actual results were just the opposite; Roosevelt got 61 percent of the vote.

From the 10 million who had a chance to express a preference, 2 million very interested persons had selected themselves. This 2 million had more than its proportional share of those who were disgruntled with Roosevelt's depression-era programs. The 2 million ballots were a biased sample; the results were not representative of the population. In fairness to the *Literary Digest,* it should be noted that it used a similar master list in 1932 and predicted the popular vote within 1 percentage point.[4]

With a nice random sample you can predict with confidence your chance of being wrong. If it is higher than you would like, you can reduce it by increasing the sample size. With a biased sample, however, you do not have a sound theoretical

---

[4] See the *Literary Digest,* August 22, 1936, and November 14, 1936.

basis for assessing your margin of error and you don't know how much confidence to put in your predictions. You may be right (as the *Literary Digest* was in 1932) or you may be wrong (the 1936 experience).

Here are some other examples of biased samples. The junior economics major in our story recognized that the sample of passersby was biased toward people taking late-afternoon classes (who perhaps had lower family incomes). The population of a city cannot be randomly sampled using a telephone book because some persons in the city do not have a telephone and others have an unlisted number. None of those people would have a chance of being selected to be in the sample. Intact groups often represent biased samples. An introductory psychology class and a cage of rats may be the handiest samples around, but, for almost any population you can name, they represent biased samples. You may get generalizable results from such samples, but you cannot be sure. The search for biased samples in someone else's research is a popular (and serious) game among researchers.

The Widespread Use of Samples

Systematic sampling is used by a very wide variety of people and organizations. The public opinion polls reported in newspapers and on television are based on carefully selected samples. ASCAP, the musicians' union, samples from the broadcasts of more than 8000 U. S. radio stations and then collects royalties for members on the basis of the sample results. The U. S. Postal Service uses sampling to make decisions about the placement of new facilities. Quality control departments in factories all over the world use sampling. The reason for this widespread use is that sampling works; you can find out about the whole thing by examining just a little of it. For a well-written introduction to sampling, one with interesting examples, see *A Sampler on Sampling* by Bill Williams (1978).

PROBLEMS

12. Explain the difference between a biased and a representative sample.

13. Can a statistic be computed on data from a biased sample?

14. Consider as a population the students at State U. whose names are listed alphabetically in the student directory. Are the following samples biased, random, or stratified?
   a. Every fifth name on the list
   b. Every member of the junior class
   c. 150 names drawn from a box containing all the names in the directory
   d. One name chosen at random from the directory
   e. 12 persons chosen at random from the freshman class, 10 chosen at random from the sophomore class, 9 at random from the junior class, and 8 at random from the senior class
   f. A random sample from those taking the required English course

15. Was the sample of light bulbs in Problem 4 random, stratified, or biased?

SAMPLING DISTRIBUTIONS

This short section is about sampling distributions in general, and the next section is about a particular sampling distribution—the sampling distribution of the mean.

Establishing these two categories in your thinking now will help you as you learn about other sampling distributions in later chapters.

You can think of a sampling distribution as a frequency distribution in which the X axis consists of the scores you get if you calculate the same statistic on each of all possible samples from a population. The Y axis is frequency. To say it again, draw many random samples from a population, calculating a statistic on each one. Then arrange all these numbers into a frequency distribution.

The sampling distributions you will use are theoretical distributions. From them you can find the probability of obtaining any particular values of the statistic.

Sampling distributions are important. As evidence of their importance, statisticians have special names for the mean and for the standard deviation of sampling distributions. The mean is called the **expected value** and the standard deviation is called the **standard error.**[5] We will not have much more to say about expected value, but if you continue your study of statistics, you will encounter it. The standard error, however, will be used many times in this text.

Every sampling distribution is for a particular statistic (such as the mean, variance, or correlation coefficient). In this chapter, you will learn only about the sampling distribution of the mean. It will serve as an introduction to sampling distributions in general, some others of which you will find out about in later chapters.

THE SAMPLING DISTRIBUTION OF THE MEAN

Remember the population of 24 self-esteem scores that you sampled from earlier in this chapter? In one of the problems, each student drew a sample ($N = 10$) and calculated the mean. For almost every sample in your class, the mean was different. Each, however, was an *estimate* of the population mean. Think about drawing 200 samples with $N = 10$ from the same population, finding their means, and making a frequency distribution of the means. If it sounds like a tedious task, let us assure you that it is! We worked on it for a while but got so bored that we asked a friend to do the problem for us on a computer.

The computer drew 200 separate random samples, each with $N = 10$, calculated the mean of each, and arranged these 200 $\bar{X}$'s into the frequency polygon seen in **Figure 6.1.** The mean, μ (a parameter), of the 24 self-esteem scores is 35.375. As you can see, most of the statistics (sample means) are fairly good estimates of that parameter. Some $\bar{X}$'s, of course, miss the mark widely, but most are pretty close. Figure 6.1 is an *empirical sampling distribution of the mean—that is, a frequency distribution of sample means.*

Notice the steps:

1. Every sample is drawn randomly from the same population.
2. The sample size (N) is the same for all samples.
3. The number of samples is very large.

[5] In this and in other statistical contexts, the term *error* means deviations or random variation. The word *error* is left over from the days of Gauss when random variation was referred to as the "normal law of error." Of course, *error* sometimes means mistake, so you will have to be alert to the context when this word appears (or you may make an error).

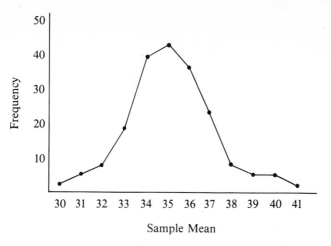

Figure 6.1 Empirical sampling distribution of the mean. The population is the self-esteem scores. For each sample, $N = 10$.

You will *never* use an empirical sampling distribution of the mean in any of your calculations; you will always use theoretical ones that come from mathematical formulas. An empirical sampling distribution of the mean is easier to understand, though, so we used it for this initial discussion.

We hope that, when you looked at Figure 6.1, you were at least suspicious that it might be the ubiquitous normal curve. It is. This puts you in the position of an educated person: what you learned in the past (how to use the normal curve) can be used to solve new and different problems—finding probabilities about sample means.

Of course, the normal curve is a theoretical curve, and we have presented you with an empirical curve that appears normal. We would like to let you prove for yourself that the form of a sampling distribution of the mean is a normal curve, but, unfortunately, that requires mathematical sophistication beyond that assumed for this course. So we will resort to a time-honored teaching technique—an appeal to authority. In this case, our authority is mathematical statistics, which has proved a theorem called the **Central Limit Theorem.** This important theorem says:

> For *any* population of scores, regardless of form, the sampling distribution of the mean will approach a normal distribution as N (sample size) gets larger. Furthermore, the sampling distribution of the mean will have a mean equal to μ and a standard deviation equal to $\sigma/\sqrt{N}$.

Our appeal to authority resulted in a lot of information. Now you know not only that sampling distributions of the mean are normal curves[6] but also that, if you know the population parameters μ and σ, you can determine the parameters of the sampling distribution. The mean of the sampling distribution of means will be the

---

[6] One qualification is that the sample size (N) must be large. How many does it take to make a large sample? The traditional answer is 30 or more, although, if the population itself is symmetrical, a sampling distribution of the mean will be normal with sample sizes much smaller than 30. If the population is severely skewed, samples with 30 (or more) may be required.

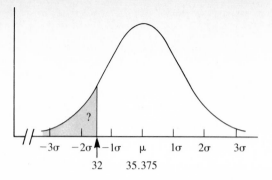

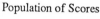

-3σ -2σ -1σ μ 1σ 2σ 3σ

32   35.375

Figure 6.2 Theoretical sampling distribution of the mean, $N = 10$

Population
mean = 35.375
standard deviation = 6.304

Sampling distribution of the mean for $N = 10$
expected value of the mean, $E(\bar{X}) = 35.375$
standard error of the mean, $(\sigma_{\bar{x}}) = 6.304/\sqrt{10} = 1.993$

Playing Cards   **Choice of Numbers**

Population of Scores

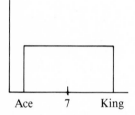

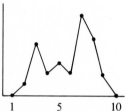

Sampling Distribution with $N = 2$

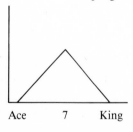

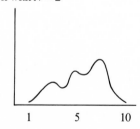

Sampling Distribution with $N = 30$

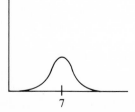

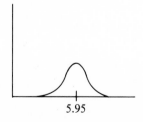

Figure 6.3 Populations of playing cards and number choices, and sampling distributions with $N = 2$ and $N = 30$

same as the population mean, μ. The standard deviation of the sampling distribution will be the standard deviation of the population (σ) divided by the square root of the sample size. One final point: the standard error of the mean is symbolized $\sigma_{\bar{X}}$ and the expected value of the mean is symbolized $E(\bar{X})$.

We will use this information to calculate the sampling distribution of the mean for the population of self-esteem scores. **Figure 6.2** shows the results.

The most remarkable thing about the Central Limit Theorem is that it works *regardless* of the form of the original population. Thus, the sampling distribution of the mean of scores coming from a rectangular or bimodal population approaches normal if N is large. Figure 6.3 illustrates this for two examples you are familiar with: the distribution of playing cards (Figure 5.1) and the distribution of choices of numbers between 1 and 10 (Figure 5.5). Examine **Figure 6.3** now.

Finally, the Central Limit Theorem *does not* apply to all statistics, but it does apply to the sampling distribution of the mean, which is a most important statistic.

Since the sampling distribution of the mean is a normal curve, you can apply what you learned in the last chapter about normally distributed *scores* to questions about sample *means*. For example, referring to Figure 6.2, what proportion of all sample means from the population of self-esteem scores would be expected to be 32 or less? You will recall that the z-score formula for an X score is $z = (X - \mu)/\sigma$. The application to a sample mean is similar:

$$z = \frac{\bar{X} - \mu}{\sigma_{\bar{X}}}$$

Thus, for $\bar{X} = 32$,

$$z = \frac{\bar{X} - \mu}{\sigma_{\bar{X}}} = \frac{32 - 35.375}{1.994} = -1.69$$

Looking in the normal-curve table (Table C in Appendix C), you can see that, for a z score of 1.69, you would *expect* a proportion of .0455 of the means to be less than 32. We have redrawn Figure 6.2 as **Figure 6.4** to summarize these findings.

The predicted proportion is .0455. Fortunately, we can check this prediction by determining the proportion of those 200 random samples that had means of 32 or less. By checking the frequency distribution from which Figure 6.1 was drawn, we

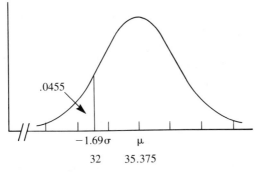

Figure 6.4   Proportion of sample means of 32 or less, $N = 10$

found the empirical proportion to be .0400. Missing by one-half of 1 percent isn't bad, and once again, you find that a theoretical normal distribution predicts an actual empirical proportion quite nicely.

What effect does sample size have on a sampling distribution? Since $\sigma_{\bar{x}} = \sigma/\sqrt{N}$, $\sigma_{\bar{x}}$ will become smaller as N gets larger. **Figure 6.5** shows some sampling distributions of the mean based on the population of 24 self-esteem scores. The sample sizes are 3, 5, 10, and 20. A sample mean of 39 is included in all four figures as a reference point. Notice that, as $\sigma_{\bar{x}}$ becomes smaller, a sample mean of 39 becomes a rarer and rarer event. A good investigator with an experiment to do will keep in

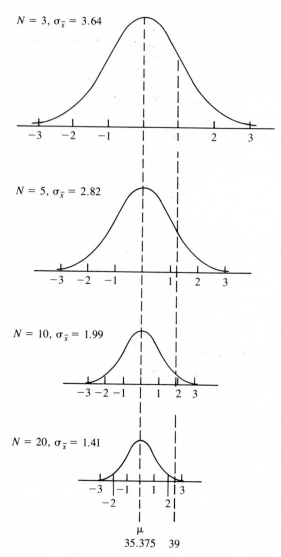

Figure 6.5   Sampling distributions of the mean for four different sample sizes. All samples are drawn from the same population. Note how a sample mean of 39 becomes more and more rare as $\sigma_{\bar{x}}$ becomes smaller.

mind what we have just demonstrated about the effect of sample size on sampling distributions and will use reasonably large samples.

PROBLEMS

16. When the population parameters are known, the standard error of the mean is $\sigma_{\bar{X}} = \sigma/\sqrt{N}$. At each intersection of the table, fill in the value for $\sigma_{\bar{X}}$.

	σ			
N	1	2	4	8
1				
4				
16				
64				

17. On the basis of the table constructed in Problem 16, write a precise verbal statement about the relationship between $\sigma_{\bar{X}}$ and N.
18. In order to reduce $\sigma_{\bar{X}}$ to one-fourth its size, you must increase N by how much?
19. Under what condition does $\sigma_{\bar{X}} = \sigma$?
20. Back in Chapter 4 you learned how to use a regression equation to predict a Y score if you were given an X score. That Y score is a statistic, so it has a sampling distribution with its own standard error. What could you conclude if the standard error was very small? Very large?

CONSTRUCTING A SAMPLING DISTRIBUTION WHEN σ IS NOT AVAILABLE

To calculate the standard error of the mean, a population parameter, σ, was given to you. In the world of empirical research, however, you rarely know population parameters. Fortunately, with a little modification of the formula and no modification of logic, you can estimate σ by calculating s from a random sample. With this estimate of σ, you can calculate an estimate of the standard error of the mean.

When you have only a sample standard deviation with which to estimate the standard error of the mean, the formula is the following:

standard error
of the mean
estimated from $s_{\bar{X}} = \dfrac{s}{\sqrt{N}}$
a sample

where $s_{\bar{X}}$ = standard error of the mean
s = standard deviation of a sample
N = sample size

The statistic $s_{\bar{X}}$ is an estimate of $\sigma_{\bar{X}}$, and $\sigma_{\bar{X}}$ is required for use of the normal curve. The larger the sample size, the more reliable $s_{\bar{X}}$ is. As a practical matter, $s_{\bar{X}}$ is considered reliable enough if $N \geq 30$. In pure mathematical statistics, though, the normal curve is appropriate only when you know μ and σ.

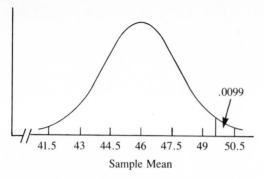

Figure 6.6   Sampling distribution of the mean for a population with a mean of 46, $N = 36$

We will apply this standard error of the mean to a problem. A math educator and an educational psychologist developed a program in which parents helped their children with arithmetic homework by supplying the answer as soon as the child worked each problem.[7] At the end of the term, the students took a standardized test. Over the years, the mean score on this test had been 46. This year, after the special homework program, the following data were obtained: $\bar{X} = 49.5$, $s = 9.0$, and $N = 36$. We calculated $s_{\bar{X}}$ and constructed the sampling distribution about the assumed parameter of 46. See **Figure 6.6.**

$$s_{\bar{X}} = \frac{s}{\sqrt{N}} = \frac{9.0}{\sqrt{36}} = 1.50$$

For the sample mean of 49.5, we can find a z score. With the z score, we can determine the probability of a mean of 49.5 or larger being drawn by chance from a population with $\mu = 46$:

$$z = \frac{\bar{X} - \mu}{s_{\bar{X}}} = \frac{49.5 - 46}{1.50} = 2.33$$

From the normal-curve table, the probability of a z score of 2.33 or larger is .0099. This means that the probability is only about 1 in 100 that a sample with a mean of 49.5 or larger would be drawn from a population with a mean of 46, *if chance is the only factor affecting the scores in the sample.*

Besides chance, what other possibilities are there? Well, perhaps the children were better than average in arithmetic to begin with. Johnston and Maertens thought of this; a pretest demonstrated that they were not. Since the study was designed so that extraneous-variable explanations were not plausible, and since chance was not a likely explanation for the difference, the authors concluded that immediate feedback produces better arithmeticians who then score higher on standardized tests.

[7] For an actual study on this topic, see Johnston and Maertens (1972).

PROBLEMS

21. A reporter covering a wildcat teachers' strike needed to know whether the strike had affected student attendance. Official records would not be available before her deadline. However, she knew that the average daily attendance for classrooms was 25.5, so she walked to the nearest school, counted the children in each of the 12 classrooms with windows next to the sidewalk, and found a mean of 23.3, $s = 3.1$.

  a. Which of the following headlines is appropriate?

    "SCHOOL ATTENDANCE SIGNIFICANTLY DOWN"

    "SCHOOL ATTENDANCE PROBABLY NOT AFFECTED BY TEACHERS' STRIKE"

    "REPORTER LAMBASTED BY EDITOR FOR BIASED SAMPLING"

  b. Assume the $\bar{X}$ and s were obtained from a random sample and that $N = 12$ is adequate. What is the probability that the strike was having *no* effect on school attendance?

22. A social worker thought his clients lacked an important skill—assertiveness. He enrolled 36 of them in an assertiveness training course. Eight weeks later, the 36 fully trained clients took a test that measured assertiveness, the national mean of which was 30. The clients produced the following statistics:

$$\Sigma X = 1188 \qquad \Sigma X^2 = 40,464$$

What is the probability of obtaining a sample with such a mean from a population with a mean of 30? Write a conclusion about the effects of the assertiveness training course.

23. Now you are in a position to return to the story of the two economics students at the beginning of this chapter. Work out, for the junior, the probability of obtaining a sample mean of $36,900 or less from a population with a mean of $38,500. The sample produced statistics of $\bar{X} = \$36,900$, $s = \$10,000$, and $N = 38$.

If the junior had done his homework as you have just done, and if he had confronted his sophomore friend with the results, the conversation might have gone like this, with the junior speaking:

"... and so, the mean of $36,900 isn't very reliable. The standard deviation is large and, with $N = 38$, samples can bounce all over the place. Why, if the real mean (parameter) is the same as the national average, $38,500, we would *expect* about one-sixth of all random samples of 38 students to have means of $36,900 or less."

"Yeah," said the sophomore, "but that *if* is a big one. What if the campus mean is really $36,900? Then the sample we got was right on the nose. After all, a sample mean is an *unbiased estimator* of the population parameter."

"I see your point. And I see mine, too. It seems like either of us could be correct. That leaves me uncertain about the real campus parameter."

"Me, too."

Not all statistical stories end with so much uncertainty. However, we said that one of the advantages of random sampling is that you can measure the uncertainty. You measured it, and there was a lot. Remember, if you agree to draw a sample, you agree to accept some uncertainty about the results.

Given such uncertainty, there is a practical method that may reduce it: increase the sample size. You have already worked some problems that showed that a fourfold increase in N will reduce the standard error by half. Applying this principle to our story, suppose the two students got a random sample of 400 students, which again produced a mean of $36,900. Now $s_{\bar{x}}$ becomes

$$\frac{s}{\sqrt{N}} = \frac{10,000}{\sqrt{400}} = 500$$

and z becomes

$$z = \frac{36,900 - 38,500}{500} = -3.20$$

compared with -0.99 when $N = 38$. The probability associated with a z score of 3.20 or larger is less than .01. Uncertainty has been reduced 16-fold. With such figures, our heroes would be rather confident that State University students have families with incomes lower than the national average.

If you are a typical student, you may still have a nagging doubt about that 1 chance in 100. Hang on to that doubt. It is real and justifiable and is a direct consequence of accepting sampling as a way to find out about parameters.

One other caution about sample size is in order. Increasing sample size will not guarantee a reduction of uncertainty. You know that any new sample will probably produce a new mean. For example, the sample with $N = 400$ might have produced a mean of $38,000.

PROBLEM 24. What is the probability that a mean of $38,000 or less would be drawn from a population where $\mu = \$38,500$? Again, let $s = \$10,000$ and $N = 400$.

CONFIDENCE INTERVALS

Statisticians divide statistical inference into two categories. The first category is called *hypothesis testing,* and the second is called *estimation.* **Hypothesis testing** means to hypothesize a value for a parameter, compare (or test) the parameter with an empirical statistic, and decide whether the parameter is reasonable. Hypothesis testing is just what you have been doing so far in this chapter. Hypothesis testing is the more popular technique of statistical inference among social and behavioral scientists.

The other kind of inferential statistics, estimation, can take two forms—point estimation and confidence intervals. **Point estimation** means that one particular number is estimated to be the parameter of the population. A **confidence interval** is a range of values bounded by a lower and an upper limit. The interval is expected, with a certain degree of confidence, to contain the parameter. These confidence intervals are based on sampling distributions.

Our purpose in this section is to describe confidence intervals. We will begin with an explanation of the concept of a confidence interval and then describe how to find the lower and upper limits of the interval. Finally, we will caution you on how to tell the story of the degree of confidence you may have that the limits contain the parameter.

The Concept of a Confidence Interval

A confidence interval is simply a range of values with a lower and an upper limit. With a certain degree of confidence (usually 95 percent or 99 percent), you can state that the interval contains the parameter. The following example shows how the size of the interval and the degree of confidence are directly related (that is, as one increases, the other increases also).

Suppose you drew a random sample of 36 scores from a population and found a mean of 76 and a standard deviation of 8. The scores ranged from a low of 60 to a high of 91. Since $\bar{X}$ is the best estimator of the parameter, μ, you might just say "I estimate that $\mu = 76$." (This is an example of a point estimate.) How confident would you be that μ is exactly 76? Not very, we hope. Although 76 is the best estimate, no one would be surprised to get a sample mean of 76 from a population with $\mu = 75$ (or 75.5, 75.8, and so on). Thus, you would not be very confident if you estimated μ with a specific point.

At the other extreme, how confident would you be if you said, "I estimate that the interval 60 to 91 contains μ"? Very, we imagine. For most purposes, however, such a wide interval would not be specific enough. The problem, as you may have surmised, is to trade off some of the specificity that goes with a point estimate for some of the confidence that comes with a wider interval (or conversely, trade off some of the confidence that goes with a wide interval for more specificity).

Fortunately, a sampling distribution can be used to establish both confidence and the interval. The result is a lower and an upper limit for the unknown population parameter.

Here is the rationale for confidence intervals. Suppose you define a population of scores. A random sample is drawn and the mean ($\bar{X}$) is calculated. Using this mean (and the techniques described in the next section), a statistic called a confidence interval is calculated. (We will use a 95 percent confidence interval in this explanation.) Now suppose that from this population many more random samples are drawn and a 95 percent confidence interval is calculated for each. For most of the samples, $\bar{X}$ will be close to μ and μ will fall within the confidence interval. Occasionally, of course, a sample will produce an $\bar{X}$ far from μ and the confidence interval about $\bar{X}$ will not contain μ. The method is such, however, that the probability of these rare events can be measured and held to an acceptable minimum like 5 percent. The result of all this is a method that produces confidence intervals, 95 percent of which contain μ.

In a real-life situation, you draw *one* sample and calculate *one* interval. You do not know whether or not μ lies between the two limits, but the method you have used makes you 95 percent confident that it does.

Calculating the Limits of a Confidence Interval

Having drawn a random sample and calculated the mean and standard error, you can find the lower limit (LL) of a 95 percent confidence interval with the formula

$$LL = \bar{X} - 1.96 s_{\bar{X}}$$

In a similar way, the upper limit (UL) may be found with

$$UL = \bar{X} + 1.96 s_{\bar{X}}$$

The number 1.96 is the z score that leaves $2\frac{1}{2}$ percent of the curve in each tail. For the problem described earlier, in which a random sample of 36 produced a mean of 76 and a standard deviation of 8, we will calculate a 95 percent confidence interval. To use the formulas for LL and UL, you need $s_{\bar{X}}$:

$$s_{\bar{X}} = \frac{s}{\sqrt{N}} = \frac{8}{\sqrt{36}} = 1.33$$

Here are the limits of the confidence interval about the sample mean of 76:

$$LL = \bar{X} - 1.96s_{\bar{X}} = 76 - 1.96(1.33) = 73.39$$
$$UL = \bar{X} + 1.96s_{\bar{X}} = 76 + 1.96(1.33) = 78.61$$

Thus, 73.39 and 78.61 are the lower and upper limits of a 95 percent confidence interval about the sample mean of 76.

A long interpretation is that 73.39 to 78.61 is one set of limits that is calculated by a method that captures μ 95 percent of the time. This leaves you 95 percent confident that 73.39–78.61 contains μ. An abbreviated interpretation is that you are 95 percent confident that the interval 73.39–78.61 contains μ.

A caution about words is in order. It is *not* proper to say that you have "95 percent confidence that μ is in the interval." This implies that μ is variable, and you want to do just the opposite: imply that the confidence interval is variable. So the proper terminology is to say that you have "95 percent confidence that the interval contains μ."

The general formulas for the limits of a confidence interval about a mean are

$$LL = \bar{X} - z(s_{\bar{X}})$$
$$UL = \bar{X} + z(s_{\bar{X}})$$

where z is chosen from the normal-curve table according to the **confidence level** you want for your interval. For a 99 percent confidence interval, a z value of 2.58 is appropriate. To use these formulas, $\bar{X}$ and $s_{\bar{X}}$ should be based on a large (30 or more) random sample from the population.

Here is a real-life situation in which confidence intervals are used. *Statistical process control* is a procedure used by manufacturing companies to reduce or eliminate "rejects" from turning up at the end of an assembly line. In a typical manufacturing process, clearly defined steps occur between raw input and finished output. The basic idea of statistical process control (SPC) is to establish a confidence interval for each step in the production process, sample products as they leave that step, and modify the process if a sample produces a mean that is outside the range of the confidence interval (Berger and Hart, 1986). An early step in SPC is to establish a confidence interval during a time when the production process is working well.

To illustrate, imagine a factory that is producing 10-inch-diameter drain pipes. During times when the assembly line is "in control," pipes are pulled out and diameters measured. Suppose a sample of 36 pipes produced the following statistics:

$$\Sigma X = 361.80$$
$$\Sigma X^2 = 3636.44$$

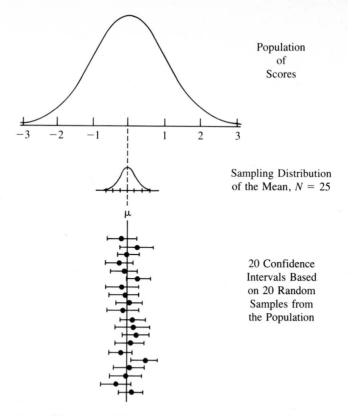

Figure 6.7 Twenty 95 percent confidence intervals, each calculated from a random sample of 25 from a normal population

We will use these measurements to construct a 99 percent confidence interval. As you can verify with your calculator, $\bar{X} = 10.05$ inches, $s = 0.100$ inch, and $s_{\bar{X}} = 0.0167$ inch.[8] The 99 percent confidence interval would be

LL = $10.05 - 2.58(0.0167) = 10.05 - 0.043 = 10.007$ inches
UL = $10.05 + 2.58(0.0167) = 10.05 + 0.043 = 10.093$ inches

Having this confidence interval, production workers can pull out samples of 36 and calculate a mean. If this sample mean is within the confidence interval, all is well.

If the sample mean is outside of the interval, two possibilities exist. The manufacturing process is fine, but a combination of rare (and random) events produced a sample mean that is out of the confidence interval. The probability of this is .01. The other possibility is that something is wrong with the manufacturing process. Adjusting or fixing is appropriate. As you can imagine, practical managers

---

[8] The statistics used in SPC are usually simpler than those in this illustration. For example, s is not calculated, it is estimated from the range, and N is smaller, usually less than 10. In addition, the confidence interval is often 3σ's wide, resulting in approximately 99.7 percent confidence.

figure that the second possibility is more likely than the first and they stop the process and look for problems to fix.

The concept of confidence intervals has been known to give students trouble. Here is another explanation of confidence intervals, this time with a picture.

Figure 6.7 (see page 135) shows the following:

1. A population of scores (top curve[9])
2. A sampling distribution of the mean when $N = 25$ (small curve)
3. Twenty 95 percent confidence intervals based on random samples ($N = 25$) from the population.

On each of the 20 horizontal lines, the endpoints represent the lower and upper limits and the filled circle shows the mean. As you can see, nearly all the confidence intervals have "captured" μ. One has not. Find it. Does it make sense to you that one out of 20 of the 95 percent confidence intervals would not contain μ?

In summary, confidence intervals for population means produce an interval statistic (lower and upper limits) that is destined to contain μ 95 percent of the time (or 99 or 90). Whether or not a particular confidence interval contains the unknowable μ is uncertain. But you have control over the degree of that uncertainty.

We will end this section with yet another caution about words. The term *confidence level* is used for problems of estimation, such as confidence intervals, and the term *significance level* is used for problems of hypothesis testing.

PROBLEMS

25. Place 95 percent confidence intervals around $\bar{X}$ for the following samples.

Sample	$\bar{X}$	s	N
a	36	2	100
b	36	10	100
c	36	10	1000

26. Here is one final problem from the data gathered by the economics students. Establish a 99 percent confidence interval about their sample mean of $36,900. Recall that $s = \$10,000$ and $N = 38$. Write a statement about the meaning of this confidence interval.

27. Place a 95 percent confidence interval about the mean arithmetic score of 49.5 made by the fifth-grade students whose parents provided immediate feedback on homework assignments ($s = 9$, $N = 36$). Tell what this interval means.

28. In Figure 6.7, the 20 lines that represent confidence intervals vary in length. What aspect of the random sample causes this variation?

29. Look at Figure 6.7. Imagine that the confidence intervals had been based on $N = 100$.
   a. How many of the 20 intervals would be expected to capture μ?
   b. Would the confidence intervals be wider or more narrow? By how much?

30. Look at Figure 6.7. Imagine that the lines were 90 percent confidence intervals rather than 95 percent confidence intervals.
   a. How many of the 20 intervals would be expected to capture μ?
   b. Would the confidence intervals be wider or more narrow?

[9] It is not necessary that the population be normal.

OTHER SAMPLING DISTRIBUTIONS

Now you have been introduced to the sampling distribution of the mean. The mean is clearly the most popular statistic among researchers. (Of all statistics, the mean is the mode.) There are times, however, when the statistic necessary to answer a researcher's question is not the mean. For example, to find the degree of relationship between two variables, you need a correlation coefficient. To determine whether a treatment causes more variable responses, you need a standard deviation. Proportions are commonly used statistics. In each of these cases (and indeed, for any statistic), the basic hypothesis-testing procedure you have just learned is often used by researchers. That procedure is:

1. Hypothesize a population parameter.
2. Draw a random sample and calculate a statistic.
3. Compare the statistic with the sampling distribution of that statistic and decide how likely it is that such a population would produce such a sample statistic.

Where do you find sampling distributions for statistics other than the mean? In the rest of this book you will encounter new statistics and new sampling distributions. Tables D–L in Appendix C represent sampling distributions from which probability figures can be obtained. In addition, some statistics have sampling distributions that are normal, which allows you to use the familiar normal curve. Finally, some statistics and their sampling distributions are covered in other books.

Along with every sampling distribution comes a standard error. Just as every statistic has its sampling distribution, every statistic has its standard error. For example, the standard error of the median is the standard deviation of the sampling distribution of the median. The standard error of the variance is the standard deviation of the sampling distribution of the variance. Worst of all, the standard error of the standard deviation is the standard deviation of the sampling distribution of the standard deviation. If you follow that sentence, you probably understand the concept of standard error quite well.

Our chapter summary is that statistics are variable things, that a picture of that variety is a sampling distribution, and that a sampling distribution can be used to obtain probability figures.

RANDOM SAMPLES—THE TRUTH REVEALED

We have used the phrase *random sample(s)* more than 25 times in this chapter. The techniques of inferential statistics that you are learning in this book are based on the assumption that a random sample has been drawn. But how often do you find random samples in actual data analysis? Seldom. Nevertheless, there are two justifications for the continued use of nonrandom samples.

In the first place, an experiment is always a practical effort. After you get an idea, you have to gather equipment or materials, arrange for subjects, conduct the experiment, analyze the results, and tell the story of what your experiment reveals.

Getting participants for your research is necessary, and it is just not practical to get a random sample of college students, first-graders, lower-middle-class workers, or people with passive-aggressive personality disorders. A nonrandom sample will have to do, so the investigator tries to obtain a representative sample, being careful to balance or eliminate as many sources of bias as possible.

In the second place, the question of generalizability can be answered empirically by finding out whether other samples produce the same conclusions. This kind of checkup is practiced continually. Usually the results based on samples that are unsystematic (but not random) are true for other samples from the same population.

Both of these justifications develop a very hollow ring, however, if someone demonstrates that one of your samples is biased and that a representative sample proves your conclusions false.

PROBLEMS
   31. Give a verbal description of the procedures you would follow to construct an empirical sampling distribution of the median.

   32. What is the name of the standard deviation of the distribution described in Problem 31?

7

DIFFERENCES BETWEEN MEANS

Objectives for Chapter 7: After studying the text and working the problems in this chapter, you should be able to:

1. Describe the design of a simple experiment
2. Explain the logic of inferential statistics
3. Define with words the null hypothesis
4. Describe a sampling distribution of mean differences in words, calculate its standard error, and use it to determine whether the two sample means came from the same population
5. Define level of significance
6. Define critical value
7. Define Type I and Type II errors
8. Define α and β and describe the relationship between them
9. Analyze the data from a two-group experiment, reject or retain the null hypothesis, and interpret the results
10. Describe the difference between a one-tailed and a two-tailed test of significance
11. Describe the factors that affect the rejection of the null hypothesis

One of the best things about statistics is that it helps you to understand experiments and the experimental method. The experimental method is probably the most powerful method we have of finding out about natural phenomena. Few *ifs*, *ands*, or *buts* or other qualifiers need to be attached to conclusions based on results from a sound experiment.

Besides being powerful, experiments can be interesting. They can answer such questions as:

1. Is there a difference between the racial attitudes of 9th- and 12th-grade students?
2. Is there a difference between the mean IQs of females and males?
3. Does the removal of 20 percent of the cortex of the brain have an effect on the memory of tasks learned before the operation?

4. Can you *reduce* people's ability to solve problems by teaching them other skills?

Our plan in this chapter is to discuss the simplest kind of experiment and then show how the statistical techniques that you learned about sampling distributions can be expanded to answer questions like those above.

A SHORT LESSON ON HOW TO DESIGN AN EXPERIMENT

The basic ideas underlying a simple two-group experiment are not very complicated.

The logic of an experiment: Start with two equivalent groups. Treat them exactly alike except for one thing. Measure both groups and attribute any statistically significant difference between the two to the one way in which they were treated differently.

This summary of an experiment is described more fully in Table 7.1 and in the following paragraph.

The fundamental question of the experiment outlined in **Table 7.1** is, What is the effect of Treatment A on a person's ability to perform Task Q? In more formal terms, the question is, For Task Q scores, is the mean of the population of those who have had Treatment A different from the mean of the population of those who have not had Treatment A? This experiment has an independent variable with two levels (Treatment A or no Treatment A) and a dependent variable (scores on Task Q). A population of subjects is defined and two random samples are drawn.[1] These

Table 7.1   Summary of a simple experiment

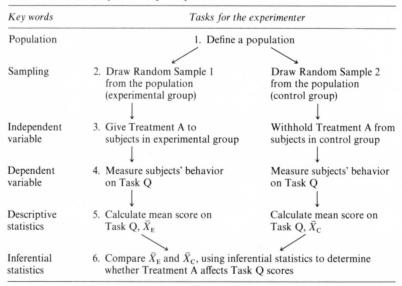

Key words	Tasks for the experimenter	
Population	1. Define a population	
Sampling	2. Draw Random Sample 1 from the population (experimental group)	Draw Random Sample 2 from the population (control group)
Independent variable	3. Give Treatment A to subjects in experimental group	Withhold Treatment A from subjects in control group
Dependent variable	4. Measure subjects' behavior on Task Q	Measure subjects' behavior on Task Q
Descriptive statistics	5. Calculate mean score on Task Q, $\bar{X}_E$	Calculate mean score on Task Q, $\bar{X}_C$
Inferential statistics	6. Compare $\bar{X}_E$ and $\bar{X}_C$, using inferential statistics to determine whether Treatment A affects Task Q scores	

[1] An equivalent statement is that there are two identical populations to begin with. One random sample is then drawn from each population.

random samples are both representative of the population and (approximately) equivalent. Treatment A is then administered to one group (commonly called the **experimental group**) but not to the other group (commonly called the **control group**). Except for Treatment A, both groups are treated in exactly the same way; that is, extraneous variables are held constant or balanced out for the two groups. Both groups perform Task Q and the mean score for each group is calculated. The two sample means almost surely will differ. The question now is whether this observed difference is caused by sampling variation (a chance difference) or by Treatment A. You can answer this question by using the techniques of inferential statistics. Table 7.1 summarizes the ideas of this paragraph.

Our generalized example uses procedures familiar to behavioral scientists. (The dependent variable was test scores.) As you know, the experimental method has broad applicability. It has been used to decide how much sugar to use in a cake recipe, what kind of tea tastes the best, whether a drug is useful in treating an illness, the effect of authoritarian parents on the political attitudes of their children, and many other things. One word that we used is recognized by all experimentalists. The word *treatment* refers to different levels of the independent variable. The experiment in Table 7.1 had two treatments.

In some experimental designs, subjects are assigned to treatments by the experimenter; in others, the experimenter uses a group of subjects who have already been "treated" (for example, being males or being children of authoritarian parents). In either of these designs, the methods of inferential statistics are the same, although the interpretation of the first kind of experiment is.usually less open to attack.[2]

The *raison d'etre* of experiments is *generalization*—generalization to the population.[3] Researchers want to tell a story about the effect of the independent variable on the dependent variable *in the population*. All the work with samples is just to provide evidence to support the story. The nature of samples, however, is that they have chance built into them. In the next section we will describe a method that allows you to use samples to get evidence to support your story about populations even though the samples are full of chance.

THE LOGIC OF INFERENTIAL STATISTICS

This section presents the logical thought processes and some terms that will be important throughout the remainder of this book. It deserves your careful attention.

A decision must be made about the population of those given Treatment A, but it must be made on the basis of sample data. Accept from the start that because of your decision to use samples, you can never know for sure whether or not Treatment A has an effect. Nothing is ever *proved* through the use of inferential statistics. You can only state probabilities, which are never exactly one or zero.

The decision-making goes like this. In a well-designed two-group experiment, all the imaginable results can be reduced to two possible outcomes: either Treatment A

[2] We are bringing up an issue that is beyond the scope of this statistics book. Courses with titles such as "Experimental Methods" and "Research Design" cover the intricacies of interpreting experiments.

[3] *Raison d'etre* (find someone who can pronounce French phrases to teach you how to say this one) means "reason for existence."

has an effect or it does not. Make a tentative assumption that Treatment A does *not* have an effect and then, using the results of the experiment for guidance, find out how probable it is that the assumption is correct. If it is not very probable, rule it out and say that Treatment A has an effect. If the assumption is probable, you are back where you began; you have the same two possibilities you started with.[4]

We put this into the language of an experiment:

1. Begin with two logical possibilities, H_0 and H_1.

   H_0: Treatment A *did not* have an effect; that is, the mean of the population of scores of those who received Treatment A is equal to the mean of the population of scores of those who did not receive Treatment A, and thus the difference between population means is zero. This possibility is symbolized H_0 (pronounced "H sub oh").

   H_1: Treatment A *did* have an effect; that is, the mean of the population of scores of those who received Treatment A is not equal to the mean of the population of scores of those who did not receive Treatment A. This possibility is symbolized H_1 (pronounced "H sub one").

2. Tentatively assume that Treatment A had no effect (that is, assume H_0). If H_0 is true, the two random samples should be alike except for the usual variations in samples. Thus, the difference in the sample means is tentatively assumed to be due to chance.

3. Determine the sampling distribution for these differences in sample means. This sampling distribution gives you an idea of the differences you can expect if only chance is at work.

4. By subtraction, obtain the actual difference between the experimental-group mean and the control-group mean.

5. Compare the difference obtained to the differences expected (from Step 3) and conclude that the difference obtained was either expected or unexpected.

   a. Expected: Differences of this size are very probable just by chance, and the most reasonable conclusion is that the difference between the experimental group and the control group may be attributed to chance. Thus, retain both possibilities in Step 1.

   b. Unexpected: Differences of this size are highly improbable and the most reasonable conclusion is that the difference between the experimental group and the control group is due to something besides chance. Thus, reject H_0 and accept H_1; that is, conclude that Treatment A had an effect.

6. Write a conclusion about the experiment using the terms of the experiment. Your conclusion should describe the relationship of the dependent variable to the independent variable for the population.

The basic idea is to assume that there is no difference between the two population means and then let the data tell you whether the assumption is reasonable. If the assumption is not reasonable, you are left with only one alternative: the populations have different means.

The assumption of no difference is so common in statistics that it has a name: the **null hypothesis,** symbolized, as you have already learned, H_0. The null hypothesis

---

[4] In courses in logic, this kind of reasoning is called the rule of negative inference (*modus tollens*).

is often stated in formal terms:

$$H_0: \mu_1 - \mu_2 = 0 \quad \text{or} \quad H_0: \mu_1 = \mu_2$$

The null hypothesis states that the mean of one population is equal to the mean of a second population.[5] We will frequently use the term *null hypothesis* and the symbol H_0. It will be important that you understand what it is.

H_1 is referred to as an **alternative hypothesis.** Actually, there is an infinite number of alternative hypotheses—that is, the existence of any difference other than zero. In practice, however, it is usual to choose one of three possible alternative hypotheses before the data are gathered:

1. $H_1: \mu_1 \neq \mu_2$. In the example of the simple experiment, this hypothesis states that Treatment A had an effect without stating whether the treatment improves or disrupts performance on Task Q. Most of the problems in this chapter use this H_1 as the alternative to H_0. If you reject H_0 and accept this H_1, *you must examine the means and decide whether Treatment A facilitated or disrupted performance on Task Q.*
2. $H_1: \mu_1 > \mu_2$. The hypothesis states that Treatment A improves performance on Task Q.
3. $H_1: \mu_1 < \mu_2$. The hypothesis states that Treatment A disrupts performance on Task Q.

The section on one-tailed and two-tailed tests that is later in this chapter will give you more information on choosing an alternative hypothesis.

A word of caution about terminology: the null hypothesis meets with one of two fates at the hands of the data. It may be rejected, which allows you to accept an alternative hypothesis, or it may be retained. If it is retained, *it is not proved as true;* it is simply retained as one among many possibilities.

Perhaps an analogy will help with this distinction about terminology. Suppose a masked man has burglarized a house and stolen all the silver. For purposes of this story, suppose there are only two people who could possibly have done the deed, HI and HO. The lawyer for HO tries to establish beyond reasonable doubt that her client was out of the state during the time of the robbery. If she can do this, it will exonerate HO (HO will be rejected, leaving only HI as a suspect). However, if she cannot establish this, the situation will revert to its original state: HI or HO could have stolen the silver away, and both are retained as suspects. So the null hypothesis can be *rejected* or *retained,*[6] but it can never be proved with certainty to be true or false by using the methods of inferential statistics. Statisticians are usually very careful with words. That is probably because they are used to mathematical symbols, which are very precise. Regardless of the reason, this distinction between *retained* and *proved,* though subtle, is important. The next section shows you how to do Steps 3 and 5 in the list of steps in an experiment.

---

[5] Actually, the concept of the null hypothesis is broader than simply the assumption of no difference, although that is the only version used in this book. Under some circumstances, a difference other than zero is the hypothesis tested.

[6] Some textbooks use the term *accepted* instead of *retained.*

<table>
<tr><td>CLUE TO
THE FUTURE</td><td>You have just completed the section that explains the reasoning that is at the heart of inferential statistics. This reasoning is a part of every test in every chapter that follows in this book (and will also turn up in many future books, articles, and discussions).</td></tr>
</table>

PROBLEMS

1. In your own words, outline the logic of an experiment.
2. In the experiment described in Table 7.1, what method was used to ensure that the two samples were equivalent before treatment?
3. In your own words, outline the logic of inferential statistics.
4. What is H_0 a symbol for?
5. Tell what the null hypothesis is.

SAMPLING DISTRIBUTION OF A DIFFERENCE BETWEEN MEANS

We have used the term *difference* a great deal in this chapter, including in the title. A difference is simply the answer in a subtraction problem. As explained in the section on the logic of inferential statistics, the difference that is of interest is the difference between two means. You evaluate the obtained difference by comparing it with a sampling distribution of differences between means (often called a sampling distribution of mean differences).

Chapter 6 contained a great deal of discussion about sampling distributions, especially the sampling distribution of the mean. Recall that a sampling distribution is a frequency distribution of sample statistics, all calculated from samples of the same size drawn from the same population; the standard deviation of that frequency distribution is called a standard error. Precisely the same logic holds for a sampling distribution of differences between means.

We can best explain a sampling distribution of differences between means by describing the procedure for generating an *empirical* sampling distribution of mean differences. Find for yourself a single population of scores. Randomly draw *two* samples, calculate the mean of each, and subtract the second mean from the first. Do this many times and then arrange all the differences into a frequency distribution. Such a distribution will consist of a number of scores, each of which is a *difference* between two sample means. Think carefully about the mean of the sampling distribution of mean differences. Stop reading and decide what the numerical value of this mean will be.

The mean of a sampling distribution of mean differences is zero because positive and negative differences occur equally often, giving you an algebraic sum close to zero. If you were successful on this problem, congratulations! Few students get it right when they first read the chapter.

This sampling distribution of mean differences has a standard deviation called the standard error of a difference between means.

In many experiments, it is obvious that there are *two* populations to begin with—for example, a population of men and a population of women. The question, however, is whether they are equal on the dependent variable. To generate a sampling distribution of differences between means in this case, assume that, on the dependent variable, the two populations have the same mean, standard deviation, and form (shape of the distribution). Then draw one sample from each population, calculate the means, and subtract one from the other. Do this many times. Arrange the differences between sample means into a frequency distribution.

The sampling distributions of differences between means that you will use will be theoretical distributions, not the empirical ones we described in the last four paragraphs. However, a description of the procedures for an empirical distribution, which is what we have just given, is usually easier to understand in the beginning.

Two things about a sampling distribution of mean differences are constant: the mean and the form. The mean is zero, and the form is normal if the sample means are based on large samples. Again, the traditional answer to the question, What is a large sample? is 30 or more.[7]

We will illustrate these ideas with an experiment your authors did (Johnston and Spatz, 1973). The question we wished to answer was, Are the racial attitudes of 9th-graders different from those of 12th-graders? The null hypothesis was that the population means were equal (H_0: $\mu_1 = \mu_2$). The alternative hypothesis was that they were not equal (H_1: $\mu_1 \neq \mu_2$). The subjects in this experiment were 9th- and 12th-grade black and white students who expressed their attitudes about persons of their own gender but different race. Higher scores represent more positive attitudes. Thus, the independent variable was grade and the dependent variable was racial attitude score. **Table 7.2** shows the summary data.

As you can quickly calculate from Table 7.2, the obtained mean difference between samples of 9th- and 12th-graders is 4.10. Now a decision must be made. Should this difference in samples be ascribed to chance (retain H_0; there is no difference between the population means)? Or should we say that such a difference is so unlikely that it is due not to chance but to the different characteristics of 9th- and 12th-grade students (reject H_0 and accept H_1; there is a difference between the populations)? Using a sampling distribution of mean differences (Figure 7.1), a decision can be made.

Figure 7.1 shows a sampling distribution of differences between means that is based on the assumption that there are no population differences between 9th- and 12th-graders—the standard assumption that any experiment would begin with.

Table 7.2 Data from an experiment
that compared the racial attitudes of 9th-
and 12th-grade students

Grade	N	$\bar{X}$	s
9	200	48.42	12.16
12	200	52.52	11.83

[7] The sampling distribution of mean differences will be normal with smaller samples if the population(s) from which they are drawn is (are) symmetrical.

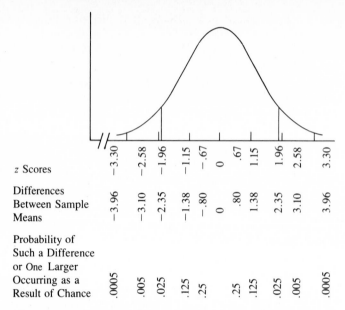

z Scores	−3.30	−2.58	−1.96	−1.15	−.67	0	.67	1.15	1.96	2.58	3.30
Differences Between Sample Means	−3.96	−3.10	−2.35	−1.38	−.80	0	.80	1.38	2.35	3.10	3.96
Probability of Such a Difference or One Larger Occurring as a Result of Chance	.0005	.005	.025	.125	.25		.25	.125	.025	.005	.0005

Figure 7.1   Sampling distribution from the racial attitudes study. It is based on chance and shows z scores, probabilities of those z scores, and differences between sample means.

(You will learn to construct sampling distributions of mean differences in later sections.) In other words, Figure 7.1 assumes H_0.

Figure 7.1 is a normal curve that shows z scores, possible differences between sample means in the racial attitudes study, and probabilities associated with the z scores and difference scores. Study Figure 7.1 carefully and notice the following characteristics:

1. The mean is zero, since it is based on the assumption of no difference between means (H_0: $\mu_1 = \mu_2$).
2. As the size of the difference between means increases, z scores increase.
3. As z scores and differences between means become larger, probabilities become smaller.
4. As before, small probabilities go with small areas of the curve.

Our obtained difference, 4.10, is not even shown on the distribution. Such events are extremely rare when the true difference is zero and only chance is at work. From Figure 7.1, you can see that a difference of 3.96 or more would be expected five times in 10,000 (.0005). Since a difference of −3.96 or greater also has a probability of .0005, we can add the two probabilities together to get .001 (more about this in the section on one- and two-tailed tests). Since our difference was 4.10 (less probable than 3.96), we can conclude that the probability of a difference of ±4.10 being due to chance is less than .001. It seems reasonable to rule out chance—that is, to reject H_0 and thus accept H_1. By examining the means of the two groups in Table 7.2, we can write a conclusion using the terms in the experiment: 12th-graders have more positive attitudes toward people of their own gender but different race than do 9th-graders.

A PROBLEM AND ITS ACCEPTED SOLUTION

The probability that populations of 9th- and 12th-grade attitude scores are the same was so small ($p < .001$) that it was easy to rule out chance as an explanation for the difference. But what if that probability had been .01, or .05, or .25, or .50? How to divide this continuum into a group of events that is "due to chance" and another that is "not due to chance"—that is the problem.

It may already be clear to you that any solution will be an arbitrary one. Breaking any continuum into two parts will leave you uncomfortable about the events close to either side of the break. Nevertheless, a solution does exist, and there are two important concepts involved: *level of significance* and *critical region*.

The generally accepted solution is to say that the .05 level of probability is the cutoff between "due to chance" and "not due to chance." The name of the cutoff point that separates "due to chance" from "not due to chance" is the **level of significance.** If an event has a probability of .05 or less (for example, $p = .03$, $p = .01$, or $p = .001$), H_0 is rejected and the event is considered *significant* (not due to chance). If an event has a probability of .051 or greater (for example, $p = .06$, $p = .50$, or $p = .99$), H_0 is retained and the event is considered not significant (may be due to chance). Here the word *significant* is not synonymous with "important." A significant event in statistics is one that is not attributed to chance.

The area of the sampling distribution that covers the events that are "not due to chance" is called the **critical region.** If an event falls in the critical region, H_0 is rejected. Figure 7.1 identifies the critical region for the .05 level of significance. As you can see, the difference in means between 9th- and 12th-grade racial attitudes (4.10) falls in the critical region, so H_0 should be rejected.

Although widely adopted, the .05 level of significance is not universal. Some investigators use the .01 level in their research. When the .01 level is used and $H_1: \mu_1 \neq \mu_2$, the critical region consists of .005 in each tail of the sampling distribution. In Figure 7.1, differences greater than -3.10 or 3.10 are required in order to reject H_0 at the .01 level.

In textbooks, a lot of lip service is paid to the .05 level of significance. In actual research, the practice is to run the experiment and report any significant differences at the smallest correct probability value. Thus, in the same report, some differences may be reported as significant at the .001 level, some at the .01 level, and some at the .05 level.

Recently there has been a trend toward reporting exact probability figures, such as $p = .032$ or $p = .008$. Regardless of how the probabilities are reported, however, researchers view .05 as an important cutoff (Nelson, Rosenthal, and Rosnow, 1986). When .10 or .20 is used as a significance level, a justification is given.

PROBLEMS

6. What does level of significance mean?
7. What events are included in the critical region?
8. What is the largest probability usually adopted as a level of significance?

HOW TO CONSTRUCT A SAMPLING
DISTRIBUTION OF DIFFERENCES BETWEEN MEANS

You already know two important characteristics of a sampling distribution of differences between means. The mean is zero, and the form is normal. When we constructed Figure 7.1, the sampling distribution of differences between the racial attitudes of 9th- and 12th-graders, we used the normal-curve table and a form of the familiar z score[8]

$$z = \frac{\bar{X}_1 - \bar{X}_2}{s_{\bar{X}_1 - \bar{X}_2}}$$

where $\bar{X}_1$ = mean of one sample
  $\bar{X}_2$ = mean of a second sample
  $s_{\bar{X}_1 - \bar{X}_2}$ = standard error of a difference

The only new term in this formula is $s_{\bar{X}_1 - \bar{X}_2}$, which is the standard error of a sampling distribution of mean differences. This longer, more descriptive name is usually shortened to that above, the **standard error of a difference.** The formula for it is

$$s_{\bar{X}_1 - \bar{X}_2} = \sqrt{s_{\bar{X}_1}^2 + s_{\bar{X}_2}^2}$$

where $s_{\bar{X}_1 - \bar{X}_2}$ = standard error of a difference between means
  $s_{\bar{X}_1}$ = standard error of the mean of Sample 1
  $s_{\bar{X}_2}$ = standard error of the mean of Sample 2

Now you are in a position to follow all the steps used in constructing Figure 7.1. First, we drew a normal curve and labeled the mean 0. Next we calculated the standard error of a difference. Remember that $s_{\bar{X}} = s/\sqrt{N}$.

$$s_{\bar{X}_1 - \bar{X}_2} = \sqrt{s_{\bar{X}_1}^2 + s_{\bar{X}_2}^2} = \sqrt{\left(\frac{s_1}{\sqrt{N_1}}\right)^2 + \left(\frac{s_2}{\sqrt{N_2}}\right)^2}$$

$$= \sqrt{\left(\frac{12.16}{\sqrt{200}}\right)^2 + \left(\frac{11.83}{\sqrt{200}}\right)^2} = \sqrt{.7393 + .6997} = 1.1996 = 1.20$$

Although the baseline of Figure 7.1 does not specifically indicate standard error units of 1.20, the hash marks on the X axis represent distances of 1.20.

Our next step was to look in Table C, column C, in Appendix C for probability figures of .25, .125, .025, .005, and .0005. We found the corresponding z scores and placed both z scores and probability figures at the appropriate places on Figure 7.1.

---

[8] The formula in the text is the "working model" of the more general formula

$$z = \frac{(\bar{X}_1 - \bar{X}_2) - (\mu_1 - \mu_2)}{s_{\bar{X}_1 - \bar{X}_2}}$$

Since our null hypothesis is that $\mu_1 - \mu_2 = 0$, the term in parentheses on the right is 0, leaving you with the "working model." This more general formula has a form you have seen before and will see again: the difference between a statistic and a parameter divided by the standard error of the statistic, where the statistic is $\bar{X}_1 - \bar{X}_2$, the parameter is $\mu_1 - \mu_2$, and the standard error is $s_{\bar{X}_1 - \bar{X}_2}$.

The final thing we needed for Figure 7.1 was the difference between sample means associated with each z score. Since

$$z = \frac{\bar{X}_1 - \bar{X}_2}{s_{\bar{X}_1 - \bar{X}_2}}$$

then

$$\bar{X}_1 - \bar{X}_2 = (z)(s_{\bar{X}_1 - \bar{X}_2})$$

Solving this formula using each z score gave us the difference between means that we placed between the z score and its probability.

There you have it, the procedures we used in constructing Figure 7.1—a picture of the differences we would expect between samples of 9th- and 12th-graders if the populations really did have the same attitudes toward race.

A SHORTCUT—TAILS OF DISTRIBUTIONS

Some students suspect that the construction of an entire sampling distribution is unnecessary. You may be such a student and, if so, you are correct. All that is needed is to determine whether the difference between sample means falls into the *critical region of the sampling distribution,* and there is a fairly easy way to determine this.

To begin with, adopt a level of significance of .05, which gives you a critical region of .025 in each tail of the distribution. Since all sampling distributions of mean differences (based on large samples) are normal curves, the z scores that mark off the critical regions will always be the same, ± 1.96.[9] Thus, any two samples that produce a z score whose *absolute value* is equal to or greater than 1.96 will fall into the critical region and lead to a rejection of the null hypothesis. If the absolute value of the z score is less than 1.96, you retain the null hypothesis. If you will glance back at Figure 7.1, you will see that the critical region is defined by z scores of ± 1.96.

This z score of 1.96 that defines the .05 critical region for every normal sampling distribution has a name: **critical value.** We mentioned earlier that the .05 level of significance is not universal; some investigators adopt a .01 level. Recall also that most investigators report significant differences at their smallest probability value (whether .05, .01, or .001). Each of these probabilities has its own critical value. Stop reading for a moment and test your understanding by deciding whether the critical values associated with critical regions of .01 and .001 are larger or smaller than 1.96.

The critical value to use with a .01 level of significance is 2.58. With a .001 level, use 3.30. You can check these values for yourself by dividing the significance levels in half (.005 and .0005), finding these values in column C of Table C, and noting the corresponding z score.

[9] This is for a two-tailed test. The z-score value for a one-tailed test is given to you later in this chapter.

Using the Shortcut—The z-Score Test

Here is the way we would analyze the data in Table 7.2 if we were starting from scratch. We would find the z score for the observed mean difference:

$$z = \frac{\bar{X}_1 - \bar{X}_2}{s_{\bar{X}_1 - \bar{X}_2}} = \frac{\bar{X}_1 - \bar{X}_2}{\sqrt{\left(\frac{s_1}{\sqrt{N_1}}\right)^2 + \left(\frac{s_2}{\sqrt{N_2}}\right)^2}} = \frac{48.42 - 52.52}{\sqrt{\left(\frac{12.16}{\sqrt{200}}\right)^2 + \left(\frac{11.83}{\sqrt{200}}\right)^2}} = -3.42$$

The result of the z-score test, $z = -3.42$, allows us to reject the null hypothesis at the .001 level of significance. It is unlikely that the two samples of racial attitudes came from the same population. By examining the means, we can write our conclusion: 12th-graders have more positive racial attitudes than do 9th-graders, $p < .001$.

The p in "$p < .001$" refers to the probability of obtaining a difference between samples as large as the one actually observed if the true difference is zero and only chance is at work. To put this in more technical language, p is the probability of obtaining the difference that was observed if the null hypothesis is true.

Here is another problem. Do males or females have higher IQ scores? The independent variable here is gender and the dependent variable is IQ. Suppose the data in **Table 7.3** were obtained.[10]

We begin by tentatively adopting H_0. In this case, H_0 is the assumption that there is no population difference between the mean IQ scores of males and females. Next we find the standard error of the difference:

$$s_{\bar{X}_1 - \bar{X}_2} = \sqrt{(s_{\bar{X}_1})^2 + (s_{\bar{X}_2})^2} = \sqrt{\left(\frac{s_1}{\sqrt{N_1}}\right)^2 + \left(\frac{s_2}{\sqrt{N_2}}\right)^2}$$

$$= \sqrt{\left(\frac{15.4}{\sqrt{72}}\right)^2 + \left(\frac{14.3}{\sqrt{57}}\right)^2} = \sqrt{6.8814} = 2.62$$

Using the z-score formula, you get

$$z = \frac{\bar{X}_1 - \bar{X}_2}{s_{\bar{X}_1 - \bar{X}_2}} = \frac{103.3 - 101.2}{2.62} = \frac{2.1}{2.62} = .80$$

Since the obtained z score is less than 1.96, we may not reject H_0. Thus, our data are not conclusive. In statistical terms, we retain H_0 and we have the same number of

Table 7.3 Hypothetical data on gender differences in IQ

	Males	Females
$\bar{X}$	103.3	101.2
s	15.4	14.3
N	72	57

[10] For a short discussion of the technical problems besetting anyone who makes this comparison on real data, see J. P. Guilford (1967, p. 403). The central problem is that an IQ is a composite score, and females do better on some parts and males do better on others. The importance assigned to a particular part can throw the results one way or the other.

hypotheses now as we had before we gathered any data—namely, H_0 and H_1. In terms of the experiment, we have no evidence that females and males differ in IQ. A terminology convention, widely adopted by researchers, is to refer to nonsignificant differences as **NS**.

PROBLEMS

9. In an experiment that compared two methods of teaching German, the holistic method was found to be superior to a lecture-discussion method (based on a test of comprehension). The article reporting the experiment included the phrase "$p < .01$." Write a description of what this phrase means. Be sure to name the event that the probability refers to.

10. What critical value would be appropriate for a .02 level of significance?

11. An educational psychologist was interested in the effect of set (previous experience) on a problem-solving task. The task was to tie together two strings (A and B) that were suspended from the ceiling 12 feet apart. Unfortunately, the subject could not reach both strings at one time. The plight is shown in the illustration. The solution to the problem is to tie a weight to string A, swing it out, and then catch it on the return. The psychologist supplied the weight (an electrical light switch) under one of two conditions. In one condition, subjects had previously wired the switch into an electrical circuit and used it to turn on a light. In the other condition, subjects had no experience with the switch until they were confronted with the two-string problem. The question was whether having used the switch as an electrical device would have any effect on the time required to solve the problem. (See Birch and Rabinowitz, 1951.)

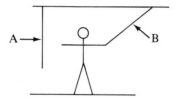

	Wired the switch into circuit	No previous experience with switch
$\bar{X}$	7.40 minutes	5.05 minutes
s	2.13 minutes	2.31 minutes
N	43	41

a. Identify the independent and dependent variables.
b. State the null hypothesis for these data.
c. Perform a z-score test.
d. Interpret the results of the probability you obtain.

12. Examine the following data and tell why you cannot test the difference between the means with a z-score test.

	Experimental group	Control group
$\bar{X}$	148	122
s	118	109
N	13	9

13. Karl Lashley (1890–1958) did many studies to find out how the brain stores memories. In one experiment, rats learned a simple maze that had one blind alley. Afterward, they were anesthetized and operated on. For half the rats, 20 percent of the cortex of the brain was removed. For the other half, the same operation was performed except that no brain tissue was removed (a sham operation). After the rats recovered, they were given 20 trials in the simple maze and the number of errors was recorded. Identify the independent variable and the dependent variable and then analyze the data in the table. Write a conclusion about the effect of a 20 percent loss of cortex on retention by rats of a simple maze task.

	Percent of cortex removed	
	0	20
ΣX	208	252
ΣX^2	1706	2212
N	40	40

14. Estimates of the incidence of asthma run from 2 to 5 percent of the population. In one experiment, a group whose attacks began before age 3 was compared with a group whose attacks began after age 6. The dependent variable was the number of months the asthma persisted. Analyze the difference between means and write a conclusion that tells what you have found.

	Onset of asthma	
	Before age 3	*After age 6*
$\bar{X}$	60	36
s	12	6
N	45	52

15. Suppose Somebody Else conducted an experiment on two methods of teaching statistics: the lecture method and the individualized approach. Students knew of the two methods and chose the method they preferred. The lecture class was taught by Professor Y and the individualized class was taught by Professor Z. For the final examination, Professor Y gave a 150-item multiple choice test and Professor Z gave an essay exam.

	Lecture	*Individualized*
$\bar{X}$	109	121
s	26	39
N	41	37

Analyze the experiment and write a conclusion.

AN ANALYSIS OF POTENTIAL MISTAKES

At first glance, the idea of adopting a significance level of 5 percent seems preposterous to some students.

"You do an experiment," they say, "and you reject H_0 at the .05 level and you conclude that there is a difference in the populations. But in your heart you are uncertain. Perhaps a rare event happened and the difference really is due to chance."

This line of reasoning is fairly common. Many thoughtful students take the next step.

"I don't have to use the .05 level. I'm going to use a level of significance of 1 in 1 million. That way I can reduce the uncertainty."

It is true that adopting the .05 level of significance leaves some room for mistaking a chance difference for a real difference and that lowering the level of significance will reduce the probability of this kind of mistake. However, it increases the probability of another kind. Uncertainty about the conclusion will remain. In this section, we will discuss the two kinds of mistakes that are possible. You will be able to pick up some hints on reducing uncertainty, but to quote a phrase in a famous statistics textbook, "If you agree to draw a sample, you agree to accept some uncertainty about the results."

Look at **Table 7.4,** which shows the two ways to make a mistake.

In Table 7.4, cell 1 shows the situation when the null hypothesis is true and you retain it—your sample data led you to a correct decision. However, if H_0 was true and your sample data led you to reject it (cell 2), you would have made a mistake called a **Type I error.** The probability of a Type I error is symbolized by α (alpha).

On the other hand, suppose H_0 is really false (the second column). If, on the basis of your sample data, you retain H_0 (cell 3), you have made a mistake—a **Type II error,** the probability of which is symbolized by β (beta). If your sample data led you to reject H_0 (cell 4), you made a correct decision.

You are already somewhat familiar with α from your study of level of significance. When the .05 level of significance is adopted, the experimenter concludes that an event with $p \leq .05$ is not due to chance. The experimenter could be wrong; if so, a Type I error has been made. The probability of a Type I error, α, is controlled by the level of significance you adopt.

A proper way to think of α and a Type I error is in terms of "in the long run." Figure 7.1 is a theoretical sampling distribution of mean differences. It is a picture of repeated sampling (that is, the long run). All those differences came from sample means that were drawn from the same population, but some differences were so large that they could be expected to occur only 5 percent of the time. In an experiment, however, you have only one difference, which is based on the two sample means. If this difference is so large that you conclude there are two populations whose means are not equal, you *may* have made a Type I error. However, the probability of such an error is not more than .05.

Table 7.4 Type I and Type II errors*

		The true situation in the population	
		H_0 true	H_0 false
The decision made on the basis of sample data	Retain H_0	1. Correct decision	3. Type II error
	Reject H_0	2. Type I error	4. Correct decision

*In this case, error means mistake.

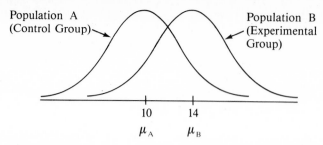

Figure 7.2   Frequency distribution of raw scores when H_0 is false

The calculation of β is a more complicated matter. For one thing, a Type II error can be committed only when the two populations have different means. Naturally, the farther apart the means are, the more likely you are to detect it, and thus the lower β is. We will discuss other factors that affect β in the last section of this chapter, "How to Reject the Null Hypothesis."

The general relationship between α and β is an inverse one. As α goes down, β goes up; that is, if you insist on a larger difference between means before you call the difference nonchance, you are less likely to detect a real nonchance difference if it is small. Figures 7.2 and 7.3 illustrate this relationship.

Figure 7.2 is a picture of two populations. Since these are populations, the "truth" is that the mean of the experimental group is four points higher than that of the control group.[11] If a sample is drawn from each population, there is only one correct decision: reject H_0. However, will the investigator make the correct decision? Suppose samples are drawn and the mean difference is exactly 4. Our samples are telling us the exact truth about the populations. But will techniques of inferential statistics lead us to a correct decision?

Our decision will be based on a sampling distribution like the one in Figure 7.3. (We constructed this particular sampling distribution so we could illustrate the following points.)

You can see in **Figure 7.3** that if the critical region covers 5 percent of the curve (all the gray areas), a mean difference of 4 points falls within it. So if α were set at .05, you would correctly reject the null hypothesis. However, if the critical region covers only 1 percent of the curve (the dark gray areas), the mean difference of 4 points is not within it. So if α was .01 you would not reject H_0. Failure to reject H_0 in this case is a Type II error.

At this point, we can return to our discussion of setting the significance level. The suggestion was, Why not reduce the significance level to 1 in 1 million? From the analysis of the potential mistakes, you can answer that when you decrease α, you increase β. So protection from one error is traded for liability to another kind of error.

Most persons who use statistics as a tool set α (usually at .05) and let β fall where it may. The actual calculation of β, though important, will not be discussed in this book. (To calculate β, see Guilford and Fruchter, 1978, pp. 176–182.)

---

[11] Such "truth" is available only in hypothetical examples in textbooks. In the real world of experimentation, you do not know population parameters. This example, however, should help you understand the relation of α to β.

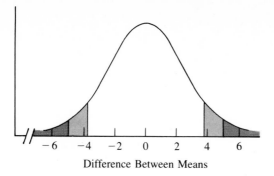

Difference Between Means

Figure 7.3 Sampling distribution of differences between means from Populations A and B of Figure 7.2, if H_0 were true

PROBLEMS

16. What is a Type I error?
17. What is a Type II error?
18. Suppose the real situation is that the treatment has an effect on scores. Is it possible to make a Type I error?
19. Distinguish among α, Type I error, and level of significance.
20. Suppose an experimenter chose to use $\alpha = .01$ rather than .05. What effect would this have on the probability of Type I and Type II errors?

ONE-TAILED AND TWO-TAILED TESTS

Earlier, we told you that it is usual to choose one of three possible alternative hypotheses before the data are gathered.

1. $H_1: \mu_1 \neq \mu_2$. This hypothesis simply says that the population means differ, but it makes no statement about the direction of the difference.
2. $H_1: \mu_1 > \mu_2$. Here, the hypothesis is made that the mean of the first population is greater than the mean of the second population.
3. $H_1: \mu_1 < \mu_2$. The mean of the first population is smaller than the mean of the second population.

So far in this chapter, you have been working with the first H_1. You have tested the null hypothesis, $\mu_1 = \mu_2$, against the alternative hypothesis, $\mu_1 \neq \mu_2$. The null hypothesis was rejected when you found large positive deviations $(\bar{X}_1 > \bar{X}_2)$ *or* large negative deviations $(\bar{X}_1 < \bar{X}_2)$. When α was .05, the .05 was divided into .025 in each tail of the sampling distribution, as seen in **Figure 7.4.** If the absolute value of the obtained z score is greater than 1.96, you can reject H_0 and accept either of the possible alternative hypotheses, $\mu_1 > \mu_2$ or $\mu_1 < \mu_2$. This is called a **two-tailed test of significance,** for reasons that should be obvious from Figure 7.4.

In some research the investigator is interested only in deviations in one direction. In formal terms, the investigator is eager to detect $\mu_1 > \mu_2$ if that is the case for the two populations. The relationship $\mu_1 < \mu_2$, however, is of no interest. In

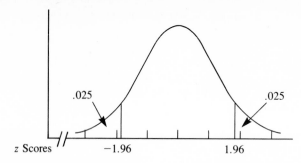

Figure 7.4 A two-tailed test of significance, with $\alpha = .05$

such a case, a **one-tailed test** is appropriate. The two ways to express a one-tailed test are:

$$H_0: \mu_1 \geq \mu_2 \qquad H_0: \mu_1 \leq \mu_2$$
$$\qquad\qquad \text{or}$$
$$H_1: \mu_1 < \mu_2 \qquad H_1: \mu_1 > \mu_2$$

Figure 7.5 is a picture of the sampling distribution for a one-tailed test for $\mu_1 > \mu_2$. In a one-tailed test, the critical region is all in one tail of the sampling distribution. The only outcome that allows you to reject H_0 is when μ_1 is so much larger than μ_2 that the z score is 1.65 or more. Notice in Figure 7.5 that if you are running a one-tailed test, there is *no way* to conclude that μ_1 is less than μ_2, even if $\bar{X}_2$ is many times the size of $\bar{X}_1$. In a one-tailed test, you are interested in only one kind of difference. One-tailed tests are generally used when an investigator knows a great deal about the particular research area or when *practical* reasons dictate an interest in establishing $\mu_1 > \mu_2$ but not $\mu_1 < \mu_2$.

Here are two examples that illustrate when one- or two-tailed tests should be used. The first is an example that calls for a two-tailed test. Jojoba (ho-ho-ba) and guayule (wy-òo-lee) are plants that grow well in semiarid regions such as the Sonoran Desert of Arizona and California (Maugh, 1977). The seeds of the jojoba can be processed into a liquid wax that has many industrial applications. At present, these needs are being met by expensive synthetics. The other plant, guayule, can be processed into rubber that is virtually identical to rubber that is now imported. Given a particular area for cultivation, an experiment could be run. The question to be answered is, Which of the two plants is more productive in this particular area?

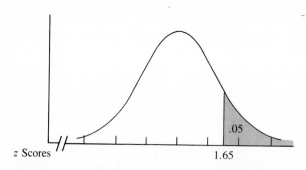

Figure 7.5 A one-tailed test of significance, with $\alpha = .05$

To answer this question, you need a two-tailed test because you want to be able to show that $\mu_1 < \mu_2$ *or* that $\mu_1 > \mu_2$.

The second example, also from agronomy, calls for a one-tailed test. Suppose a newly developed variety of rice is being compared with an old, established variety. For the established variety, much is already known: yield, resistance to disease, optimal schedules for flooding, and the like. Research on the new variety will continue only if the new variety shows promise of a greater yield. Now the question is, Is the new variety more productive? and the appropriate statistical test for an experiment is one-tailed. If the new variety is equal to the old standard, or if it is worse, research will stop. The experimenter is only interested in being able to detect a difference such as $\mu_1 > \mu_2$.

There is some controversy among behavioral scientists about the use of one- or two-tailed tests. When in doubt, most behavioral scientists use a two-tailed test. The decision to use a one-tailed or a two-tailed test should be made before the data are gathered. The answers to the problems in this text are based on two-tailed tests unless a one-tailed test is specified. (For a collection of articles discussing one- and two-tailed tests, see Kirk, 1972.)

PROBLEMS

21. Two sisters who attended different midwestern universities began a friendly discussion of the relative intellectual capacities of their fellow students. Each was sure that she had to compete with brighter students than her sister. Though both were majoring in literature, each had an appreciation for quantitative thinking, so they decided to settle their discussion with ACT admission scores made by entering freshmen at the two schools. Suppose they bring you the following data and ask for your analysis. The N's represent all the freshmen for one year.

	The U.	State U.
$\bar{X}$	21.4	21.5
s	3	3
N	8000	8000

The sisters agreed to set $\alpha = .05$. Begin by deciding whether to run a one-tailed or a two-tailed test. Run the test and interpret your results. Be sure to carry several decimal places in your work.

22. What z score should be used for a one-tailed test with $\alpha = .01$?

23. A computer salesman is promoting Brand Z by telling a researcher that his brand is much faster than Brand Y, the computer on the researcher's desk. The researcher works some problems on both computers and records the amount of time required (seconds).
    a. Should a one-tailed or a two-tailed test be used?
    b. What advice would you give?

	Time scores		
	ΣX	ΣX^2	N
Brand Z	1216	30,830	54
Brand Y	984	19,390	54

24. A class of nursing students, interested in preventive medicine, set up a blood-pressure monitoring station in a gym where many students and faculty engaged in a noontime aerobics program. Some of those who had their blood pressure measured were veterans of the aerobics program and some were newcomers to it. Identify the independent and dependent variables, analyze the following data, and write a sentence about the effect of participating in the aerobics program.

	Veterans	Newcomers
$\bar{X}$	116	125
s	15	19
N	36	36

WHY .05?

The story of how science and other disciplines came to adopt .05 as the arbitrary point separating differences attributed to chance from those not attributed to chance is not usually covered in textbooks. The story takes place in England at the turn of the century.

It appears that the earliest explicit rule about α was in a 1910 article by Wood and Stratton, "The Interpretation of Experimental Results," which was published in *The Journal of Agricultural Science*. Thomas B. Wood, the principal editor of this journal, advised researchers to take "30:1 as the lowest odds which can be accepted as giving practical certainty that a difference... is significant" (Wood and Stratton, 1910). "Practical certainty" meant "enough certainty for a practical farmer." A consideration of the circumstances of Wood's journal provides a plausible explanation of his advice to researchers.

At the turn of the century, farmers in England were doing quite well. For example, Biffen (1905) reported that wheat production averaged 30 bushels per acre compared to 14 in the United States, 10 in Russia, and 7 in Argentina. Politically, of course, England was near the peak of its influence.

In addition, a science of agriculture was beginning to be established. Practical scientists with an eye on increased production and reduced costs conducted studies of weight gains in cattle, pigs, and sheep, amounts of dry matter in milk and mangels (a kind of beet), and yields of many kinds of grain. In 1905, *The Journal of Agricultural Science* was founded so these results could be shared around the world. Conclusions were reached after calculating probabilities, but the researchers avoided or ignored the problem created by probabilities between .01 and .25.

Wood's recommendation in his "how to interpret experiments" article was to adopt an α level that would provide a great deal of protection against a Type I error. (Odds of 30:1 against the null hypothesis convert to $p = .0323$.) A Type I error in this context would be a recommendation that, when implemented, did not produce improvement. Your authors think that Wood and his colleagues wanted to be *sure* that when they made a recommendation, it would be correct.

A Type II error (failing to recommend an improvement) would not cause much damage; farmers would just continue to do well (though not as well as they might). A Type I error, however, would at best result in no agricultural improvement *and* in a loss of credibility and support for the fledgling institution of agriculture science.

To summarize our argument, it made sense for the agricultural scientists to protect themselves and their institution against Type I errors. They did this by not making a recommendation unless the odds were 30:1 against a Type I error.

When tables were published as an aid to researchers, probabilities (.10, .05, .02, and .01) replaced odds. So 30:1 ($p = .0323$) gave way to .05. Those tables were used by researchers in many areas besides agriculture; they appear to have adopted .05 as the most important dividing line. (See Nelson, Rosenthal, and Rosnow, 1986, for a survey of psychologists.)

SIGNIFICANT RESULTS AND IMPORTANT RESULTS

Once, when one of us was a rookie instructor, we did some research with an undergraduate student, David Cervone, who was interested in hypnosis. We got permission to use two rooms in the library to conduct the experiment. When the results were analyzed, two of the groups were significantly different. One day, the librarian asked how the experiment had come out.

"Oh, we got some significant results," I said.

"I imagine David thought they were more significant than you did," was the reply.

At first I was confused. Why would David think they were more significant than I would? Point oh-five was point oh-five. Then I realized that the librarian and I were using the word *significant* in two quite different ways. He meant "important" and I meant "not due to chance."

By **statistically significant,** I meant that the difference was a *reliable* one that would be expected to occur again if the study was run again. That is the meaning we have had in mind in this chapter. However, let's pursue the librarian's meaning and ask about the *importance* of a difference. A statistically significant difference may not be an important difference. When you worked Problem 21, you found that the difference in the college admissions scores between two schools was significant at the .05 level. However, the difference was only one-tenth of a point, which doesn't seem important at all.

When you worked Problem 14, you found that the amount of time a person was subject to asthma attacks depended on whether the first attack occurred before age 3 or after age 6. Such a difference would be important if you wanted to tell parents what to expect.

Arthur Irion (1976) captured the two meanings of the word *significant* when he reported that his study "reveals that there is significance among the differences although, of course, it doesn't reveal what significance the differences have."

The basic point we are making is that a study that has statistically significant results may or may not have important results. You have to decide about the importance without the help of inferential statistics.

HOW TO REJECT THE NULL HYPOTHESIS—THE TOPIC OF POWER

From your reading of this text and other material about experiments, you may have the impression that researchers are excited when they reject H_0 and unhappy when they don't. That impression is correct. It is also reasonable. To reject H_0 is to be left with only one alternative, H_1, from which a conclusion can be drawn. To retain H_0 is to be left up in the air. You don't know whether the null hypothesis is really true or whether it is false and you just failed to detect it. So if you are going to design and run an experiment, you should maximize your chances of rejecting H_0.[12] There are three factors to consider: actual difference, standard error, and α.

In order to get this discussion out of the realm of the hypothetical and into the realm of the practical, consider the following problem. Suppose you must do an experiment for a class. You are free to choose any project, and you want to reject H_0. You decide to try to show that widgets are different from controls. Accept for a moment the idea that widgets *are* different—that H_0 *should* be rejected. What are the factors that determine whether you will conclude from your experiment that widgets are different?

1. *Actual difference.* The larger the actual difference between widgets and controls, the more likely you are to reject H_0. There is a practical limit, though. If the difference is too large, other people, including your instructor, will call your experiment trivial, saying that it demonstrates the obvious and that anyone can see that widgets are different. On the other hand, small differences can be difficult to detect. Pre-experiment estimations of actual differences are usually based on your own experience.[13]

2. *The standard error of a difference.* Look at the formula for z on page 150. You can see that as $s_{\bar{X}_1 - \bar{X}_2}$ gets smaller, z gets larger, and you are more likely to reject H_0. This is true, of course, *only if widgets are really different from controls.* Here are two ways you can reduce the size of $s_{\bar{X}_1 - \bar{X}_2}$.

   a. *Sample size.* The larger the sample, the smaller the standard error of the difference. Figure 6.5 shows that the larger the sample size, the smaller the standard error of the mean. The same relationship is true for the standard error of a difference. Some texts (for example, Guilford and Fruchter, 1978, pp. 182–186) show you how to calculate the sample size required to reject H_0. In order to do this calculation, you must make assumptions about the size of the actual difference. Many times, the size of the sample is dictated by practical consideration—time, money, or the availability of widgets.

   b. *Sample variability.* Reducing the variability in the sample will produce a smaller $s_{\bar{X}_1 - \bar{X}_2}$. You can reduce variability by using reliable measuring instruments, recording data correctly, and, in short, reducing the "noise" or random error in your experiment.

---

[12] Some of you may object to this philosophy of deliberately setting out to reject H_0. You might argue that a scientist should try to discover the "true situation," which certainly may be that H_0 is true. Unfortunately, inferential statistics are poor tools for establishing the truth of H_0.

[13] There is some danger in trusting your experience. Otto Loewi received a Nobel Prize in 1936 for demonstrating the chemical nature of nerve transmission in 1921. The idea for the experiment came to him during the night. He got out of bed, went to his laboratory, and performed an experiment that involved isolated frog hearts and some Ringer solution. Later he wrote, "If I had carefully considered it in the daytime, I would undoubtedly have rejected the kind of experiment I performed." Loewi goes on to explain that it seems improbable that the actual difference between the Ringer solution with the chemical and that without could be detected by the method he used. "It was good fortune that at the moment of the hunch I did not think but acted" (Loewi, 1963).

3. *Alpha.* The larger α is, the more likely you are to reject H_0. The limit to this factor is your colleagues' sneer when you report that widgets are "significantly different at the .40 level." Everyone believes that such differences should be attributed to chance. Sometimes practical considerations may permit the use of $\alpha = .10$. If both widgets and controls could be used to treat a deadly illness and both have the same side effects, but "widgets are significantly better at the .10 level," then widgets will be used. (Also, more data will then be gathered [sample size increased] to see whether the difference between widgets and controls is reliable.)

These three factors are discussed by statisticians under the topic *power*.[14] The power of a statistical test is defined as $1 - \beta$. The more powerful the test, the more likely it is to detect any actual difference between widgets and controls.

We will close this section on power by asking you to imagine that you are a researcher directing a project that could make a Big Difference. (Since you are imagining this, the difference can be in anything you would like to imagine—the health of millions, the destiny of nations, your bank account, whatever.) Now suppose that the success or failure of the project hinges on one final statistical test. One of your assistants comes to you with the question, "How much power do you want for this last test?"

"All I can get," you answer.

If you examine the list of factors that influence power, you will find that there is only one item that you have some control over, and that is the standard error of a difference. Of the factors that affect the size of the standard error of a difference, the one that most researchers can best control is N.

Our advice to you is to allocate plenty of power to important statistical tests—use large Ns.

PROBLEMS

25. In your own words, distinguish between a significant difference and an important difference.

26. List the four factors that determine the power of an experiment. (If you can recall them without looking at the text, you will know that you have been reading actively and effectively.)

27. It's again time for one of those *integrative* questions. As we did at the end of Chapter 4, we suggest that you plan to devote at least 30 minutes to this question (even if you cannot do it right now). Write an essay on inferential statistics. We suggest that your essay not contain any reference to the normal curve because, as you will see, you can do inferential statistics without using the normal curve. Once again, a good procedure is to write down from memory the things you would like to include. Then go back over these last three chapters and make more notes of the facts, considerations, or organizational ideas. Write the essay and revise it. Rest. Revise again and write your final answer.

---

[14] Texts that devote a separate chapter to power are Howell (1987), Loftus and Loftus (1988), and Minium and Clarke (1982).

Transition Page

The techniques you have learned so far require the use of the normal distribution to assess probabilities. These probabilities will be accurate if you have used σ in your calculations and fairly accurate if N is so large that s is a reliable estimator of σ. In the next chapter, you will learn about the t distribution, which statisticians use, regardless of sample size, when they do not know σ.

The logic you have used will be used again; that is, you assume the null hypothesis, draw random samples, introduce the independent variable, and calculate a mean difference on the dependent variable. If this difference cannot be attributed to chance, reject the null hypothesis and interpret the results.

At this point in their studies, most students suspect that the normal curve is an indispensable part of modern statistical living. Up until now in this book, it has been. However, in the next five chapters you will encounter several sampling distributions, none of which is normal, but all of which can be used to determine the probability that a particular event occurred by chance. Deciding which distribution to use is not a difficult task, but it does require some practice. Remember that a theoretical distribution is accurate if the assumptions on which it is based are true for the data from the experiment. By knowing the assumptions a distribution requires and the nature of your data, you can pick an appropriate distribution.

8

THE *t* DISTRIBUTION AND THE *t* TEST

Objectives for Chapter 8: After studying the text and working the problems in this chapter, you should be able to:

1. Contrast the *t* distribution with the normal distribution
2. Use the *t* distribution to determine whether a sample mean came from a particular population
3. Distinguish between independent-samples designs and correlated-samples designs
4. Calculate *t*-test values for both independent-samples designs and correlated-samples designs and write interpretations
5. Calculate confidence intervals for both independent-samples designs and correlated-samples designs and write interpretations
6. List and explain the assumptions for the *t* test
7. Use the *t* distribution to determine whether a Pearson product-moment correlation coefficient is significantly different from .00

This chapter is about a theoretical distribution called the **t distribution.** The *t* distribution is used when σ is not known and sample sizes are too small to ensure that *s* is a reliable estimate of σ.

In this chapter, the *t* distribution will be used to find answers to these four kinds of questions:

1. Did a sample with a mean $\bar{X}$ come from a population with a mean μ?
2. Did two samples with means $\bar{X}_1$ and $\bar{X}_2$ come from the same population?
3. What is the confidence interval about the difference between two sample means?
4. Did a Pearson product-moment correlation coefficient, based on sample data, come from a population with a true correlation of .00 for the two variables?

Questions 1, 2, and 4 are problems of hypothesis testing. Question 3 requires the establishment of a confidence interval. Although the *t* distribution is new to you and requires explanation, the logic of hypothesis testing and confidence intervals is not.

Traditionally, the *t* distribution is referred to with a lowercase *t*. A capital *T* is used for other distributions (see Chapter 12) and as a standardized test score. Because some computer programs do not print lowercase letters, however, *t* becomes *T* on some printouts (and often in text based on that printout). Be alert.

The story of the man who invented (discovered?) the *t* distribution is an interesting one that tells something about the motivation of statisticians. W. S. Gosset (1876–1937) was educated at Oxford in chemistry and mathematics.[1] In 1899, he went to work for Arthur Guinness, Son & Company, a brewery in Dublin, Ireland, where the management had just begun a policy of employing scientists. Gosset's task was to make recommendations about various brewing processes, but he had limited amounts of data to work with.

Gosset was familiar with the normal curve, a distribution based on σ. Unfortunately, the samples he had were small and Gosset was confident that such small-sample *s* values were not accurate estimators of σ. Thus, the normal-curve model was not appropriate.

Gosset's solution was to work out the mathematics of curves based on *s*. He found that the theoretical distribution depended on sample size, with a different distribution for each *N*. These distributions make up a family of curves that have come to be called the *t* distribution.

In Gosset's work, you again see how a practical question forced the development of a statistical tool. (Remember that Francis Galton invented the concept of the correlation coefficient in order to assess the degree to which characteristics of fathers are found in their sons.) In Gosset's case, an example of a practical question was, Will this new strain of barley, developed by the botanical scientists, have a greater yield than our old standard? Such questions were answered with data from experiments carried out on the ten farms maintained by the Guinness Company in the principal barley-growing regions of Ireland. A typical experiment might involve two 1-acre plots (one planted with the old barley, one with the new) on each of the ten farms. Gosset then was confronted with ten 1-acre yields for the old barley and ten for the new. Was the difference in yields caused by sampling fluctuation, or was it a reliable difference between the two strains? He made the decision using his newly derived *t* distribution.

Gosset wanted to publish his work on the *t* distribution in *Biometrika*, a journal founded in 1901 by Francis Galton, Karl Pearson, and W. R. F. Weldon. However, at the Guinness Company there was a rule that employees could not publish (the rule there, apparently, was publish *and* perish). Because the rule was designed to keep brewing secrets from escaping, there was no particular ferment within the company when Gosset, in 1908, published his new mathematical statistics under the pseudonym "Student." The *t* distribution then came to be known as "Student's *t*." (No one seems to know why the letter *t* was chosen. E. S. Pearson surmises that *t* was simply a "free letter"; that is, no one had yet used *t* to designate a statistic.) Since he worked for the Guinness Company all his life, Gosset continued to use the pseudonym "Student" for his publications in mathematical statistics. Gosset was very devoted to his company, working hard and rising through the ranks. He was appointed head brewer a few months before his death in 1937.

[1] For more information, see "Gosset, W. S.," in *Dictionary of National Biography, 1931–40,* London: Oxford University Press, 1949, or McMullen and Pearson (1939, pp. 205–253).

In this chapter, we will describe some characteristics of the *t* distribution and compare it to the normal distribution. Then you will do some hypothesis testing: one section on samples that are independent of each other and one on samples that are correlated. Next you will use the *t* distribution to establish confidence intervals about a mean difference. Following confidence intervals you will learn the assumptions that are required if you choose to use a *t* test to analyze your data. Finally, you will learn how to determine whether a correlation coefficient is statistically significant. Problems 1–4, mentioned at the beginning of the chapter, will be dealt with in order.

THE *t* DISTRIBUTION

Rather than just one *t* distribution, there are many *t* distributions. In fact, there is a *t* distribution for each sample size from 2 to ∞. These different *t* distributions are described as having different **degrees of freedom,** and there is a different *t* distribution for each degree of freedom. *Degrees of freedom* is abbreviated *df*. For now we will define *degrees of freedom* as sample size minus one, and provide you with a more thorough explanation later. Thus, $df = N - 1$. If the sample consists of 12 members, $df = 11$.

Figure 8.1 is a picture of four of these *t* distributions, each based on a different number of degrees of freedom. You can see that, with fewer degrees of freedom, a larger proportion of the curve is contained in the tails.

You know from your work with the normal curve that a theoretical distribution is used to determine a probability and that, on the basis of the probability, the null hypothesis is retained or rejected. You will be glad to learn that the logic of using the *t* distribution to make a decision is just like the logic of using the normal distribution. Recall that

$$z = \frac{\bar{X} - \mu}{\sigma_{\bar{X}}}$$

and that *z* is normally distributed. You probably also recall that at the critical *z* value

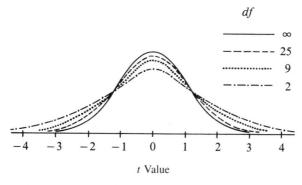

Figure 8.1 Four different *t* distributions. The difference between $df = 25$ and $df = \infty$ is exaggerated here so that the curves will print distinctively.

of ± 1.96, the chances are only 5 in 100 that the mean $\bar{X}$ came from the population with mean μ.

In a similar way, if the samples are small, you can perform a **t test** with the formula

$$t = \frac{\bar{X} - \mu}{s_{\bar{X}}}$$

The number of degrees of freedom (df) determines which t distribution is appropriate, and from it you can find a t value that would be expected to occur by chance 5 times in 100. Figure 8.2 separates the t distributions of Figure 8.1. The critical values in **Figure 8.2** are those associated with the interval that contains 95 percent of the cases, leaving 2.5 percent in each tail. Look at each of the four curves.

If you looked at Figure 8.2 carefully, you may have been suspicious that the t distribution for $df = \infty$ is a normal curve. It is. As df approaches ∞, the t distribution approaches the normal distribution. When $df = 30$, the t distribution is almost normal. Now you understand why we repeatedly cautioned, in chapters that used the normal curve, that N must be at least 30 (unless you know σ or that the distribution of the population is symmetrical). Even when $N = 30$, the t distribu-

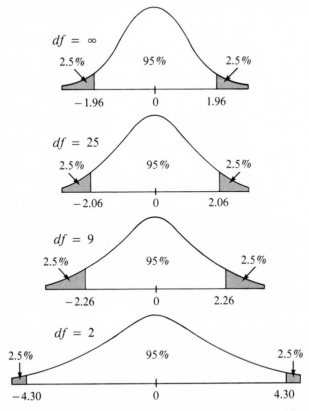

Figure 8.2   Four t distributions showing the t values that enclose 95 percent of the cases

tion is more accurate than the normal distribution for assessing probabilities; so in most research studies (that use samples), t is used rather than z.

A reasonable question now is, Where did those critical values of ± 4.30, ± 2.26, ± 2.06, and ± 1.96 come from? The answer is Table D in Appendix C. Table D is really a condensed version of 34 t distributions. Turn to **Table D** and note that there are 34 different degrees of freedom in the left-hand column. Across the top under "α Levels for Two-Tailed Test" you will see six selected probability figures, .20, .10, .05, .02, .01, and .001. Follow the .05 column down to $df = 2, 9, 25$, and ∞, and you will find critical values of 4.30, 2.26, 2.06, and 1.96.

Table D differs in several ways from the normal-curve table. In the normal-curve table, the body of the table contains probability figures, which range from .00 to .50. z scores are found in the margin. In Table D, the body contains t values and the probability figures are found on the top margin. And, instead of hundreds of probability values, there are only six. These are the six that are commonly chosen as α levels by experimenters. Finally, for a one-tailed t test, use the α levels shown under that heading at the top of Table D.

Note that the one-tailed figures are half those for a two-tailed test. You might draw a t distribution, put in values for a two-tailed test, and see for yourself that reducing the probability figure by half is appropriate for a one-tailed test.

As a general rule, researchers run two-tailed tests. If a one-tailed test is used, a justification is usually given. In this text we will routinely use two-tailed tests.

We'll use Student's t distribution to decide whether a particular sample mean came from a particular population.

A Belgian, Adolphe Quételet ('Ka-tle) (1796–1874), is regarded as the first person to recognize that social and biological measurements may be distributed according to the "normal law of error" (the normal distribution). Quételet made this discovery while developing actuarial (life expectancy) tables for a Brussels life insurance company. Later, he examined anthropometric (body) measurements and, in 1836, he developed Quételet's Index (QI), a ratio in which weight in grams is divided by height in centimeters. This index was supposed to permit the evaluation of a person's nutritional status: very large numbers indicated obesity and very small numbers indicated starvation.

Suppose a present-day anthropologist read that Quételet had found a mean QI value of 375 on the entire population of French army conscripts. No standard deviation was given because it had not yet been invented. Our anthropologist, wondering if there has been a change during the last 100 years, obtains a random sample of 20 present-day Frenchmen who have just been inducted into the army. He finds a mean of 400 and a standard deviation of 60. One now familiar question remains, Should this mean increase of 25 QI points be attributed to chance or not? To answer this question, we will perform a ***t*** **test**. As usual, we will require $p \leq .05$ to reject chance as an explanation. Since

$$t = \frac{\bar{X} - \mu}{s_{\bar{X}}}$$

and

$$s_{\bar{X}} = \frac{s}{\sqrt{N}} = \frac{60}{\sqrt{20}} = 13.42$$

you get

$$t = \frac{400 - 375}{13.42} = 1.86$$

and

$$df = N - 1 = 20 - 1 = 19$$

If you look in Table D under the column for a two-tailed test with $\alpha = .05$ at the row for 19 df, you will find a critical value of 2.09. Our anthropologist's t is less than 2.09, so the null hypothesis should be retained. The difference between present-day soldiers and those of old may be due to chance.

Quételet's Index is not currently used by anthropologists. There were several later attempts to develop a more reliable index of nutrition and most of those attempts were successful.

Quételet's promotion of statistics, however, was widely influential in the 19th century. Florence Nightingale, a pioneer in using statistical analysis to understand health care, said that Quételet was the founder of "the most important science in the whole earth" (Cohen, 1984). Quételet's contribution to criminology is documented by Beirne (1987). Quételet's work suggested to Francis Galton that genius could be treated mathematically, an idea that led to the concept of correlation.[2]

PROBLEMS

1. Fill in the blanks in the statements below with "t" or "normal."
   a. There is just one _____ distribution, but there are many _____ distributions.
   b. Given a value of 2.00 on the horizontal axis, a _____ distribution has a greater proportion of the curve to the right than does a _____ distribution.
   c. The _____ distribution must be used if the samples are small.
2. Would the standard deviation of a t distribution with $df = 3$ be larger than, smaller than, or equal to the standard deviation of the normal curve?
3. Determine the df for the following samples.
   a. $N = 25$ b. $N = 4$ c. $N = 42$
4. Who invented the t distribution? Why?
5. Find the t value for the following means and standard errors of the mean, given that $\mu = 29.00$.
   a. $\bar{X} = 31.50$; $s_{\bar{x}} = 2.50$ b. $\bar{X} = 25.00$; $s_{\bar{x}} = 12.00$
   c. $\bar{X} = 29.60$; $s_{\bar{x}} = 0.15$ d. $\bar{X} = 38.00$; $s_{\bar{x}} = 18.00$
   e. $\bar{X} = 23.00$; $s_{\bar{x}} = 1.33$
6. In the following problems, the investigator thought that μ might be 81.00. Your task is to calculate t values for each of the five problems, find the appropriate critical values in Table D, and assign each sample to one of the following categories: $p > .05$, $p = .05$, $.01 < p < .05$, $p = .01$, $p < .01$. This table should help you.

---

[2] Quételet was widely respected during his lifetime. A statue was erected in his honor in Brussels and he was the first foreign member of the American Statistical Association. A short intellectual biography of Quételet is given in the *International Encyclopedia of the Social Sciences*.

If $t_{ob} < t_{.05}$,   then $p > .05$ Retain H_0

If $t_{ob} = t_{.05}$ then $p = .05$ Reject H_0

If $t_{.01} > t_{ob} > t_{.05}$,   then $.01 < p < .05$ Reject H_0

If $t_{ob} > t_{.01}$,   then $p < .01$ Reject H_0

where t_{ob} is the value calculated from the data (observations), and $t_{.05}$ and $t_{.01}$ are t values found in Table D.

a. $\bar{X} = 84.92$; $s = 7.75$, $N = 15$ **b.** $\bar{X} = 76.59$; $s = 5.56$, $N = 7$

c. $\bar{X} = 76.59$; $s = 10.29$, $N = 24$ **d.** $\bar{X} = 132.21$; $s = 82.49$, $N = 21$

e. $\bar{X} = 80.014$; $s = 0.45$, $N = 5$

7. You determined a number of probabilities in Problem 6. What event do these probabilities refer to?

DEGREES OF FREEDOM

You have been determining degrees of freedom by a rule-of-thumb technique: $N - 1$. Now it is time for us to explain the concept more thoroughly in order to prepare you for statistical techniques in which $df \neq N - 1$.

It is somewhat difficult to obtain an intuitive understanding of the concept of degrees of freedom without the use of mathematics. If the following explanation leaves you scratching your head, you might read Helen Walker's excellent article in the *Journal of Educational Psychology* (1940).

The "freedom" in *degrees of freedom* refers to the freedom of a number to have any possible value. If you were asked to pick two numbers and there were no restrictions, both numbers would be free to vary (take any value) and you would have two degrees of freedom. If a restriction is imposed—say, that $\Sigma X = 0$, then one degree of freedom is lost because of that restriction; that is, when you now pick the two numbers, only one of them is free to vary. As an example, if you choose 3 for the first number, the second number *must be* -3. The second number is not free to vary because of the restriction that $\Sigma X = 0$. In a similar way, if you were to pick five numbers with a restriction that $\Sigma X = 0$, you would have four degrees of freedom. Once four numbers are chosen (say, -5, 3, 16, and 8), the last number (-22) is determined.

The restriction that $\Sigma X = 0$ may seem to you to be an "out-of-the-blue" example and unrelated to your earlier work in statistics, but some of the statistics you have calculated have had such a restriction built in. For example, when you found $s_{\bar{X}}$, as required in the formula for t, you used some algebraic version of

$$s_{\bar{X}} = \frac{s}{\sqrt{N}} = \frac{\sqrt{\dfrac{\Sigma(X - \bar{X})^2}{N - 1}}}{\sqrt{N}}$$

The restriction that is built in is that $\Sigma(X - \bar{X})$ is always zero and, in order to meet that requirement, one of the X's is determined. All X's are free to vary except one, and the degrees of freedom for $s_{\bar{x}}$ is $N - 1$. Thus, for the problem of using the t distribution to determine whether a sample came from a population with a mean μ, $df = N - 1$. Walker (1940) summarizes this reasoning by stating: "A universal rule holds: The number of degrees of freedom is always equal to the number of observations minus the number of necessary relations obtaining among these observations." A necessary relationship for $s_{\bar{x}}$ is that $\Sigma(X - X) = 0$. Another way of stating this rule is that the number of degrees of freedom is equal to the number of original observations minus the number of parameters estimated from the observations. In the case of $s_{\bar{x}}$, one degree of freedom is subtracted because $\bar{X}$ is used as an estimate of μ.

INDEPENDENT-SAMPLES AND CORRELATED-SAMPLES DESIGNS

Now we switch from the question of whether a sample came from a population with a mean, μ, to the more common question of whether two samples came from populations with identical means. To answer the question the mean of one sample is compared with the mean of another sample, and the difference is attributed either to chance (null hypothesis retained) or to a treatment (null hypothesis rejected).

There are two kinds of two-group designs. With an **independent-samples design,** the subjects serve in only one of the two groups, and there is no reason to believe that there is any correlation between the scores of the two groups. With a **correlated-samples design,** subjects may serve in both groups or in just one group, but in either case, there is a correlational relationship between the scores of the two groups. The difference between these designs is important because the calculation of the t value for independent samples is different from the calculation of the t value for correlated samples. You may not be able to tell which design has been used just by looking at the numbers or by identifying the independent and dependent variable; instead, you must examine the description of the procedures of the experiment.

Although the t formulas appear to be different, the purpose of both designs is the same: to determine the probability that two such samples could have a common population mean.

CLUE TO THE FUTURE

> Most of the rest of this chapter is organized around independent-samples and correlated-samples designs. Three-fourths of Chapter 12 is also organized around these two designs. In Chapters 9 and 10, the procedures you will learn are appropriate only for independent samples.

Correlated-Samples Design

In a correlated-samples design, there is a logical reason to pair a score in one group with one particular score in the second group.[3] Correlated-samples designs always consist of *pairs* of scores.

---

[3] Some texts use the term *dependent samples* instead of *correlated samples*.

A correlated-samples design may come about in a number of ways. Fortunately, the actual arithmetic in calculating a *t* value is the same for any of three correlated-samples designs. The three types of designs are **natural pairs, matched pairs,** and **repeated measures.**

Natural Pairs. In a natural-pairs investigation, the experimenter does not assign the subjects to one group or the other; the pairing occurs prior to the investigation. **Table 8.1** identifies one way in which natural pairs may occur—father and son. In such an investigation, you might ask whether fathers are shorter than their sons (or more religious, or more racially prejudiced, or whatever). Notice, though, that it is easy to decide that these are correlated-samples data: there is a logical way to pair up the scores in the two groups.

Did you notice that Table 8.1 is the same as Table 4.2, which outlined the basic requirements for the calculation of a correlation coefficient? As you will soon see, that correlation coefficient is a part of determining a *t* value.

Matched Pairs. In some situations, the experimenter has control over the ways pairs are formed and a match can be arranged. One method is for two subjects to be paired on the basis of similar scores on a pretest that is related to the dependent variable. For example, a hypnotic susceptibility test might be given to a group of subjects. Subjects with similar scores could be paired and then one member of each pair randomly assigned to either the experimental group or the control group. The result is two groups equivalent in hypnotizability.

Another variation of matched pairs is the split-litter technique used with nonhuman animals. A pair from a litter is split and one is put in each group. In this way, the genetics of one group is matched with that of the other. The same technique has been used in human experiments with twins or siblings. Student's barley experiments are examples of starting with two similar subjects (adjacent plots of ground) and assigning them at random to one of two treatments.

Still another example of the matched-pairs technique is the treatment of each member of the control group according to what happens to its paired member in the experimental group. Because of the forced correspondence, this is called a *yoked-control* design.

The difference between the matched-pairs design and a natural-pairs design is that, with the matched pairs, the investigator can randomly assign one member of the pair to a treatment. In the natural-pairs design, the investigator has no control over assignment. Although the statistics are the same, the natural-pairs design is usually open to more interpretations than the matched-pairs design.

Table 8.1 Illustration of a correlated-samples design

Father	Height (in.) X	Son	Height (in.) Y
Michael Smith	64	Mike, Jr.	64
Matthew Johnson	66	Matt, Jr.	66
Christopher Williams	68	Chris, Jr.	68
Brian Brown	70	Brian, Jr.	70
David Jones	72	Dave, Jr,	72
Adam Miller	74	Adam, Jr.	74

Repeated Measures. A third kind of correlated-samples design is called a repeated-measures design because more than one measure is taken on each subject. This design often takes the form of a before-and-after experiment. A pretest is given, some treatment is administered, and a post-test is given. The mean of the scores on the post-test is compared with the mean of the scores on the pretest to determine the effectiveness of the treatment. Clearly, there are two scores that should be paired: the pretest and the post-test scores of each subject. In such an experiment, each person is said to serve as his or her own control.

All three of these methods of forming groups have one thing in common: a meaningful correlation may be calculated for the data. The name *correlated samples* comes from this fact. With a correlated-samples design, one variable is designated X, the other Y.

Independent-Samples Design

In an independent-samples design,[4] there is no reason to pair a score in one group with a particular score in the second group. Often in this design, the whole pool of subjects is assigned in a random fashion to the groups.

The designs you analyzed in Chapter 7 were all independent-samples designs. This design is outlined in Table 7.1, which you might want to review now.

One caution is in order. Both of these designs use random sampling, but with an independent-samples design, the subjects are randomly selected from a population of *individuals*. In a correlated-samples design, *pairs* are randomly selected from a population of pairs.

Finally, by way of comparison, both these designs have an independent variable that has two levels. Both have a dependent variable on which every subject has a score. The basic difference is that with the correlated-samples design there is a logical reason to pair up scores from the two groups, and in an independent-samples design there is not.

PROBLEMS   Identify each of the following experiments as an independent-samples design or a correlated-samples design. Work all six problems before checking your answers.

8. An investigator gathered as many case histories as she could of situations in which identical twins were raised apart—one in a "good" environment and one in a "bad" environment. The group raised in the "good" environment was compared with that raised in the "bad" environment on attitude toward education.

9. Alaskans with seasonal affective disorder (depression) spent 2 hours a day under bright artificial light. Before treatment the mean depression score was 20.0, and at the end of 1 week of treatment the mean depression score was 6.4. (See Hellekson, Kline, and Rosenthal, 1986.)

10. A researcher counted the number of aggressive encounters among children who were playing with parts of toys. Then he lifted a screen to reveal other children playing with whole toys. The children with the parts of toys watched for a while, and then the researcher lowered the screen. The children resumed playing, and the researcher counted

---

[4] Some texts use the terms *noncorrelated* or *uncorrelated* for this design.

the number of aggressive encounters. The procedure was completed for six groups of children, and the investigator compared the number of aggressive encounters within the groups before and after the screen was lifted. (For a similar experiment, see Barker, Dembo, and Lewin, 1941.)

11. To study the effect of sleep deprivation, investigators placed a rat in an aquarium. Half of a circular platform extended into the aquarium above the water line. By staying on the half-circle, the rat could avoid the water. If the rat fell asleep, the platform began to turn, forcing the rat to wake up and walk in order to stay out of the water. On the other half of the platform was a second rat, faced with the same problem of staying out of the water in its separate aquarium. The amount of activity for the two rats was obviously the same, but the second rat could sleep whenever the first rat was awake. Thus, one group of rats was much more sleep-deprived than the other. The dependent variables were physiological measures. (See Rechtschaffen et al., 1983.)

12. A college dean faced a problem that suggested an experiment. Thirty-two freshmen had applied for the sophomore honors course. Only 16 could be accepted, so the dean flipped a coin for each applicant, with the result that 16 were selected. At graduation, she compared the mean grade point average of the "winners" (those who had taken the course) with that of the "losers" to see if the sophomore honors course had any effect on GPA.

13. A social psychologist was interested in the effect of birth order on IQ. He obtained data from many families with two children and then compared the IQs of the firstborn with those of the second-born. (See Zajonc, 1975.)

USING THE *t* DISTRIBUTION FOR INDEPENDENT-SAMPLES DESIGNS

Using the *t* distribution to test a hypothesis is very similar to using the normal distribution. The null hypothesis is that the two populations have the same mean, and thus any difference between the two sample means is due to chance. The *t* distribution tells you the probability that the difference you observe is due to chance if the null hypothesis is true. You simply establish an α level, and if your observed difference is less probable than α, reject the null hypothesis and conclude that the two means came from populations with different means. If your observed difference is more probable than α, retain the null hypothesis. Does this sound familiar? We hope so.

The way to find the probability of the observed difference is to use a *t* test. The probability of the resulting *t* value can be found in Table D in Appendix C. For an independent-samples design, the formula for the *t* test is

$$t = \frac{\bar{X}_1 - \bar{X}_2}{s_{\bar{X}_1 - \bar{X}_2}}$$

This formula is the "working formula" for the more general case in which the numerator is $(\bar{X}_1 - \bar{X}_2) - (\mu_1 - \mu_2)$. For the cases in this book, the hypothesized value of $\mu_1 - \mu_2$ is zero, which reduces the general case to the working formula above. Thus, the *t* test, like many other statistical tests, consists of a difference between a statistic and a parameter divided by the standard error of the statistic.

Table 8.2 shows two formulas for calculating $s_{\bar{X}_1 - \bar{X}_2}$. Use the formula in the top half of the table when the two samples have an unequal number of scores.

Table 8.2   Formulas for $s_{\bar{X}_1 - \bar{X}_2}$ for independent-samples
t tests

If $N_1 \neq N_2$:

$$s_{\bar{X}_1 - \bar{X}_2} = \sqrt{\left(\frac{\Sigma X_1^2 - \dfrac{(\Sigma X_1)^2}{N_1} + \Sigma X_2^2 - \dfrac{(\Sigma X_2)^2}{N_2}}{N_1 + N_2 - 2}\right)\left(\frac{1}{N_1} + \frac{1}{N_2}\right)}$$

If $N_1 = N_2 = N$:

$$s_{\bar{X}_1 - \bar{X}_2} = \sqrt{s_{\bar{X}_1}^2 + s_{\bar{X}_2}^2} = \sqrt{\frac{\Sigma X_1^2 - \dfrac{(\Sigma X_1)^2}{N} + \Sigma X_2^2 - \dfrac{(\Sigma X_2)^2}{N}}{N(N - 1)}}$$

In the special situation where $N_1 = N_2$, the formula simplifies to that shown on the bottom of Table 8.2.

The formula for degrees of freedom for independent samples is $df = N_1 + N_2 - 2$. Here is the reasoning. For each sample, the number of degrees of freedom is $N - 1$, since, for each sample, a mean has been calculated with the restriction that $\Sigma(X - \bar{X}) = 0$. Thus, the total degrees of freedom is $(N_1 - 1) + (N_2 - 1) = N_1 + N_2 - 2$.

Here is an example of an experiment in which the results were analyzed with an independent-samples *t* test. Thirteen monkeys were randomly assigned to either a drug group or a control group (placebo).[5] The drug group ($N = 7$) received pills for 8 days, while the placebo group ($N = 6$) was given an inert substance. After 8 days of pills, training began on a complex problem-solving task. Training and pills were continued for 6 days, after which the number of errors was tabulated. The number of errors each animal made and the *t* test are presented in **Table 8.3.** The null hypothesis is that the drug made no difference—that the difference obtained was due just to chance. Since the *N*'s are unequal for the two samples, the longer formula for the standard error must be used. Consulting Table D for 11 *df*, you'll find that $t = 2.20$ is required in order to reject the null hypothesis with $\alpha = .05$. Since the obtained $t = -2.99$, reject the null hypothesis. The final (and perhaps most important) step is to interpret the results. Since the drug group, on the average, made fewer errors (39.71 vs. 57.33), we may conclude that the drug treatment *facilitated* learning. We will often express tabled *t* values as $t_{.05}$ (11 *df*) = 2.20. This gives you the critical value of *t* (2.20) for a particular *df* (11) and level of significance ($\alpha = .05$).

Notice that the absolute value of the obtained *t* ($|t| = |-2.99| = 2.99$) is *larger* than the tabled *t* (2.20). In order to reject the null hypothesis, the absolute value of the obtained *t* must be as great as, or greater than, the tabled *t*. The larger the obtained $|t|$, the smaller the probability that the difference between means occurred by chance. **Figure 8.3** should help you see why this is so. Notice in Figure 8.3 that, as the values of $|t|$ become larger, less and less of the area of the curve remains in the tails of the distribution. Remember that the area under the curve is a probability.

---

[5] Monkey research is very expensive, so experiments are carried out with small *N*'s. Thus, small-sample statistical techniques are a must.

Table 8.3   Number of errors for each monkey during 6 days of training

Drug group, X_1	Placebo group, X_2
34	39
52	57
26	68
47	74
42	49
37	57
40	
ΣX 278	344
ΣX^2 11,478	20,520
N 7	6
$\bar{X}$ 39.71	57.33

Applying the t test,

$$t = \frac{\bar{X}_1 - \bar{X}_2}{s_{\bar{x}_1 - \bar{x}_2}} = \frac{\bar{X}_1 - \bar{X}_2}{\sqrt{\dfrac{\Sigma X_1^2 - \dfrac{(\Sigma X_1)^2}{N_1} + \Sigma X_2^2 - \dfrac{(\Sigma X_2)^2}{N_2}}{N_1 + N_2 - 2}\left(\dfrac{1}{N_1} + \dfrac{1}{N_2}\right)}}$$

$$= \frac{39.71 - 57.33}{\sqrt{\left(\dfrac{11,478 - \dfrac{(278)^2}{7} + 20,520 - \dfrac{(344)^2}{6}}{7 + 6 - 2}\right)\left(\dfrac{1}{7} + \dfrac{1}{6}\right)}}$$

$$= \frac{-17.62}{\sqrt{\left(\dfrac{437.43 + 797.33}{11}\right)(0.31)}} = \frac{-17.62}{5.90} = -2.99$$

$df = N_1 + N_2 - 2 = 7 + 6 - 2 = 11$

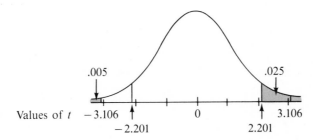

Values of t -3.106 0 3.106

-2.201 2.201

.005   .025

Figure 8.3   Values of t at the .05 and .01 levels of significance when $df = 11$

Recall that we have been conducting a two-tailed test; that is, the probability figure for a particular t value is the probability of $+t$ or larger plus the probability of $-t$ or smaller. In Figure 8.3, $t_{.05}$ (11 df) = 2.201. This means that, if the null hypothesis is true, a t value of $+2.201$ or larger would occur $2\frac{1}{2}$ percent of the time and a t value of -2.201 or smaller would occur $2\frac{1}{2}$ percent of the time.

PROBLEMS   For each problem in this section, decide whether a one-tailed or two-tailed test is appropriate. Analyze the data and write a conclusion.

14. A French teacher was interested in the effect of "immediate, concrete experience" on learning vocabulary. He decided to conduct an experiment using volunteers from his beginning French class. The college was 30 miles from a city that students often visited. As he drove to the city one day, he made a tape describing in French the terrain, signs, distances, and so forth. ("L' auto est sur le pont. Carrefour prochain est dangereux.") Some students listened to the trip tape while driving (immediate, concrete experience) and some listened to the trip tape in the language laboratory. Students then took a vocabulary test and the number of errors was recorded. Analyze the data in the table with an independent-samples t test. Write a conclusion about immediate, concrete experience.

Listened to tape while in laboratory	Listened to tape while in car
14	6
9	8
18	5
11	2
	9

15. The table lists the data from the hypothetical experiment described in Problem 12. Reread Problem 12, analyze the summary data in the table, and write a conclusion.

	"Winners"	"Losers"
ΣX	55.36	56.32
ΣX^2	193.38	199.60
N	16	16

16. The general manager of a large office force is considering a new word processing package. To compare it to the one in use, she picked 20 "new hires" and randomly assigned them to training on the old package or the new package. After 2 weeks of training and work, each employee completed the same "test procedure." The number of minutes required is shown in the table. N was 17, since two of the workers were absent several times and one had been fired. Analyze the data and draw a conclusion.

New package	Old package
4.3	6.3
4.7	5.8
6.4	7.2
5.2	8.1
3.9	6.3
5.8	7.5
5.2	6.0
5.5	5.6
4.9	

USING THE *t* DISTRIBUTION FOR CORRELATED-SAMPLES DESIGNS

The formula for *t* when the data come from correlated samples has a familiar theme: a difference between means divided by the standard error of the difference. The standard error of the difference between means of correlated samples is symbolized $s_{\bar{D}}$.

One formula for a *t* test between correlated samples is

$$t = \frac{\bar{X} - \bar{Y}}{s_{\bar{D}}}$$

where $s_{\bar{D}} = \sqrt{s_{\bar{X}}^2 + s_{\bar{Y}}^2 - 2r_{XY}(s_{\bar{X}})(s_{\bar{Y}})}$

$df = N - 1$, where N = number of pairs

The number of degrees of freedom in a correlated-samples case is the number of *pairs* minus one. Although each pair has two values, once one value is determined, the other is restricted to a similar value. (After all, they are called *correlated* samples.) In addition, another degree of freedom is subtracted when $s_{\bar{D}}$ is calculated. This loss is similar to the loss of 1 *df* when $s_{\bar{X}}$ is calculated.

As you can see by comparing the denominator of the correlated-samples *t* test with that of the *t* test on page 174 for independent samples (when $N_1 = N_2$), the difference lies in the term $2r_{XY}(s_{\bar{X}})(s_{\bar{Y}})$. Of course, when $r_{XY} = 0$, this term drops out of the formula, and the standard error is the same as for independent samples.

Also notice what happens to the standard error term in the correlated-samples case where $r > 0$: the standard error is reduced. Such a reduction will increase the size of *t*. Whether this reduction will increase the likelihood of rejecting the null hypothesis depends on how much *t* is increased, since there are fewer degrees of freedom in a correlated-samples design than in the independent-samples design.

The formula $s_{\bar{D}} = \sqrt{s_{\bar{X}}^2 + s_{\bar{Y}}^2 - 2r_{XY}(s_{\bar{X}})(s_{\bar{Y}})}$ is used only for illustration purposes. There is an algebraically equivalent but arithmetically easier calculation called the *direct-difference method*, which does not require you to calculate *r*. To find $s_{\bar{D}}$ by the direct-difference method, find the difference between each pair of scores, calculate the standard deviation of these difference scores, and divide the standard deviation by the square root of the number of pairs.

Thus,

$$t = \frac{\bar{X} - \bar{Y}}{s_{\bar{D}}} = \frac{\bar{X} - \bar{Y}}{s_D / \sqrt{N}}$$

where $s_D = \sqrt{\dfrac{\sum D^2 - \dfrac{(\sum D)^2}{N}}{N - 1}}$

$D = X - Y$

N = number of *pairs* of scores

Here is an example of a correlated-samples design and a *t*-test analysis. Suppose you were interested in the effects of interracial contact on racial attitudes. You have a fairly reliable test of racial attitudes in which high scores indicate more positive attitudes. You administer the test one Monday morning to a biracial group of 14 12-year-old girls who do not know each other but who have signed up for a week-long community day camp. The campers then spend the next week taking nature walks, playing ball, eating lunch, swimming, making things, and doing the kinds of things that camp directors dream up to keep 12-year-old girls busy. On Saturday morning, the girls are again given the racial attitude test. Thus, the data consist of 14 pairs of before-and-after scores. The null hypothesis is that the mean of the population of "before" scores is equal to the mean of the population of "after" scores or, in terms of the specific experiment, that a week of interracial contact has no effect on racial attitudes.

Suppose the data in **Table 8.4** were obtained. Using the sum of the D and D^2 columns in Table 8.4, we can find s_D:

$$s_D = \sqrt{\frac{\Sigma D^2 - \frac{(\Sigma D)^2}{N}}{N-1}} = \sqrt{\frac{897 - \frac{(-81)^2}{14}}{13}} = \sqrt{32.95} = 5.74$$

$$s_{\bar{D}} = \frac{s_D}{\sqrt{N}} = \frac{5.74}{\sqrt{14}} = 1.53$$

Thus,

$$t = \frac{\bar{X} - \bar{Y}}{s_{\bar{D}}} = \frac{28.36 - 34.14}{1.53} = \frac{-5.78}{1.53} = -3.78$$

$$df = N - 1 = 14 - 1 = 13$$

Table 8.4   Hypothetical data from a racial attitudes study

| | Racial attitude scores | | | |
| | Before day camp | After day camp | | |
Name*	X	Y	D	D^2
Jennifer	34	38	−4	16
Sarah	22	19	3	9
Jessica	25	36	−11	121
Amanda	31	40	−9	81
Nicole	27	36	−9	81
Ashley	32	31	1	1
Megan	38	43	−5	25
Melissa	37	36	1	1
Katherine	30	30	0	0
Stephanie	26	31	−5	25
Elizabeth	16	34	−18	324
Lindsay	24	31	−7	49
Rebecca	26	36	−10	100
Erin	29	37	−8	64
Sum	397	478	−81	897
Mean	28.36	34.14		

* The girls' names are the 14 most common in the United States (Dunkling and Gosling, 1983).

Since $t_{.01}$ (13 df) = 3.01, this difference is significant beyond the .01 level; that is, $p < .01$. The "after" mean was larger than the "before" mean; therefore, we may conclude that, after the week of camp, racial attitudes were significantly more positive than before.

You might note that $\bar{X} - \bar{Y} = \bar{D}$, the mean of the difference scores. In this problem, $\Sigma D = -81$ and $N = 14$, so $\bar{D} = \Sigma D/N = -81/14 = -5.78$.

Gosset preferred the correlated-samples design. In his agriculture experiments, there were significant correlations between the yields of the old barley and the new barley grown on adjacent plots. This correlation reduced the standard-error term in the denominator of the t test, making the correlated-samples design more sensitive than the independent-samples design for detecting a difference between means.

PROBLEMS

17. Give a formula and definition for each of the following symbols.

    a. s_D **b.** D

    c. $s_{\bar{D}}$ **d.** t (verbal definition)

    e. $\bar{Y}$

18. The table shows data based on the experiment described in Problem 10. Reread Problem 10 and analyze the scores in this table. Write a conclusion.

Before	After
16	18
10	11
17	19
4	6
9	10
12	14

19. Here are some data from a consulting job that one of your authors worked on. He was asked by a lawyer to analyze salary data for employees at a large rehabilitation center to determine whether there was any evidence of sex discrimination. For all employees, men made about 25 percent more than women did. In order to rule out explanations like "more educated" and "more experienced," a subsample of employees with bachelor's degrees was separated out. From this group, men and women with equal experience were paired, resulting in an N of 16 pairs. For the data in the table, compare the mean annual salaries of women and men. In your conclusion (which you can direct to the judge), explain whether education, experience, or chance could account for the observed difference. *Note:* You may have to do some thinking before you can set this problem up. Start by looking at the data you have.

	Women, X	Men, Y
ΣX or ΣY	$204,516	$251,732
ΣX^2 or ΣY^2	2,697,647,000	4,194,750,000
$\Sigma X Y$		3,314,659,000

20. Decide whether the following study is an independent or correlated-samples experiment and analyze the data. An experimental methods class was randomly divided into two groups. Six participants found their reaction time (RT) to an auditory "go" signal and then their RT to a visual "go" signal. Five participants found their RT first to the visual signal and then to the auditory signal. The scores for each person are arranged in rows for the 11 members of the class.

Auditory RT (seconds)	Visual RT (seconds)
0.16	0.17
0.19	0.23
0.23	0.20
0.14	0.19
0.19	0.19
0.18	0.22
0.21	0.23
0.18	0.16
0.17	0.21
0.16	0.19
0.17	0.18

21. Two groups were chosen randomly from a large sociology class to study the effects of primacy versus recency. Both groups were given a two-page description of a person at work, with a paragraph telling how the person had been particularly helpful to a new employee. For half of the descriptions, the "helping" paragraph was near the beginning (primacy), and for the other half the "helping" paragraph was near the end (recency). One group read the primacy description and the other the recency description. Each subject then wrote a page summary of what the worker would do during leisure time. The dependent variable was the number of positive adjectives (words like *good, exciting, cheerful*) in the page summary of leisure activities. From the data, write a conclusion about primacy and recency.

Recency	Primacy
13	10
14	17
4	9
8	11
11	12
6	5
6	10
10	16
14	14

USING THE *t* DISTRIBUTION TO ESTABLISH A CONFIDENCE INTERVAL ABOUT A MEAN DIFFERENCE

As you probably recall from Chapter 6, a confidence interval is a range of values that is expected to contain a parameter. A confidence interval is established for a specified degree of confidence, usually 95 percent or 99 percent.

In this section, you will learn how to establish a confidence interval about a mean difference. The problems here are similar to those dealt with in Chapter 6, except for the following:

1. Probabilities will be established with the *t* distribution rather than with the normal distribution.
2. The parameter of interest is a *difference* between two population means rather than a population mean.

The first point can be dispensed with rather quickly. You have already practiced using the *t* distribution to establish probabilities; you will use Table D in this section, too.

The second point requires a little more explanation. The questions you have been answering so far in this chapter have been *hypothesis-testing* questions of the form, Does $\mu_1 - \mu_2 = 0$? You answered each question by drawing two samples, calculating the means, and finding the difference. If the probability of the difference was very small, the hypothesis $H_0: \mu_1 - \mu_2 = 0$, was rejected. Suppose you have rejected the null hypothesis, but someone wants more information and asks, What is the real difference between μ_1 and μ_2? The person recognizes that the real difference is not zero and wonders what it is. You are being asked to make an estimate of $\mu_1 - \mu_2$. If you establish a confidence interval about the difference between $\bar{X}_1$ and $\bar{X}_2$ or $\bar{X}$ and $\bar{Y}$, you can state with a specified degree of confidence that the interval contains $\mu_1 - \mu_2$.

Confidence Intervals for Independent Samples

The sampling distribution of $\bar{X}_1 - \bar{X}_2$ is a *t* distribution with $N_1 + N_2 - 2$ degrees of freedom. The lower and upper limits of the confidence interval about a mean difference are found with the following formulas:

$$LL = (\bar{X}_1 - \bar{X}_2) - t_\alpha(s_{\bar{X}_1 - \bar{X}_2})$$
$$UL = (\bar{X}_1 - \bar{X}_2) + t_\alpha(s_{\bar{X}_1 - \bar{X}_2})$$

For a 95 percent confidence interval, use the two-tailed *t* value in Table D associated with $\alpha = .05$. For 99 percent confidence, change α to .01.

For an example, we will use the calculations you worked up in Problem 16 on the time required to do the test procedure with the two different word processing packages. We will establish a 95 percent confidence interval about the difference found. As your calculations revealed,

$$\bar{X}_1 - \bar{X}_2 = 6.6 - 5.1 = 1.5 \text{ minutes} \qquad s_{\bar{X}_1 - \bar{X}_2} = 0.40$$
$$N_1 = 8 \qquad N_2 = 9 \qquad df = 8 + 9 - 2 = 15$$

From Table D, $t_{.05}(15\ df) = 2.13$ and

$$LL = (\bar{X}_1 - \bar{X}_2) - t_\alpha(s_{\bar{X}_1 - \bar{X}_2}) = 1.5 - 2.13(0.40) = 0.65$$
$$UL = (\bar{X}_1 - \bar{X}_2) + t_\alpha(s_{\bar{X}_1 - \bar{X}_2}) = 1.5 + 2.13(0.40) = 2.35$$

Thus, 0.65 and 2.35 minutes are the lower and upper limits of a 95 percent confidence interval for the mean difference between the two packages.

One of the benefits of establishing a confidence interval about a mean difference is that you also test the null hypothesis, $\mu_1 - \mu_2 = 0$, in the process. If 0 is outside the confidence interval, then the null hypothesis would be rejected using hypothesis-testing procedures. In the example we just worked, the confidence interval was

0.65 to 2.35 minutes; a value of 0 falls outside this interval. Thus, you can reject $H_0: \mu_1 - \mu_2 = 0$, at the .05 level. (You confirmed this in working Problem 16.)

In some situations, the extra information you get from a confidence interval can be very useful. For example, the new word processing package is faster, but as you might expect, it is also more expensive. Is it a better buy? Suppose a cost-benefit analysis shows that a time reduction of 0.5 minute per procedure would be required to recover the additional cost. Given the confidence interval you just worked out, the manager can be confident that the new package is the better buy. (The lower limit of the confidence interval, 0.65 minute, is more than the required reduction of 0.50 minute.)

Confidence Intervals for Correlated Samples

The sampling distribution of $\bar{X} - \bar{Y}$ is also a t distribution. The number of degrees of freedom is $N - 1$. As in the section on hypothesis testing of correlated samples, N is the number of *pairs* of scores. The lower and upper limits of the confidence interval about a mean difference between correlated samples are

$$\text{LL} = (\bar{X} - \bar{Y}) - t_\alpha(s_{\bar{D}}) \quad \text{and} \quad \text{UL} = (\bar{X} - \bar{Y}) + t_\alpha(s_{\bar{D}})$$

The interpretation of a confidence interval about a difference between means is very similar to the interpretation you made of confidence intervals about a sample mean. Again, the method is such that repeated sampling from two populations will produce a series of confidence intervals, 95 (or 99) percent of which will contain the true difference between the population means. You have sampled only once, so the proper interpretation is that you are 95 (or 99) percent confident that your lower and upper limits capture the true difference. It would probably be helpful for you to reread the material on interpreting a confidence interval about a mean, which is found in the section "The Concept of a Confidence Interval" near the end of Chapter 6.

ERROR DETECTION
> For confidence intervals for either independent or correlated samples, use a t value from Table D, not one calculated from the data.

PROBLEMS

22. Suppose $s_{\bar{X}_1 - \bar{X}_2} = 0.04$ for the problem on word processing packages. Establish a 95 percent confidence interval about the mean difference.

23. Based on the results of Problem 22, what decision should the manager make about the new package if a difference of 1.7 minutes is required to justify the extra expense?

24. Suppose a difference in time per problem of 0.25 minute is required to justify the new package. What decision should the manager make?

25. Establish a 99 percent confidence interval about the difference you found in Problem 20.

26. For an experiment on the effects of sleep on memory, eight volunteers were randomly divided into two groups. Everyone learned a list of ten nonsense syllables (later called consonant-vowel-consonants—CVCs). One group then slept for 4 hours, while the other group engaged in daytime activities. Then each subject recalled the CVCs. Two days later, everyone learned a new list. The group that had slept now engaged in daytime activities,

and the group that had been active slept. After 4 hours, each person recalled CVCs. The scores in the table represent the number of CVCs recalled by each person under the two conditions. Establish a 90 percent confidence interval about the mean difference. (This problem is modeled after a 1924 study by Jenkins and Dallenbach.)

Awake	Asleep
2	3
0	2
4	4
0	3
1	2
2	5
1	2
0	4

27. Interpret the confidence interval established in Problem 26.
28. This problem requires you to establish a 95 percent confidence interval about a mean rather than about a mean difference. This is just what you did in Chapter 6, but this time use the *t* distribution rather than the normal distribution. A researcher had a graph that showed a generally upward trend but with a dip in the middle (see illustration). To illustrate that the dip is not a chance phenomenon, calculate a 95 percent confidence interval about each mean. Redraw the graph and place a vertical line the length of the confidence interval on each mean. This will give you an idea of the sampling variation you would expect for each mean. By looking at the variation around the means for Trials 3, 4, and 5, you can determine whether the dip at Trial 4 can be attributed to chance.

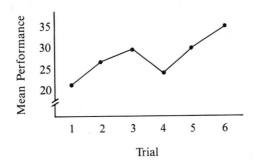

	Trials					
	1	2	3	4	5	6
$\bar{X}$	21	27	30	25	30	36
s	3.3	2.7	3.0	3.6	5.1	4.5
N	17	17	17	17	17	17

ASSUMPTIONS WHEN USING THE *t* DISTRIBUTION

You can perform a *t* test on the difference between means on any two-group data you have or any that you can beg, borrow, buy, or steal. No doubt about it, you can easily come up with a *t* value using

$$t = \frac{\bar{X}_1 - \bar{X}_2}{s_{\bar{X}_1 - \bar{X}_2}}$$

Once you have obtained a *t* value from your data, you can leap to the conclusion that the *t* distribution will give you an accurate probability figure. By consulting Table D you can *get* a probability figure and, using the .05 rule, retain or reject the null hypothesis.

In a similar way, you can calculate a confidence interval about the difference between means in any two-group experiment. If you then leap to the conclusion that the *t* distribution would be an accurate model, you can say you are "99 percent confident that the interval contains the true difference between the population means."

Should you leap to the conclusion that the *t* distribution will be accurate? *Will* the *t* distribution reflect actual empirical probabilities?

The *t* distribution will give correct results when the assumptions it is based on are true for the populations being analyzed. The *t* distribution, like the normal curve, is a theoretical distribution. In deriving the *t* distribution, mathematical statisticians make three assumptions:

1. The dependent-variable scores for both populations are normally distributed.
2. The variances of the dependent-variable scores for the two populations are equal.
3. The scores on the dependent variable are random samples from the population.

Assumption 3 requires three explanations. First, in a correlated-samples design, the *pairs* of scores should be random samples from the population you are interested in.

Second, assumption 3 ensures that any sampling errors will fall equally into both groups and that you may generalize from sample to population. Many times it is a physical impossibility to sample randomly from the population. In these cases, you should randomly assign the available subjects to one of the two groups. This will randomize errors, but your generalization to the population will be on less secure grounds than if you had obtained a truly random sample.

Third, assumption 3 ensures the *independence* of the scores; that is, knowing one score within a group does not help you predict other scores in that same group. Either random sampling from the population or random assignment of subjects to groups will serve to achieve this independence.

Now we can return to the major question of this section, When will the *t* distribution produce accurate probabilities? The answer is, When random samples are obtained from populations that are normally distributed and have equal variances.

This may appear to be a tall order. It is, and in practice no one is able to demonstrate these characteristics exactly. The next question becomes, Suppose I am not sure that my data have these characteristics. Am I likely to reach the wrong conclusion if I use Table D?

We don't have a simple answer to your reasonable question. In the past, several studies have suggested that the answer is no, because the *t* test is a robust test. (*Robust* means that a test gives you fairly accurate probabilities even when the data do not meet the assumptions on which it is based.) However, Bradley (1978) points out that robustness has not been defined quantitatively and, furthermore, that several studies (largely ignored by textbooks) show sizable departures from accuracy in a variety of situations. (See, for example, Blair and Higgins, 1985.) In actual practice, however, many researchers routinely use a *t* test unless one or more of these assumptions is clearly not justified.

USING THE *t* DISTRIBUTION TO TEST THE SIGNIFICANCE OF A CORRELATION COEFFICIENT

In Chapter 4, you learned to calculate Pearson product-moment correlation coefficients. This section is on testing the statistical significance of these coefficients. The question is whether an obtained *r*, based on a sample, could have come from a population of pairs of scores for which the parameter correlation is .00. The answer to this question is based on the size of a *t* value that is calculated from the correlation coefficient. The *t* value is found using the formula[6]

$$t = (r)\sqrt{\frac{N - 2}{1 - r^2}}$$

$df = N - 2$,   where N = number of pairs

The null hypothesis is that the population correlation is .00. A sample is drawn and an *r* is calculated. The *t* distribution is then used to determine whether the obtained *r* is significantly different from .00.

As an example, suppose you had obtained $r = .40$ with 22 pairs of scores. Does such a correlation indicate a significant relationship between the two variables, or should it be attributed to chance? Applying the formula for *t*,

$$t = (r)\sqrt{\frac{N - 2}{1 - r^2}} = (.40)\sqrt{\frac{22 - 2}{1 - (.40)^2}} = 1.95$$

$df = N - 2 = 22 - 2 = 20$

Table D shows that, for 20 *df*, a *t* value of 2.09 is required to reject the null hypothesis. The obtained *t* for $r = .40$, where $N = 22$, is less than the tabled *t*, so the null hypothesis is retained. That is, a coefficient of .40 would be expected by chance alone more than 5 times in 100 from a population in which the true correlation is zero.

In fact, for $N = 22$, $r = .43$ is required for significance at the .05 level and $r = .54$ for the .01 level. As you can see, even medium-sized correlations can be expected by chance alone for samples as small as 22. Most researchers strive for *N*'s of 30 or more

---

[6] This formula is an algebraic manipulation of $t = (r - 0)/s_r$, where $s_r = \sqrt{(1 - r^2)/(N - 2)}$. The form $(r - 0)/s_r$ should be familiar to you now: the deviation of a statistic from a parameter divided by the standard error of the statistic.

for correlation problems.

Sometimes you may wish to determine whether the *difference* between two correlations is statistically significant. Intermediate-level texts discuss this test. (See Ferguson, 1981, p. 196, or Howell, 1987, p. 240.)

PROBLEMS

29. A novice researcher developed a questionnaire on attitudes toward "consumerism." Responses were obtained from a random sample of males and females. A *t* test showed no significant differences between genders. The researcher then went back to the same random sample, administered the questionnaire a second time, pooled the data with those obtained earlier, and calculated a new *t* value, which was significant. Which of the assumptions of the *t* test was violated?

30. Determine whether the following Pearson product-moment correlation coefficients are significantly different from .00.
   a. $r = .62$; $N = 10$ **b.** $r = -.19$; $N = 122$
   c. $r = .50$; $N = 15$ **d.** $r = -.34$; $N = 64$

31. This is a think problem. Suppose you had a summer job in an industrial research laboratory testing the statistical significance of hundreds of correlation coefficients based on varying sample sizes. An α level of .05 had been adopted by the management and, for each coefficient, your task was to test whether it was "significant" or "not significant" (NS). How could you construct a table of your own, using Table D and the *t* formula for testing the significance of a correlation coefficient, that would allow you to label each coefficient without working out a *t* value for each?

The table you mentally designed in Problem 31 already exists. One version is reproduced in Table A in Appendix C. The α values of .10, .05, .02, .01, and .001 are included there. Use **Table A** in the future to determine whether a Pearson product-moment correlation coefficient is significantly different from .00, but remember that this table is based on the *t* distribution.

In summary, you have learned about a new sampling distribution in this chapter. You can now use the *t* distribution to determine the following:

1. The probability that a sample came from a population with a specific mean μ.
2. The probability that two samples (either *independent* or *correlated*) have a common population mean.
3. The probability that a sample correlation coefficient came from a population with a correlation of .00.
4. A confidence interval about a *mean* or a *mean difference*.

Transition Page

The t test, at which you are now skilled, is a very efficient way to analyze the results from an experiment that is designed to compare two treatments. The next step is to learn a technique you can use to analyze the results of an experiment that has more than two treatments.

This technique is called the **analysis of variance** (ANOVA for short, pronounced uh-ǹove-uh). ANOVA was invented by Sir Ronald Fisher, an English biologist and statistician, and it is appropriate for both small and large samples, just as t is. In fact, ANOVA is a close relative of t.

So the transition this time is from a t test and its sampling distribution, the t distribution, to ANOVA and its sampling distribution, the F distribution. Chapter 9 will show you how to use ANOVA to make comparisons among two or more groups. Chapter 10 will show you that the ANOVA technique can be extended to the analysis of experiments in which there are *two* independent variables, each of which may have two or more levels of treatment.

The analysis of variance is one of the most widely used statistical techniques, and Chapters 9 and 10 are devoted to an introduction to its more elementary forms. Intermediate and advanced books explain more sophisticated (and complicated) analysis-of-variance designs.

9 ANALYSIS OF VARIANCE: ONE-WAY CLASSIFICATION

Objectives for Chapter 9: After studying the text and working the problems in this chapter, you should be able to:

1. Identify the independent and dependent variables in a one-way ANOVA
2. Explain the rationale of ANOVA
3. Define F and explain its relationship to t
4. Compute sums of squares, degrees of freedom, mean squares, and F for an ANOVA
5. Interpret an F value obtained in an experiment
6. Construct a summary table of ANOVA results
7. Distinguish between *a priori* and *post hoc* tests
8. Use Tukey's Honestly Significant Difference test to make all pairwise comparisons following an ANOVA
9. List and explain the assumptions of ANOVA

In this chapter, you will work with the simplest analysis-of-variance design, a design called **one-way ANOVA**. One-way ANOVA is used to find out if there are any statistically significant differences among three or more sample means.[1] Experiments with more than two treatment levels are common in all disciplines that use statistics. Here are some examples:

1. Samples of lower-, middle-, and upper-class persons were compared on attitudes toward religion.
2. Four groups learned a task. A different schedule of reinforcement was used for each group. Afterward, response persistence was measured.
3. Suicide rates were compared for countries that represented low, medium, and high degrees of modernization.
4. The effect of strychnine on memory was assessed. Control groups received saline injections or no injection.

Except for the fact that these experiments have more than two groups, they are like those you analyzed with an *independent-samples t test* in Chapter 8. There is

---

[1] As will become apparent, ANOVA can be used also when there are just two groups.

one independent and one dependent variable. The null hypothesis is that the population means are the same for all groups. The subjects in each group are independent of subjects in the other groups. The only difference is that the independent variable has more than two levels.[2] To analyze such designs, use a one-way ANOVA.[3]

In example 1, the independent variable is social class, and it has three levels. The dependent variable is attitudes toward religion. The null hypothesis is that the religious attitudes are the same in all three populations of social classes; that is, $H_0: \mu_{\text{lower}} = \mu_{\text{middle}} = \mu_{\text{upper}}$.

PROBLEM   **1.** For examples 2, 3, and 4, identify the independent variable, the number of levels of the independent variable, the dependent variable, and the null hypothesis.

The task, as we stated, is to find any statistically significant differences among several sample means. A common first reaction to this problem is to run t tests on all possible combinations. For three means $(\bar{X}_1, \bar{X}_2, \bar{X}_3)$, three t tests would be required $(\bar{X}_1$ vs. $\bar{X}_2, \bar{X}_1$ vs. $\bar{X}_3$, and $\bar{X}_2$ vs. $\bar{X}_3)$. For four means, six tests are needed.[4] This multiple t-test approach *will not work* (and not just because doing lots of t tests is tedious).

Here is the reasoning: Suppose you had 15 samples all drawn from the same population. (Since there is just one population, the null hypothesis is clearly true.) These 15 sample means will vary from one another as a result of chance factors. Now suppose you ran every possible t test (all 105 of them), retaining or rejecting each null hypothesis at the .05 level. How many times would you reject the null hypothesis? About five; that is, if a t test has a .05 probability of making a Type I error, and you run about 100 tests on samples all drawn from the same population, you'll have about five Type I errors when you finish.

Now, let's move from this theoretical analysis to the reporting of an experiment. Suppose you conducted a 15-group experiment, ran 105 t tests, and found five significant differences. If you then pulled out those five and said they were reliable differences, people who understand the preceding paragraph would laugh at you (and recommend that your manuscript not be published). You can protect yourself from being laughed at if you use a test that allows you to compare the several means of your experiment, while keeping α at an acceptable level (like .05 or .01).

Sir Ronald A. Fisher (1890–1962) developed such a test. Fisher's brilliant work in genetics has been overshadowed by his fundamental work in statistics. In genetics, it was Fisher who showed that Mendelian genetics were compatible with Darwinian evolution. (For a while after 1900, genetics and evolution were in opposing camps.) And, among other contributions, Fisher showed how a recessive gene could become established in a population.

---

[2] Some writers prefer to call this a *completely randomized design.*

[3] Note that this book will not cover the ANOVA technique for analyzing a correlated-samples design. In the analysis of variance, such designs are referred to as *repeated measures designs.* To understand the analysis of repeated measures designs, work through Chapter 10 in this book and then read one of the following references: Ferguson (1981, chap. 19), Christensen and Stoup (1986, chap. 14), Porter and Hamm (1986, chap. 14), or Howell (1987, chap. 14).

[4] The formula for the number of combinations of n things taken two at a time is $[n(n-1)]/2$.

In statistics, it was Fisher who developed ANOVA, the topic of this chapter and the next. He also discovered the exact sampling distribution of r (1915), developed a general theory of estimation, and wrote *the* book on statistics. (*Statistical Methods for Research Workers* was first published in 1925 and went through 14 editions and several translations by 1973.)

Before getting into biology and statistics in such a big way, Fisher worked for an investment company for 2 years and taught in a public school for 4 years. (See Greene, 1966, or any encyclopedia.)

RATIONALE OF ANOVA

The question to be answered by ANOVA is whether the samples all came from populations with the same mean μ or whether at least one of the samples came from a population with a different mean.

Fisher, who was a friend of Gosset, came to understand that Student's t used a principle that was also applicable to experiments that have more than two groups. *The principle is that of dividing one estimate of the population variability by another.*[5] (See Box, 1981, Fisher's daughter, for an article on the correspondence between Fisher and Gosset.)

The sampling distribution that Fisher derived is the **F distribution.** As will be shown, F values that make up the F distribution are obtained by dividing one estimate of the population variance by a second estimate. Thus,

$$F = \frac{\text{estimate of } \sigma^2}{\text{estimate of } \sigma^2}$$

These two estimates of σ^2 are obtained by different methods. The numerator is obtained by a method that accurately estimates σ^2 only when H_0 is true. If H_0 is false, the estimate of σ^2 in the numerator will be too large.

The denominator is obtained by a method that is unaffected by the truth or falsity of H_0. Thus, when the null hypothesis is true, the expected value of F is about 1.00, since both methods are good estimators of σ^2, and $\sigma^2/\sigma^2 \cong 1.00$. Values somewhat larger and smaller than 1.00 are to be expected because of sampling fluctuation, but if an F value is too *large*, there is cause to suspect that H_0 is false.

We'll take these two estimates of σ^2 one at a time and discuss them. The estimate of σ^2 in the numerator is obtained from the two or more *sample means*. The conceptual steps follow. (Computational steps will come later.)

1. Find the standard deviation of the two or more sample means. This standard deviation of sample means is an old friend of yours, the standard error of the mean, $s_{\bar{X}}$.
2. Since $s_{\bar{X}} = s/\sqrt{N}$, squaring both sides gives $s_{\bar{X}}^2 = s^2/N$. Multiplying both sides by N and rearranging, you get $s^2 = N s_{\bar{X}}^2$.
3. s^2 is, of course, an estimate of σ^2.

---

[5] In the case of t,

$$t = \frac{\bar{X}_1 - \bar{X}_2}{s_{\bar{X}_1 - \bar{X}_2}} = \frac{\text{a range}}{\text{standard error of the difference between means}} = \frac{\text{a measure of variability}}{\text{a measure of variability}}$$

Thus, to find this s^2, you need to multiply the sample size (N) by the variance of the sample means, both of which you can calculate. This estimate of σ^2 is called the *between-means estimate* (also called the between-groups estimate). Notice that this between-means estimate of σ^2 is accurate only if the sample means are all drawn from the same population. If one or more means come from a population with a larger or smaller mean, the variance of the sample means will be larger.

The other estimate of σ^2 (the denominator of the F ratio) is obtained from the variability within each of the samples. Each sample variance is an independent estimate of σ^2, so, by averaging them, an even better estimate can be made. This estimate is called the *within-groups estimate,* and it is an unbiased estimate even if the null hypothesis is false. Once calculated, the two estimates can be compared. If the between-means estimate is much larger than the within-groups estimate, the null hypothesis is rejected.

We'll express these same ideas with pictures. **Figure 9.1** illustrates the situation when the null hypothesis is true ($\mu_A = \mu_B = \mu_C = \mu_D$). Four samples have been drawn from identical populations and a mean calculated for each sample. (See where the means are projected onto the dependent-variable line.) The means are fairly close together so the variability of these four means (the between-means estimate) will be small.

Figure 9.2 illustrates one situation in which the null hypothesis is false (one group comes from a population with a larger μ). The projection of the means this time shows that $\bar{X}_D$ will greatly increase the variability of the four means.

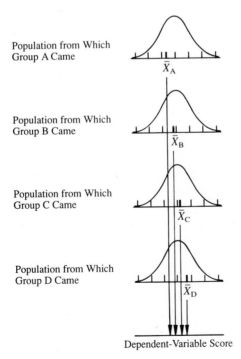

Population from Which
Group A Came

$\bar{X}_A$

Population from Which
Group B Came

$\bar{X}_B$

Population from Which
Group C Came

$\bar{X}_C$

Population from Which
Group D Came

$\bar{X}_D$

Dependent-Variable Score

Figure 9.1 H_0 is true. All four groups are drawn from identical populations. Not surprisingly, the sample means are close together.

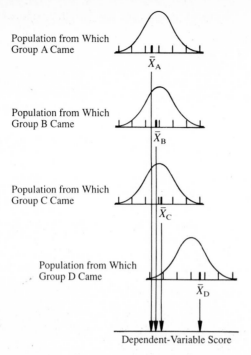

Population from Which Group A Came — $\bar{X}_A$

Population from Which Group B Came — $\bar{X}_B$

Population from Which Group C Came — $\bar{X}_C$

Population from Which Group D Came — $\bar{X}_D$

Dependent-Variable Score

Figure 9.2 H_0 is false. Three groups come from identical populations. A fourth group comes from a population with a larger mean. The sample means show more variability than those in Figure 9.1.

Study Figures 9.1 and 9.2. They illustrate how the between-means estimate is larger when the null hypothesis is false. So, if you have a small amount of variability between means, retain H_0. If you have a large amount of variability between means, reject H_0. Small and large, however, are relative terms and, in this case, they are relative to the population variance. A comparison of **Figures 9.3** and 9.1 illustrates how the amount of between-means variability *depends* on the population variance. In both Figures 9.1 and 9.3, the null hypothesis is true, but notice the projection of the sample means onto the dependent-variable line. There is more variability among the means that come from populations with greater variability. Figure 9.3, then, shows a large between-means estimate that is the result of large population variances and not the result of a false null hypothesis.

In order to decide whether a large between-means estimate is due to a false null hypothesis or to a population variance, you need another estimate of the population variance. The best such estimate is the average of the sample variances.

All of this discussion brings us back to the principle that Fisher found Gosset to be using in the *t* test: dividing one estimate of the population variability by another. In the case of ANOVA, if the null hypothesis is true, the two estimates should be very similar, and dividing one by the other should produce a value close to 1.0. If the null hypothesis is false, dividing the between-means estimate by the within-groups estimate will produce a value somewhat greater than 1.0.

What numbers qualify as "somewhat greater than 1.0"? You won't be surprised to learn that there is a table in Appendix C of this text that gives you critical values of

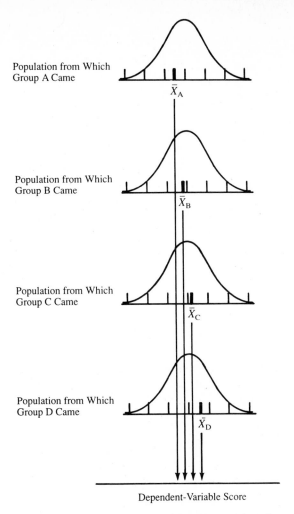

Figure 9.3 H_0 is true. All four groups are drawn from identical populations. Because the populations have larger variances than those in Figure 9.1, the sample means show more variability.

F. F values are in **Table F.** The first version of this table was compiled by George W. Snedecor of Iowa State University in 1934. At that time Snedecor named the variance ratio *F* in honor of Fisher.

Table F allows you to compare an *F* value from an experiment with those listed in the table. *If the F value from the data is as large as or larger than the tabled value, the null hypothesis is rejected.* If the *F* value from the data is not as large as the tabled value, the null hypothesis must be retained.

Figure 9.4 is a picture of the *F* distribution when the variance estimate of the numerator has 9 degrees of freedom and the variance estimate of the denominator has 15 *df*. If such an experiment produced *F* = 2.59, the null hypothesis would be rejected.

The *F* distribution and the *t* distribution are closely related. Like *t*, the *F* distribution is a family of curves. The shape of an *F* curve depends on the number of

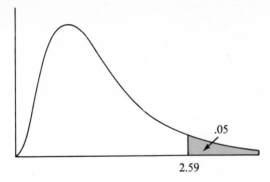

.05

2.59

Value of F

Figure 9.4   Form of the F distribution for $df_1 = 9$ and $df_2 = 15$. Adapted from Kirk (1982, p. 54), with permission

degrees of freedom in the numerator *and* the number in the denominator. All *t* curves are symmetrical, but *F* curves are positively skewed. As the number of degrees of freedom approaches infinity, the *F* curve approaches a normal distribution (a statement that is true for *t*, too). Finally, the mathematical relationship is $t^2 = F$. For example, the square of a *t* value on 15 *df* is equal to *F* on 1, 15 *df*.

Be sure you understand the rationale of ANOVA. That rationale is basic to the procedures explained in this chapter and the next. Soon you will learn how to actually compute an *F* value and interpret it. First, however, work the following problems and then learn the new terms used in ANOVA.

PROBLEMS

2. Tell what the two designs described here have in common and why the ANOVA technique discussed in this chapter is *not* appropriate.
   a. An experimenter studied the effect of 0, 2, 4, and 8 ounces of alcohol on hand steadiness. Nine subjects participated for 4 days, each taking a steadiness test after consuming 0, 2, 4, or 8 ounces in 1 hour. Thus, the mean for each level of the independent variable was based on the same nine subjects.
   b. An experimenter set up a design to study the effect of three types of propaganda on opinion change. In order to make sure the family background was equivalent for the three groups, the subjects were sets of three siblings from the same family. One sibling was assigned to each of the three groups.

3. Suppose three means came from a population with $\mu = 100$ and a fourth mean from a population with $\mu = 50$. Will the between-means estimate of variability be larger or smaller than the between-means estimate of variability of four means drawn from a population with $\mu = 100$?

4. How can estimates of a population variance be obtained?

5. Define *F*.

6. Interpret the meaning of the *F* values given here for the experiments described in examples 1 and 3 on the first page of this chapter.
   a. A very small *F* value for example 1
   b. A very large *F* value for example 3

MORE NEW TERMS

Sum of squares In the "Clue to the Future" on page 58, we pointed out that, in the computation of the standard deviation, certain values were obtained that would be important in future chapters. That future is now. The term Σx^2 (the numerator of the basic formula for the standard deviation) is called the **sum of squares** (abbreviated *SS*). So, $SS = \Sigma x^2 = \Sigma(X - \bar{X})^2$. A more descriptive name for sum of squares is "sum of the squared deviations."

Mean square **Mean square** (*MS*) is the ANOVA term for a variance s^2. The mean square is a sum of squares divided by its degrees of freedom.

Grand mean The **grand mean** is the mean of all scores; it is computed without regard for the fact that the scores come from different groups (samples).

tot The subscript *tot* after a symbol makes the symbol stand for all such numbers in the experiment; for example, ΣX_{tot} is the sum of all scores.

g The subscript *g* after a symbol means that the symbol applies to a group; for example, $\Sigma(\Sigma X_g)^2$ tells you to sum the scores in each group, square each sum, and then sum these squared values.

K *K* is the number of groups in the experiment. This is the same as the number of levels of the independent variable.

CLUE TO THE FUTURE

You will soon work problems with several sets of numbers. For each set you will need the sum, the sum of squares, and other values. Calculators with memories can get these sums simultaneously with just one entry of each number. If you know (or learn) how to exploit your calculator's capabilities, you can spend less time on statistics problems.

SUMS OF SQUARES

The technique you are learning about is called analysis of variance because the variability of all the scores in an experiment can be divided among two or more sources. In the case of simple analysis of variance, just two sources contribute all the variability to the scores. One source is the variability between groups, and the other source is the variability within each group. The sum of these two sources is equal to the total variability.

We will illustrate this using the sum of squares as a measure of variability. Calculations will be made on the data in **Table 9.1**, which shows fictional data from an experiment in which there were three groups of mental patients. The independent variable was drug therapy. Three levels of the independent variable were used— Drug A, Drug B, and Drug C. The dependent variable was the number of psychotic episodes observed for each patient during the therapy period. The experimenter believed that the use of Drug C would reduce the number of psychotic episodes more than would Drug A or B, which were in common use.

Table 9.1   Computation of sums of squares of fictional data from a drug study

	Drug A (1)		Drug B (2)		Drug C (3)	
	X_1	X_1^2	X_2	X_2^2	X_3	X_3^2
	9	81	9	81	4	16
	8	64	7	49	3	9
	7	49	6	36	1	1
	5	25	5	25	1	1
Σ	29	219	27	191	9	27
$\bar{X}$	7.25		6.75		2.25	

$\Sigma X_{tot} = 29 + 27 + 9 = 65$

$\Sigma X_{tot}^2 = 219 + 191 + 27 = 437$

$\bar{X}_{tot} = 5.42$

$$SS_{tot} = \Sigma X_{tot}^2 - \frac{(\Sigma X_{tot})^2}{N_{tot}} = 437 - \frac{65^2}{12} = 84.92$$

$$SS_{drugs} = \Sigma\left[\frac{(\Sigma X_g)^2}{N_g}\right] - \frac{(\Sigma X_{tot})^2}{N_{tot}}$$

$$= \frac{29^2}{4} + \frac{27^2}{4} + \frac{9^2}{4} - \frac{65^2}{12}$$

$$= 210.25 + 182.25 + 20.25 - 352.08 = 60.67$$

$$SS_{wg} = \Sigma\left[\Sigma X_g^2 - \frac{(\Sigma X_g)^2}{N_g}\right] = \left(219 - \frac{29^2}{4}\right) + \left(191 - \frac{27^2}{4}\right) + \left(27 - \frac{9^2}{4}\right)$$

$$= 8.75 + 8.75 + 6.75 = 24.25$$

Check: $SS_{tot} = SS_{bg} + SS_{wg} = 84.92 = 60.67 + 24.25$

This experimenter's belief is an alternative hypothesis (sometimes called a private hypothesis or experimenter's hypothesis), but it is not the alternative hypothesis being tested in ANOVA. The alternative hypothesis in ANOVA is that there are one or more differences among the population means. No direction of differences is specified. The null hypothesis is that $\mu_1 = \mu_2 = \mu_3$ or, in the terms of the experiment, that any differences among the psychotic episodes of the three groups are due to sampling fluctuation and not to differences caused by the three drugs.

First, we'll focus on the total variability as measured by the total sum of squares. Actually, as you will see, you are already familiar with the total sum of squares (SS_{tot}). To find SS_{tot}, subtract the grand mean from each score. Square these deviation scores and sum them up:

$$SS_{tot} = \Sigma(X - \bar{X}_{tot})^2$$

Computationally, SS_{tot} is more readily (and accurately) obtained using the algebraically equivalent raw-score formula

$$SS_{tot} = \Sigma X_{tot}^2 - \frac{(\Sigma X_{tot})^2}{N_{tot}}$$

which you may recognize as the numerator of the raw-score formula for s. This formula is equivalent to Σx^2. Its computation, as illustrated in Table 9.1, requires you to square each score and sum the squared values to obtain ΣX_{tot}^2. Next, the scores are summed and the sum squared. That squared value is divided by the total number of scores to obtain $(\Sigma X_{tot})^2/N_{tot}$. Subtraction of $(\Sigma X_{tot})^2/N_{tot}$ from ΣX_{tot}^2 yields the total sum of squares. For the data in Table 9.1,

$$SS_{tot} = \Sigma X_{tot}^2 - \frac{(\Sigma X_{tot})^2}{N_{tot}} = 437 - \frac{65^2}{12} = 84.92$$

Thus, the total variability of all the scores in Table 9.1 is 84.92 when measured by the sum of squares. This total comes from two sources: the between-groups sum of squares and the within-groups sum of squares. Each of these can be computed separately.

The *between-groups sum of squares* (SS_{bg}) is the variability of the group means from the grand mean of the experiment, weighted by the size of the group:

$$SS_{bg} = \Sigma[N_g(\bar{X}_g - \bar{X}_{tot})^2]$$

SS_{bg} is more easily computed by the raw-score formula:

$$SS_{bg} = \Sigma\left[\frac{(\Sigma X_g)^2}{N_g}\right] - \frac{(\Sigma X_{tot})^2}{N_{tot}}$$

This formula tells you to sum the scores for each group and then square the sum. Each squared sum is then divided by the number of scores in that group. These values (one for each group) are then summed, giving you $\Sigma[(\Sigma X_g)^2/N_g]$. From this sum is subtracted the value $(\Sigma X_{tot})^2/N_{tot}$, which was obtained in the computation of SS_{tot}.

When describing experiments in general, the term SS_{bg} is used. In a specific analysis, *between groups* is changed to a summary word for the independent variable. For the experiment in Table 9.1, that word is *drugs*. Thus,

$$SS_{drugs} = \Sigma\left[\frac{(\Sigma X_g)^2}{N_g}\right] - \frac{(\Sigma X_{tot})^2}{N_{tot}}$$
$$= \frac{29^2}{4} + \frac{27^2}{4} + \frac{9^2}{4} - \frac{65^2}{12} = 60.67$$

The other source of variability is the *within-groups sum of squares* (SS_{wg}), which is the sum of the variability in each of the groups. SS_{wg} is defined as

$$SS_{wg} = \Sigma(X_1 - \bar{X}_1)^2 + \Sigma(X_2 - \bar{X}_2)^2 + \cdots + \Sigma(X_K - \bar{X}_K)^2$$

or the sum of the squared deviations of each score from the mean of its group added to the sum of the squared deviations from all other groups for the experiment. As with the other SS values, there is an arrangement of the arithmetic that is easiest. For SS_{wg}, it is

$$SS_{wg} = \Sigma\left[\Sigma X_g^2 - \frac{(\Sigma X_g)^2}{N_g}\right]$$

This formula tells you to square each score in a group and sum them (ΣX_g^2). Subtract from this a value that you obtain by summing the scores, squaring the sum, and dividing by the number of scores in the group: $(\Sigma X_g)^2/N_g$. For each group, a value is calculated, and these values are summed to get SS_{wg}. For the data in Table 9.1, you get

$$SS_{wg} = \Sigma \left[\Sigma X_g^2 - \frac{(\Sigma X_g)^2}{N_g} \right]$$

$$= \left(219 - \frac{29^2}{4} \right) + \left(191 - \frac{27^2}{4} \right) + \left(27 - \frac{9^2}{4} \right) = 24.25$$

We mentioned that, if you work with SS as a measure of variability, the total variability is the sum of the variability of the parts. Thus,

$$SS_{tot} = SS_{bg} + SS_{wg}$$
$$84.92 = 60.67 + 24.25$$

ERROR DETECTION

$SS_{tot} = SS_{bg} + SS_{wg}$. All sums of squares are always zero or positive, never negative. This check will not catch errors made in summing the scores (ΣX) or in summing squared scores (ΣX^2).

PROBLEMS

7. Show that $SS_{tot} = SS_{bg} + SS_{wg}$ for the data in the table.

X_1	X_2	X_3	X_4
9	2	5	7
5	6	5	6
3	7	9	7
6	8	2	3
5	3	4	2

8. Emile Durkheim (1858—1917), a founder of sociology, believed that modernization produced social problems like suicide. He compiled data from several European countries and published *Suicide* (1897), a book that helped establish quantification in social science disciplines. The data in this problem were created so that the conclusion you reach will mimic that of Durkheim (and present-day sociologists). Using variables like electricity consumption, newspaper circulation, and gross national product, 13 countries were classified as to their degree of modernization (low, medium, or high). The dependent variable was the suicide rate per 100,000 persons. For the data in the table, compute SS_{tot}, SS_{bg}, and SS_{wg}.

Degree of modernization		
Low	*Medium*	*High*
4	17	20
8	10	22
7	9	19
5	12	9
		14

9. Thirty undergraduate students enrolled in an introductory statistics course were randomly assigned to three groups. Group 1 students were taught by the traditional lecture method. Group 2 students were also taught by the lecture method but, in addition, had a 2-hour laboratory each week during which they worked on statistics problems. Group 3 students used a book designed for individualized teaching. They attended no lectures but worked problems on their own and took short tests when they felt they were ready. At the end of the term, all students took the same comprehensive statistics test. The number of errors made by each student is recorded in the table. Compute SS_{tot}, SS_{bg}, and SS_{wg}.

Group 1 (lecture)	Group 2 (lecture and lab)	Group 3 (individualized)
19	13	12
18	12	11
17	10	11
15	10	10
15	9	9
14	9	8
12	8	7
11	6	5
9	6	5
7	5	4

MEAN SQUARES AND DEGREES OF FREEDOM

The next step in an analysis of variance is to find the mean squares. A mean square is simply a sum of squares divided by its degrees of freedom. It is an estimate of the population variance, σ^2.

Each sum of squares has a particular number of degrees of freedom associated with it. Thus, in a one-way classification, the df are df_{tot}, df_{bg}, and df_{wg}. Also, like SS,

$$df_{tot} = df_{bg} + df_{wg}$$

The df_{tot} is $N_{tot} - 1$. The df_{bg} is the number of groups minus one ($K - 1$). The df_{wg} is the sum of the degrees of freedom for each group $[(N_1 - 1) + (N_2 - 1) + \cdots + (N_K - 1)]$. If there are equal numbers of scores in the K groups, the formula for df_{wg} reduces to $K(N_g - 1)$. A little algebra will reduce this still further:

$$df_{wg} = K(N_g - 1) = KN_g - K$$

But $KN_g = N_{tot}$, so $df_{wg} = N_{tot} - K$. This formula for df_{wg} works whether the number in each group is the same or not.

ERROR DETECTION

$df_{tot} = df_{bg} + df_{wg}$. Degrees of freedom are always positive.

Mean squares, then, can be found using the following formulas:[6]

$$MS_{bg} = \frac{SS_{bg}}{df_{bg}} \qquad \text{where } df_{bg} = K - 1$$

$$MS_{wg} = \frac{SS_{wg}}{df_{wg}} \qquad \text{where } df_{wg} = N_{tot} - K$$

For the data in Table 9.1,

$$df_{drugs} = K - 1 = 3 - 1 = 2$$
$$df_{wg} = N_{tot} - K = 12 - 3 = 9$$

Check: $df_{tot} = df_{bg} + df_{wg} = 2 + 9 = 11$

$$MS_{drugs} = \frac{SS_{drugs}}{df_{bg}} = \frac{60.67}{2} = 30.333$$

$$MS_{wg} = \frac{SS_{wg}}{df_{wg}} = \frac{24.25}{9} = 2.694$$

Notice that, although $SS_{tot} = SS_{bg} + SS_{wg}$ and $df_{tot} = df_{bg} + df_{wg}$, mean squares are *not* additive.

CALCULATION AND INTERPRETATION OF F VALUES USING THE F DISTRIBUTION

We said earlier that F is a ratio of two estimates of the population variance. MS_{bg} is an estimate based on the variability between means. MS_{wg} is an estimate based on the sample variances. An **F test** consists of dividing MS_{bg} by MS_{wg} to obtain an F value:

$$F = \frac{MS_{bg}}{MS_{wg}}$$

The null hypothesis is $H_0: \mu_1 = \mu_2 = \cdots = \mu_K$. When the null hypothesis is true, the expected value of F is about 1.00. Of course, sampling error will produce some F's greater than 1.00 and some less than 1.00.

There are two different degrees of freedom associated with any F value. They are the df associated with $MS_{bg}(K - 1)$ and the df associated with $MS_{wg}(N_{tot} - K)$. For the data in Table 9.1,

$$F = \frac{MS_{drugs}}{MS_{wg}} = \frac{30.333}{2.694} = 11.26 \qquad df = 2, 9$$

The next question is, What is the probability of obtaining $F = 11.26$ if all three samples come from populations with the same mean? As before, if the probability is less than α, reject the null hypothesis.

Turn now to **Table F** in Appendix C, which gives the critical values of F when $\alpha = .05$ and when $\alpha = .01$. Across the top of the table are degrees of freedom

[6] MS_{tot} is not used in ANOVA; only MS_{bg} and MS_{wg} are calculated.

Table 9.2 Summary table of ANOVA for data in Table 9.1

Source	df	SS	MS	F	p
Between drugs	2	60.67	30.33	11.26	<.01
Within groups	9	24.25	2.69		
Total	11	84.92			

$$F_{.01}\,(2, 9\ df) = 8.02$$

associated with the numerator or MS_{bg}. For Table 9.1, $df_{bg} = 2$, so 2 is the column you want. Along the left side of the table are degrees of freedom associated with the denominator or MS_{wg}. In this case, $df_{wg} = 9$, so look for 9 along the side. The tabled value for 2 and 9 df is 4.26 at the .05 level (lightface type) and 8.02 at the .01 level (boldface type). Our obtained F is 11.26. Therefore, if the three samples were drawn from populations with the same mean (the null hypothesis), $F = 11.26$ would occur less than 1 percent of the time. Thus, reject the null hypothesis and conclude that the three samples do not have a common population mean. At least two drugs had different effects on the number of psychotic episodes.

At this point, your interpretation must stop. An ANOVA does not tell you which of the population means is greater than or less than the others. Such an interpretation requires more statistical analysis, which is the topic of the last part of this chapter.

It is customary to summarize the results of an ANOVA in a *summary table*. **Table 9.2** is an example. The values in the table are those we have calculated. Look at the right side of the table under p (for probability). The notation "$p < .01$" is shorthand for "the probability of $F = 11.26$ or greater occurring as a result of chance fluctuations is less than one in 100."

Sometimes Table F does not contain an F value for the df in your problem. For example, an F with 2, 35 df or 4, 90 df is not tabled. When this happens, be conservative; use the F value that is given for *fewer df* than you have. Thus, the proper F values for those two examples would be based on 2, 34 df and 4, 80 df, respectively.[7]

SCHEDULES OF REINFORCEMENT—A LESSON IN PERSISTENCE

Persistence is when you keep on trying even though the rewards are scarce or nonexistent. It is something that seems to vary a great deal from person to person and from task to task. What causes this variation? What has happened to lead to such differences in people?

Persistence can often be explained if you know how frequently reinforcement (rewards) occurred in the past. We will illustrate with the data produced by pigeons,

[7] The F distribution may also be used to test hypotheses about variances as well as hypotheses about means. To determine the probability that two sample variances came from the same population (or from populations with equal variances), form a ratio with the larger sample variance in the numerator. The resulting F value can be interpreted with Table F. The proper df are $N_1 - 1$ and $N_2 - 1$ for the numerator and denominator, respectively. For more information, see Kirk (1984, p. 307) or Ferguson (1981, p. 189).

Table 9.3 Partial analysis of the data on the number of minutes to extinction after four schedules of reinforcement during learning

Schedule of reinforcement during learning			
crf	FR2	FR4	FR8
3	5	8	10
5	7	12	14
6	9	13	15
2	8	11	13
5	11	10	11
	10		
	6		

ΣX 21   56   54   63

ΣX^2 99   476   598   811

$\bar{X}$ 4.2   8.0   10.8   12.6

$\Sigma X_{tot} = 21 + 56 + 54 + 63 = 194$

$\Sigma X_{tot}^2 = 99 + 476 + 598 + 811 = 1984$

$$SS_{tot} = \Sigma X_{tot}^2 - \frac{(\Sigma X_{tot})^2}{N_{tot}} = 1984 - \frac{(194)^2}{22} = 1984 - 1710.73 = 273.27$$

$$SS_{schedules} = \Sigma\left[\frac{(\Sigma X_g)^2}{N_g}\right] - \frac{(\Sigma X_{tot})^2}{N_{tot}} = \frac{(21)^2}{5} + \frac{(56)^2}{7} + \frac{(54)^2}{5} + \frac{(63)^2}{5} - \frac{(194)^2}{22} = 202.47$$

$$SS_{wg} = \Sigma\left[\Sigma X_g^2 - \frac{(\Sigma X_g)^2}{N_g}\right]$$

$$= \left(99 - \frac{(21)^2}{5}\right) + \left(476 - \frac{(56)^2}{7}\right) + \left(598 - \frac{(54)^2}{5}\right) + \left(811 - \frac{(63)^2}{5}\right) = 70.80$$

$$df_{tot} = N_{tot} - 1 = 22 - 1 = 21$$

$$df_{schedules} = K - 1 = 4 - 1 = 3$$

$$df_{wg} = N_{tot} - K = 22 - 4 = 18$$

but the principles they illustrate are true for many other forms of life, including students and professors.

The data in **Table 9.3** represent typical results from a *schedule of reinforcement* study. A hungry pigeon is taught to peck at a disk on the wall. A peck produces food according to a schedule the experimenter has set up. For some pigeons every peck produces food—a continuous reinforcement schedule (crf). For some birds every other peck produces food—a fixed-ratio schedule of 2:1 (FR2). A third group of pigeons gets food after every fourth peck—an FR4 schedule. Finally, for a fourth group, eight pecks are required to produce food—an FR8 schedule. After all groups receive 100 reinforcements, no more food is given (extinction begins). Under such conditions pigeons will continue to peck for a while and then stop. The dependent variable is the number of minutes a bird continues to peck (persist) after the food stops. As you can see in Table 9.3, three groups had five birds and one had seven.

Table 9.3 shows the raw data and the steps required to get the three *SS* figures and the three *df* figures. Work through that table now. When you finish, examine the summary table **(Table 9.4),** which continues the analysis by giving the mean squares, *F* value, and the probability of such an *F*.

Table 9.4 Summary table of the ANOVA analysis of the schedules
of reinforcement study

Source	df	SS	MS	F	p
Between schedules	3	202.47	67.49	17.16	<.01
Within groups	18	70.80	3.93		
Total	21	273.27			

$$F_{.05}\,(3, 18) = 3.16 \qquad F_{.01}\,(3, 18) = 5.09$$

When an experiment produces an F with a probability value less than .01, a conclusion is called for. Here it is: The schedule of reinforcement used during learning has a significant effect on persistence of responding during extinction. It is unlikely that the four samples have a common population mean.

You might feel unsatisfied by that conclusion. You might ask which groups differ from the others, or which groups are causing that "significant difference." Good questions. You will find answers in the next section, but first here are a few problems to reinforce what you learned in this section.

PROBLEMS

10. Calculate the F value for the data in Problem 7 and compose a summary table. Determine the critical value of F and retain or reject the null hypothesis. Tell what the analysis has shown.

11. Same as Problem 10 but use the data in Problem 8.

12. Same as Problem 10 but use the data in Problem 9.

13. Suppose you obtained $F = 2.56$ with 5, 75 df. How many groups are in the experiment? What are the critical values for the .05 and .01 levels? What conclusion should be reached?

COMPARISONS AMONG MEANS

A significant F by itself tells you only that it is not likely that the samples have a common population mean. Usually you want to know more, but exactly what you want to know often depends on the particular problem you are working on. For example, in the schedules of reinforcement study, many would ask which schedules are significantly different from others (all pairwise comparisons). Another question might be whether there are sets of schedules that do not differ among themselves but do differ from other sets. In the drug study, the interest is on the new experimental drug. Is it better than either of the standard drugs or is it better than the average of the standards? This is just a *sampling* of the kinds of questions that can be asked.

Unfortunately, there isn't one **multiple comparison** method that is satisfactory for all questions. In fact, there aren't even three or four methods that will handle all questions satisfactorily.

The reason there are so many methods is that specific questions are best answered by tailor-made tests. Over the years, statisticians have developed many tests for specific situations and each test has something to recommend it. For an excellent chapter on the tests and their rationale, see Howell (1987, "Multiple Comparisons Among Treatment Means").

With this background in place, here is our plan for this elementary statistics book. We will discuss the two major categories of tests subsequent to ANOVA (*a priori* and *post hoc*) and then cover one *post hoc* test (Tukey's HSD test).

A Priori and Post Hoc Tests

A priori **tests** require that you *plan* a limited number of tests before gathering the data. *A priori* means "based on reasoning, not on immediate observation." Thus, for *a priori* tests, you must choose them during the design stage of the experiment or certainly before you look at the scores on the dependent variable.

Post hoc **tests** can be chosen after the data are gathered and you have examined the means. Any and all differences for which a story can be told are fair game for *post hoc tests*. *Post hoc* means "after this" or "after the fact."

The necessity for two kinds of tests will be evident to you if you think again about the 15 samples that were drawn from the same population. Before the samples are drawn, what would be your expectation that any two randomly chosen means would be significantly different if tested with an independent-measures *t* test? About .05, we would suppose. But what about after the data are drawn and the means compiled? If you then pick out the largest and the smallest means and test them with a *t* test, what is your expectation that the two would be significantly different? Much higher, we would suppose. Thus, if you want to make comparisons after the data are gathered (keeping α at an acceptable level like .05), then larger differences should be required before a difference is declared "not due to chance."

The preceding paragraph should give you a start on understanding that there is a good reason for more than one statistical test to answer questions subsequent to ANOVA. For this text, however, we will confine ourselves to just one test, a *post hoc* test named Tukey's Honestly Significant Difference.

Tukey's Honestly Significant Difference

Suppose an ANOVA produces a significant *F*. For many research problems, the next question is, Which sample means are significantly different from the others? Tukey's Honestly Significant Difference (HSD) is designed to test the significance of every pairwise difference when *N* is the same for all samples.

The formula for the HSD is

$$HSD = \frac{\bar{X}_1 - \bar{X}_2}{s_{\bar{X}}}$$

where $\quad s_{\bar{X}} = \sqrt{\dfrac{MS_{wg}}{N}}$

N = the number in each group

The critical value against which the HSD is compared is found in Table G in Appendix C. Turn to **Table G.** Critical values for HSD$_{.05}$ are on the first page; those for HSD$_{.01}$ are on the second. After you have chosen $\alpha = .05$ or $.01$ and selected the correct page, enter the row that gives the *df* for the MS_{wg} from the ANOVA. Go over to the column that gives K for the experiment. If the HSD you calculated from the data is larger than the number at the intersection, reject the null hypothesis and conclude that the two means are significantly different. Describe that difference using the names of the independent and dependent variables.

We will illustrate Tukey's HSD by completing the analysis on the drug experiment (Tables 9.1 and 9.2). Since the one-way ANOVA produced a significant F value, pairwise comparisons are permissible.

We could calculate the three pairwise comparisons in any order, but there is a strategy that may save us from some arithmetic. The trick is to pick a mean difference of intermediate size and test it. If this difference proves significant, then all *larger* differences will also be significant. On the other hand, if the difference proves to be not significant, then all *smaller* differences also will be not significant.

In the drug study, there are three means and so there will be three comparisons. It will be easy to choose "a mean difference of intermediate size" because there is only one possibility, which is the difference between the experimental drug (C) and the standard drug (B):

$$HSD = \frac{\bar{X}_2 - \bar{X}_3}{s_{\bar{X}}} = \frac{6.75 - 2.25}{\sqrt{\dfrac{2.69}{4}}} = \frac{4.50}{0.820} = 5.49$$

Because 5.49 is larger than the tabled value for HSD$_{.01}$(5.43), you can conclude that the experimental drug (C) is significantly better than Drug B at reducing psychotic episodes ($p < .01$). Because the difference between Drug C and Drug A is even greater than the difference between Drugs C and B, you can also conclude that the experimental drug is significantly better than Drug A in reducing psychotic episodes ($p < .01$).

One comparison remains, the comparison between the two standard drugs, A and B:

$$HSD = \frac{7.25 - 6.75}{\sqrt{\dfrac{2.69}{4}}} = 0.61$$

Since HSD$_{.05} = 3.95$, you can conclude that this study does not provide evidence of the superiority of Drug B over Drug A.

Tukey's HSD When *N*'s Are Not Equal.   Tukey's HSD was developed for studies in which N's are equal for all groups. Sometimes, despite the best of intentions, unequal N's occur. A modification in the denominator of HSD will produce a statistic that can be tested with the values in Table G. The modification recommended by Kramer (1956) and shown to be accurate (Smith, 1971) is

$$s_{\bar{X}} = \sqrt{\frac{MS_{wg}}{2}\left(\frac{1}{N_1} + \frac{1}{N_2}\right)}$$

14. Distinguish between the two categories of tests subsequent to ANOVA, *a priori* and *post hoc.*

15. For Durkheim's modernization and suicide data, make all possible pairwise comparisons and write a conclusion telling what the data show.

16. For the schedules of reinforcement data, make all possible pairwise comparisons and write a conclusion telling what the data show.

ASSUMPTIONS OF THE ANALYSIS OF VARIANCE

In developing the analysis of variance and the *F* test, Fisher started with the following assumptions about the dependent variable:

1. *Normality.* It is assumed that the dependent variable is normally distributed in the populations from which samples are drawn. It is often difficult or impossible to demonstrate normality or lack of normality in the parent populations. Such a demonstration usually occurs only with very large samples. On the other hand, because of extensive research, some populations are known to be skewed, and researchers in those fields may decide that ANOVA is not appropriate for their data analysis. Unless there is a reason to suspect that populations depart severely from normality, the inferences made from the *F* test will probably not be affected. ANOVA is robust. (It results in correct probabilities even when the populations are not exactly normal.) Where there is suspicion of a severe departure from normality, however, use the nonparametric method explained in Chapter 12.

2. *Homogeneity of variance.* This means that the two or more population variances are equal. In ANOVA, the variances of the dependent-variable scores for each of the populations sampled are assumed to be equal. Figures 9.1 and 9.2, which we used to illustrate the rationale of ANOVA, show populations with equal variances. Several methods for testing this assumption are presented in advanced texts, such as Kirk (1982), Keppel (1982), and Winer (1971). An alternative is to use a nonparametric method for comparing all pairs. This will be discussed in Chapter 12.

3. *Random sampling.* Every care should be taken to assure that sampling is random and that assignment to groups is also random, so that the measurements are all independent of one another.

We hope these assumptions have a familiar ring to you. They are the same as those you learned for the *t* distribution. This makes sense; *t* is a special case of *F*. If you need a reminder of the status of the issue, How do I know if the data meet the assumptions?, reread "Assumptions When Using the *t* Distribution," near the end of Chapter 8.

Here is the final set of problems on one-way ANOVA. Careful work here will facilitate your understanding of the more complex experiments in the next chapter.

17. List the three assumptions of the analysis of variance.

18. In about 1920 Hermann Rorschach began to get an inkling of an idea. Before long he had devised a test. After people had responded to his test, he counted the number of responses each gave. (Today this measure is symbolized R by those in the know.) The data in the table were obtained from three different groups—normals, schizophrenics, and depressives. Analyze the data with an ANOVA and, if appropriate, a set of Tukey's HSDs. Write a conclusion about Rorschach's test.

Normals	Schizophrenics	Depressives
23	17	16
18	20	11
31	22	13
14	19	10
22	21	12
28	14	10
20	25	
22		

19. A researcher interested in the effects of stimulants on memory gave three groups of rats 20 trials of training to avoid a shock. Immediately after learning, one group of rats was injected with a small amount of strychnine (a stimulant), a second group was injected with saline as a control for the injection, and a third group was not injected. Seven days later, the researcher tested the rats' memory by training them until they reached a criterion of avoiding the shock five trials in a row. The number of trials to reach this criterion was recorded for each rat. (See McGaugh and Petrinovich, 1965). Perform an ANOVA and write a sentence summary.

	Strychnine	Saline	No injection
ΣX	121	138	152
ΣX^2	1608	2464	3000
N	10	8	8

20. An agricultural experimenter wanted to know which of three varieties of soybeans would produce the highest yield in a particular type of soil. He divided a field with that type of soil into 15 plots of equal size and then randomly assigned the three varieties to five plots each. The dependent variable is the number of bushels harvested from each plot. Perform an ANOVA. Construct a summary table. If appropriate, calculate Tukey's HSD on the three pairwise comparisons.

Variety of soybean		
X_1	X_2	X_3
9	13	11
7	12	10
7	12	9
6	10	9
6	9	8

10 ANALYSIS OF VARIANCE: FACTORIAL DESIGN

Objectives for Chapter 10: After studying the text and working the problems in this chapter, you should be able to:

1. Define the terms *factorial design, factor, level,* and *cell*
2. Identify the sources of variance in a factorial design
3. Compute sums of squares, degrees of freedom, mean squares, and *F* values in a factorial design
4. Determine whether *F* values are significant or not
5. Describe the concept of interaction
6. Interpret interactions and main effects
7. Make pairwise comparisons between means with Tukey's HSD test

In Chapter 9, you learned the basic concepts and procedures for conducting a simple one-way ANOVA. In this chapter, we will extend those concepts and procedures to a popular and more complex design, the factorial ANOVA. For many problems, a factorial ANOVA will give you a special kind of information that has not yet been discussed in this book. First, however, we have new terms for you and a review of the progression of designs that lead up to a factorial ANOVA.

In this chapter, we introduce the term *factor*. **Factor** is just a shorter word for independent variable. The **levels** *of a factor* are simply the different treatments that make up the independent variable. To illustrate, in Chapter 9 you analyzed data from a four-group study on persistence. The factor in that study was schedule of reinforcement and there were four levels. In Chapter 8, you analyzed data from a two-group study of wage discrimination. The factor was gender and there were two levels of the factor, male and female. We will continue our summarizing by describing the designs you learned to analyze in the two previous chapters.

In Chapter 8, you learned to analyze data from a two-group experiment, schematically shown in **Table 10.1**. In Table 10.1, there is *one independent variable*, Factor *A*, with data from two levels, A_1 and A_2. For the *t* test, the question was whether $\bar{X}_{A_1}$ was significantly different from $\bar{X}_{A_2}$. You learned to determine this whether the samples were independent or correlated.

Table 10.1   Illustration of a two-group design that can be analyzed with a *t* test

Factor A	
A_1	A_2
Scores on the dependent variable	Scores on the dependent variable
$\bar{X}_{A_1}$	$\bar{X}_{A_2}$

Table 10.2   Illustration of a four-group design that can be analyzed with an *F* test

Factor A			
A_1	A_2	A_3	A_4
Scores on the dependent variable	Scores on the dependent variable	Scores on the dependent variable	Scores on the dependent variable
$\bar{X}_{A_1}$	$\bar{X}_{A_2}$	$\bar{X}_{A_3}$	$\bar{X}_{A_4}$

In Chapter 9, you learned to analyze data from a two-or-more-group experiment, schematically shown in **Table 10.2** for a four-group design. In Table 10.2, there is *one independent variable,* Factor *A*, with data from four levels, A_1, A_2, A_3, and A_4. In the ANOVA of Chapter 9, the question was whether the four sample means represent populations that have a common mean. You learned to determine this for independent but not for correlated samples.

FACTORIAL DESIGN AND INTERACTION

In this chapter, you will learn to analyze data from a design in which there are *two independent variables* (factors), each of which may have two or more levels. **Table 10.3** illustrates an example of this design with one factor (Factor *A*) having three levels (A_1, A_2, and A_3) and another factor (Factor *B*) having two levels (B_1 and B_2). Such a design is called a **factorial design.**

Table 10.3   Illustration of a 2 × 3 factorial design that can be analyzed with *F* tests

		Factor A			
		A_1	A_2	A_3	
Factor B	B_1	$A_1 B_1$ Scores on the dependent variable	$A_2 B_1$ Scores on the dependent variable	$A_3 B_1$ Scores on the dependent variable	$\bar{X}_{B_1}$
	B_2	$A_1 B_2$ Scores on the dependent variable	$A_2 B_2$ Scores on the dependent variable	$A_3 B_2$ Scores on the dependent variable	$\bar{X}_{B_2}$
		$\bar{X}_{A_1}$	$\bar{X}_{A_2}$	$\bar{X}_{A_3}$	

A factorial design is one that has *two or more* independent variables. Each level of a factor is paired with all levels of the other factors. In this chapter, we will cover only two-factor designs.[1] These two-factor designs will be for independent samples and not for correlated samples.

Factorial designs are identified with a shorthand notation such as "2 × 3" or "3 × 5." The general term is $R \times C$ (Rows × Columns). The first number tells you the number of levels of one factor; the second number tells you the number of levels of the other factor. The design in Table 10.3 is a 2 × 3 design. Assignment of a factor to a row or column is arbitrary; we could just as well have made Table 10.3 a 3 × 2 table.

In Table 10.3, there are six cells. Each **cell** represents a different way to treat subjects. A subject in the upper left cell is given treatment A_1 *and* treatment B_1. That cell is, therefore, identified as Cell $A_1 B_1$. Subjects in the lower right cell are given treatment A_3 and treatment B_2, and that cell is called Cell $A_3 B_2$.

You are now in the paragraph that tells you what information a factorial ANOVA gives you.[2] In the first place, a factorial ANOVA will give you the same information that you can get from two separate one-way ANOVAs. Look again at Table 10.3. A factorial ANOVA will determine the probability that $\bar{X}_{A_1}$, $\bar{X}_{A_2}$, and $\bar{X}_{A_3}$ have a common population mean (a one-way ANOVA on three groups). The same factorial ANOVA will determine the probability that $\bar{X}_{B_1}$ and $\bar{X}_{B_2}$ have a common population mean (a second one-way ANOVA). In addition, a factorial ANOVA will help you decide whether the two factors *interact* on the dependent variable. The interaction test is the special kind of information that a factorial design provides. An interaction between two factors means that the effect that one factor has on the dependent variable depends on which level of the other factor is being administered. There would be an interaction in the factorial design in Table 10.3 if the difference between Level B_1 and Level B_2 depended on whether you looked at the A_1 column, the A_2 column, or the A_3 column. To expand on this, each of the three columns represents an experiment that compares B_1 to B_2. If there is *no* interaction (that is, Factor A does not interact with Factor B), then the difference between B_1 and B_2 will be the same for all three experiments.

If there *is* an interaction, however, the difference between B_1 and B_2 will *depend* on whether you are examining the experiment at A_1, at A_2, or at A_3. [The effect of Factor B (B_1 or B_2) depends on Factor A (A_1, A_2, or A_3).]

Perhaps a couple of examples of interactions will help at this point. Suppose a group of friends were sitting in a dormitory lounge one Monday discussing the weather of the previous weekend. What would you need to know to predict each person's rating of the weather? The first thing you probably want to know is what the weather was actually like. A second important variable is the activity each person had planned for the weekend. For purposes of this little illustration, suppose that weather comes in one of two varieties, snow or no snow, and that our subjects had planned only one of two activities, camping or skiing. Now we have the ingredients for an interaction. We have two independent variables (weather and plans) and a dependent variable (rating of the weather).

---

[1] Advanced textbooks discuss the analysis of three-or-more-factor designs. See Howell (1987), Kirk (1982), Loftus and Loftus (1988), Keppel (1982), Edwards (1972), or Winer (1971).

[2] See Kirk (1982, pp. 422–423) for both the advantages and the disadvantages of factorial ANOVA.

Table 10.4   Illustration of a significant interaction—
good and bad refer to ratings of the weather

		Kind of weather	
		Snow	No snow
	Camping	Bad	Good
Plans for weekend	Skiing	Good	Bad

If plans called for camping, "no snow" is rated good, but if plans called for skiing, "no snow" is rated bad. To complete the possibilities, campers rated snow bad and skiers rated it good. Thus, the effect of the weather (Factor A) depends on plans (Factor B). **Table 10.4** summarizes this paragraph. Study it before going on.

Here is a similar example in which there is *no* interaction. Suppose you wanted to know how people would rate the weather, which again could be snow or no snow. This time, however, the people are divided into camping enthusiasts and rock-climbing enthusiasts. For both groups, snow would rate as bad weather. You might make up a version of Table 10.4 that describes this second example.

In this example of no interaction, the rating of the weather does not depend on a person's plans for the weekend. A change from snow to no snow brings joy to the hearts of both groups. You can see this in the table you constructed.

MAIN EFFECTS AND INTERACTION

In a factorial ANOVA, the comparison of the levels of Factor A is called a **main effect.** Likewise, the comparison of the levels of Factor B is a main effect. The extent to which scores on Factor A depend on Factor B is the **interaction.** Thus, comparisons for main effects are like one-way ANOVAs, and information about the interaction is a bonus that comes with the factorial design.

Table 10.5 shows a 2×3 factorial ANOVA. The numbers in each of the six cells are cell means. Examine those six means and see for yourself what effect changing from A_1 to A_2 to A_3 has. Make the same examination for B_1 to B_2. (Incidentally, this kind of preliminary examination is very valuable for any set of data.)

Table 10.5   A 2×3 factorial design (The number in each cell represents the mean for all the subjects in the cell. In this example, N is the same for each cell.)

		Factor A			
		A_1	A_2	A_3	Factor B means
Factor B	B_1	10	20	60	30
	B_2	50	60	100	70
	Factor A means	30	40	80	Grand mean 50

We will use Table 10.5 to show you what goes on in a factorial ANOVA's analysis of the main effects and the interaction. Look at the comparison between the mean of B_1 (30) and the mean of B_2 (70). A factorial ANOVA will give the probability that the two means came from populations with identical means. Thus, with this probability you can decide whether the null hypothesis

$$H_0: \mu_{B_1} = \mu_{B_2}$$

should be rejected or not. A factorial ANOVA will also give you the probability of getting means of 30 (A_1), 40 (A_2), and 80 (A_3) from populations with identical means if chance only is at work. Thus, with the same factorial ANOVA you can decide whether the null hypothesis

$$H_0: \mu_{A_1} = \mu_{A_2} = \mu_{A_3}$$

should be rejected or not.

Notice that a comparison of B_1 and B_2 satisfies the requirements of an experiment (page 140) in that, except for getting either B_1 or B_2, the groups were treated alike. That is, Group B_1 and Group B_2 are alike in that both groups have equal numbers of subjects who received A_1. Whatever the effect is of receiving A_1, it will occur as much in the B_1 group as it does in the B_2 group; the only way B_1 and B_2 differ is in their levels of Factor B. It would be worthwhile for you to apply this same line of reasoning to the question of whether comparisons of A_1, A_2, and A_3 satisfy the requirement for an experiment.

In Table 10.5 there is *no* interaction. The effect of changing from B_1 to B_2 is to increase the mean score 40 points and this is true at *each* level of A. A similar constancy is seen in the rows. The effect of changing from A_1 to A_2 is to increase the mean 10 points, *regardless* of whether you are looking at the B_1 row or the B_2 row. In a similar way, the effect of moving from A_2 to A_3 is to increase scores by 40. Factor B has no effect on the change from A_2 to A_3; it is 40 for both levels. In Table 10.5 there is no interaction between Factor A and Factor B.

It is common to display an interaction or lack of one with a graph (with the dependent-variable scores on the ordinate). There is a good reason for this. A graph is the best way of arriving at a clear interpretation of an interaction. We urge you always to draw a graph of the interaction on the factorial problems you work. **Figure 10.1** is one graph of the data in Table 10.5. The result is two parallel curves. Parallel curves mean that there is no interaction between two factors.

We can also graph the data in Table 10.5 with each curve representing a level of A. **Figure 10.2** is the result. Again, the parallel lines indicate that there is no interaction.

Table 10.6 shows a 2×3 factorial design in which there *is* an interaction between the two independent variables. The main effect of Factor A is indicated by the overall means along the bottom. The average effect of a change from A_1 to A_2 to A_3 is to *reduce* the mean score by 10 (main effect). But look at the cells. For B_1, the effect of changing from A_1 to A_2 to A_3 is to *increase* the mean score by 10 points. For B_2, the effect is to *decrease* the score by 30 points. These data illustrate an interaction because the effect of one factor *depends* on which level of the other factor you administer.

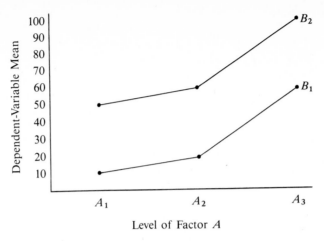

Figure 10.1 Graphic representation of data in Table 10.5; no interaction exists

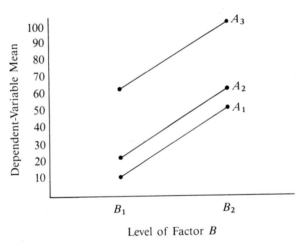

Figure 10.2 Graphic representation of data in Table 10.5; no interaction exists

Table 10.6 A 2 × 3 factorial design with an interaction between factors (The numbers represent the means of all scores within each cell.)

		Factor A			
		A_1	A_2	A_3	Factor B means
Factor B	B_1	10	20	30	20
	B_2	100	70	40	70
	Factor A means	55	45	35	Grand mean 45

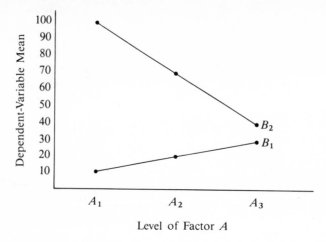

Figure 10.3   Graphic representation of the interaction of Factors A and B from Table 10.6

We will describe this interaction in Table 10.6 another way. Look at the difference between Cell A_1B_1 and Cell A_2B_1—a difference of -10. If there were no interaction, we would predict this same difference (-10) for the difference between Cells A_1B_2 and A_2B_2. But this latter difference is $+30$; it is in the opposite direction. Something about B_2 reverses the effect of changing from A_1 to A_2 that was found under the condition B_1.[3]

This interaction is illustrated graphically in **Figure 10.3.** You can see that B_1 increases across the levels of Factor A. B_2, however, decreases as you go from A_1 to A_2 to A_3. The lines for the two levels of B are not parallel.

Finally, we will graph in **Figure 10.4,** the example of rating the weather by skiers and campers. Again, the lines are not parallel.

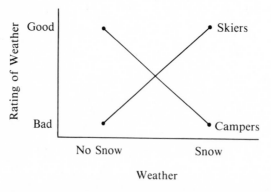

Figure 10.4   Graphic representation of the "rating of the weather" example indicating an interaction. Whether snow or no snow is rated higher depends on whether the ratings came from skiers or campers.

[3] Students often have trouble with the concept of interaction. Usually, having the same idea presented in different words facilitates understanding. Two good references are Christensen and Stoup (1986, pp. 394–398) and Ferguson (1981, p. 256).

RESTRICTIONS AND LIMITATIONS

We have tried to emphasize throughout this book the limitations that go with each statistical test you learn. For the factorial analysis of variance presented in this chapter, the restrictions include the three for one-way ANOVA. If you cannot recall those three, refresh your memory by rereading the section at the end of Chapter 9, "Assumptions of the Analysis of Variance." In addition to the basic assumptions of ANOVA, the factorial ANOVA presented in this chapter requires the following:

1. The number of scores in each cell must be equal. For techniques dealing with unequal N's, see Kirk (1982), Howell (1987), or Ferguson (1981).
2. The cells must be independent. This is usually accomplished by randomly assigning a subject to only one of the cells. This restriction means that these techniques should not be used with any type of correlated-samples design. For factorial designs that use correlated samples, see Howell (1987), Loftus and Loftus (1988), Keppel (1982), or Winer (1971).
3. The levels of both factors are chosen by the experimenter. The alternative is that the levels of one or both factors are chosen at random from several possible levels of the factor. The techniques of this chapter are used when the levels are *fixed* by the experimenter and not chosen randomly. For a discussion of fixed and random models of ANOVA, see Howell (1987), Winer (1971), or Ferguson (1981).

PROBLEMS

1. Identify the following factorial designs, using $R \times C$ notation.
   a. Three methods of teaching Spanish were evaluated using either a grade report or a credit/no credit report.
   b. Four strains of mice were injected with three types of virus.
   c. Males and females worked on a complex problem at either 7 A.M. or 7 P.M.
   d. This one is a little different. Four different species of pine were grown with three different kinds of fertilizer in three different soil types.
2. What is a main effect?
3. What is an interaction?
4. In a *t* test, are you testing for a main effect or for an interaction?
5. In a one-way ANOVA, are you testing for a main effect or for an interaction?
6. The following five tables present cell means for some ANOVAs. For each of them, do the following.
   a. Graph the means.
   b. Decide whether an interaction appears to be present. In these problems, you may state that there appears to be *no* interaction or that there appears to be one. Interactions, like main effects, are subject to sampling variations.
   c. Write a statement of explanation of the interaction or lack of one.
   d. Decide whether any main effects appear to be present.
   e. Write a statement of explanation of the main effects or lack of them.

   i. The dependent variable for this ANOVA is attitude scores toward playing games of house or baseball (see table that follows on page 216).

Factor A (gender)

		A_1 (boys)	A_2 (girls)
Factor B	B_1 (house)	50	75
(games)	B_2 (baseball)	75	50

ii. This is a study of the effect of dosage of a new drug on three types of schizophrenics. The dependent variable is scores on a "happiness" test.

A (diagnosis)

		A_1 (disorganized)	A_2 (catatonic)	A_3 (paranoid)
	B_1 (small)	5	10	70
B	B_2 (medium)	20	25	85
(dose)	B_3 (large)	35	40	100

iii. The dependent variable in this one is the number of cars sold by Overseas Motor Corporation in each of the three cities during a 90-day period.

A (cities)

		A_1 (Atlanta)	A_2 (Omaha)	A_3 (Tulsa)
B	B_1 (without)	15	60	30
(air-conditioning)	B_2 (with)	75	120	90

iv. In this problem, the dependent variable is the number of months a piece of "indoor-outdoor" carpeting lasted.

A (brands)

		A_1 (Brand X)	A_2 (Brand Y)	A_3 (Brand Z)
B	B_1 (outdoors)	10	20	30
(location)	B_2 (indoors)	5	35	65

v. Here, the dependent variable is attitude toward lowering the taxes on profits from investments.

A (socioeconomic status)

		A_1 (low)	A_2 (middle)	A_3 (high)
B	B_1 (males)	10	35	60
(gender)	B_2 (females)	15	35	55

7. Name the six requirements that must be met if a factorial ANOVA is to be analyzed using techniques outlined in this chapter.

A SIMPLE EXAMPLE OF A FACTORIAL DESIGN

As you read the following story, try to pick out the two factors and identify the levels of each factor.

Two groups of hunters, six squirrel hunters and six quail hunters, met in a bar. An argument soon began over the marksmanship required for the two kinds of hunting.

"Squirrel hunters are just better shots," barked a biased squirrel hunter.

"Poot, poot, and balderdash!" heartily swore a logic-oriented quail hunter. "It takes a lot better eye to hit a moving bird than it does to hit a still squirrel."

"Hold it a minute, you guys," demanded an empirically minded squirrel hunter. "We can settle this easily enough on our hunting-club target range. We'll just see if you six quail hunters can hit the target as often as we can."

"OK," agreed a quail hunter. "What kind of trap throwers do you have out there?"

"What kind of what? Oh, you mean those gadgets that throw clay pigeons into the air? Gee, yeah, there are some out there, but we never use them."

"Well, if you want to shoot against us, you will have to use them this time," the quail hunter insisted. "It's one thing to hit a still target, but hitting a target flying through the air above you is something else. We'll target shoot against you guys, but let's do it the fair way. Three of us and three of you will shoot at still targets and the other six will shoot at clay pigeons."

"Fair enough," the squirrel hunters agreed; and all 12 men took up their shotguns and headed for the target range.

This yarn establishes conditions for a 2×2 factorial ANOVA with three scores per cell. The cell means and computation of the sums of squares are illustrated in **Table 10.7.** The dependent variable is the number of hits made by each man out of 15 shots. Factor A is the type of target and it has two levels: A_1 for still targets and A_2 for moving targets. Factor B is the type of hunter and it also has two levels: B_1 for squirrel hunters and B_2 for quail hunters. Before going to the next section of the table, which shows calculations, simply *look* at the data in Table 10.7. Compare the mean number of hits for moving ($\bar{X}_{A_2}$) and still ($\bar{X}_{A_1}$) targets. Compare the mean hits for the quail hunters ($\bar{X}_{B_2}$) with that for the squirrel hunters ($\bar{X}_{B_1}$). Look at the cell means. Does there appear to be an interaction? This kind of preliminary inspection of the data will improve your understanding of a study and will help you catch any gross computational errors. For example, in Table 10.7, the difference between targets is larger than the difference between hunters. This will be reflected in a larger F value for Factor A than for Factor B.

Sources of Variance and Sums of Squares

Remember that in Chapter 9 you identified three sources of variance in the one-way analysis of variance. They were (1) the total variance, (2) the between-groups variance, and (3) the within-groups variance. In a factorial design with two factors, the same sources of variance can be identified. However, the between-groups variance may now be partitioned into three components. These are the two *main effects* and the *interaction*. Thus, of the variability among the means of the four

Table 10.7 Cell means and computation of sums of squares for the hunters' contest

		Targets (Factor A)			
		Still (A_1)		Moving (A_2)	
		X	X^2	X	X^2
Hunters (Factor B)	Squirrel (B_1)	12	144	8	64
		11	121	7	49
		10	100	7	49
		$\Sigma X_g = 33$		$\Sigma X_g = 22$		$\Sigma X_{B_1} = 55$
		$\Sigma X_g^2 = 365$		$\Sigma X_g^2 = 162$		$\Sigma X_{B_1}^2 = 527$
		$\bar{X}_g = 11.00$		$\bar{X}_g = 7.3333$		$\bar{X}_{B_1} = 9.1667$
	Quail (B_2)	12	144	10	100
		10	100	9	81
		9	81	8	64
		$\Sigma X_g = 31$		$\Sigma X_g = 27$		$\Sigma X_{B_2} = 58$
		$\Sigma X_g^2 = 325$		$\Sigma X_g^2 = 245$		$\Sigma X_{B_2}^2 = 570$
		$\bar{X}_g = 10.3333$		$\bar{X}_g = 9.00$		$\bar{X}_{B_2} = 9.6667$
		$\Sigma X_{A_1} = 64$		$\Sigma X_{A_2} = 49$		$\Sigma X_{tot} = 113$
		$\Sigma X_{A_1}^2 = 690$		$\Sigma X_{A_2}^2 = 407$		$\Sigma X_{tot}^2 = 1097$
		$\bar{X}_{A_1} = 10.6667$		$\bar{X}_{A_2} = 8.1667$		$\bar{X}_{tot} = 9.4167$

$$SS_{tot} = \Sigma X_{tot}^2 - \frac{(\Sigma X_{tot})^2}{N_{tot}} = 1097 - \frac{(113)^2}{12} = 1097 - 1064.0833 = 32.9167$$

$$SS_{bg} = \Sigma\left[\frac{(\Sigma X_g)^2}{N_g}\right] - \frac{(\Sigma X_{tot})^2}{N_{tot}} = \frac{(33)^2}{3} + \frac{(22)^2}{3} + \frac{(31)^2}{3} + \frac{(27)^2}{3} - \frac{(113)^2}{12}$$
$$= 363 + 161.3333 + 320.3333 + 243 - 1064.0833 = 23.5833$$

$$SS_{targets} = \frac{(\Sigma X_{A_1})^2}{N_{A_1}} + \frac{(\Sigma X_{A_2})^2}{N_{A_2}} - \frac{(\Sigma X_{tot})^2}{N_{tot}} = \frac{(64)^2}{6} + \frac{(49)^2}{6} - \frac{(113)^2}{12}$$
$$= 682.6667 + 400.1667 - 1064.0833 = 1082.8334 - 1064.0833 = 18.7501$$

$$SS_{hunters} = \frac{(\Sigma X_{B_1})^2}{N_{B_1}} + \frac{(\Sigma X_{B_2})^2}{N_{B_2}} - \frac{(\Sigma X_{tot})^2}{N_{tot}} = \frac{(55)^2}{6} + \frac{(58)^2}{6} - \frac{(113)^2}{12}$$
$$= 504.1667 + 560.6667 - 1064.0833 = 1064.8334 - 1064.0833 = 0.7501$$

$$SS_{AB} = N_g[(\bar{X}_{A_1B_1} - \bar{X}_{A_1} - \bar{X}_{B_1} + \bar{X}_{tot})^2 + (\bar{X}_{A_2B_1} - \bar{X}_{A_2} - \bar{X}_{B_1} + \bar{X}_{tot})^2$$
$$+ (\bar{X}_{A_1B_2} - \bar{X}_{A_1} - \bar{X}_{B_2} + \bar{X}_{tot})^2 + (\bar{X}_{A_2B_2} - \bar{X}_{A_2} - \bar{X}_{B_2} + \bar{X}_{tot})^2]$$
$$= 3[(11.00 - 10.6667 - 9.1667 + 9.4167)^2 + (7.3333 - 8.1667 - 9.1667 + 9.4167)^2$$
$$+ (10.3333 - 10.6667 - 9.6667 + 9.4167)^2 + (9.00 - 8.1667 - 9.6667 + 9.4167)^2]$$
$$= 4.0836$$

Check: $SS_{AB} = SS_{bg} - SS_A - SS_B = 23.5833 - 18.7501 - 0.7501 = 4.0831$

$$SS_{wg} = \Sigma\left[\Sigma X_g^2 - \frac{(\Sigma X_g)^2}{N_g}\right]$$
$$= \left(365 - \frac{(33)^2}{3}\right) + \left(162 - \frac{(22)^2}{3}\right) + \left(325 - \frac{(31)^2}{3}\right) + \left(245 - \frac{(27)^2}{3}\right)$$
$$= (365 - 363) + (162 - 161.3333) + (325 - 320.3333) + (245 - 243)$$
$$= 2 + 0.6667 + 4.6667 + 2 = 9.3333$$

Check: $SS_{tot} = SS_{bg} + SS_{wg}$
$32.9167 = 23.5833 + 9.3333$

groups in Table 10.7, some can be attributed to the *A* main effect, some to the *B* main effect, and the rest to the interaction.

Total Sum of Squares.   This calculation will be easy for you, since it is the same as SS_{tot} in the one-way analysis. It is defined as $\Sigma(X - \bar{X}_{tot})^2$, or the sum of the squared deviations of all the scores in the experiment from the grand mean of the experiment. To actually compute SS_{tot}, use the formula

$$SS_{tot} = \Sigma X_{tot}^2 - \frac{(\Sigma X_{tot})^2}{N_{tot}}$$

For the hunters' contest, you get

$$SS_{tot} = 1097 - \frac{(113)^2}{12} = 32.9167$$

Between-Groups Sum of Squares.[4]   In order to find the main effects and interaction, you should first find the between-groups variability and then partition it into its component parts. As in the one-way design, SS_{bg} is defined $\Sigma[N_g(\bar{X}_g - \bar{X}_{tot})^2]$. A "group" in this context is a group of participants treated alike; therefore, for example, squirrel hunters shooting at still targets constitute a group. In other words, a group is composed of those scores in the same cell.

The computational formula for SS_{bg} is the same as that for one-way analysis:

$$SS_{bg} = \Sigma\left[\frac{(\Sigma X_g)^2}{N_g}\right] - \frac{(\Sigma X_{tot})^2}{N_{tot}}$$

For the hunters' contest, you get

$$SS_{bg} = \frac{(33)^2}{3} + \frac{(22)^2}{3} + \frac{(31)^2}{3} + \frac{(27)^2}{3} - \frac{(113)^2}{12} = 23.5833$$

After SS_{bg} is obtained, it is partitioned into its three components: the *A* main effect, the *B* main effect, and the interaction.

The sum of squares for each main effect is somewhat like a one-way ANOVA. The sum of squares for Factor *A* ignores the existence of Factor *B* and considers the deviations of the Factor *A* means from the grand mean. Thus,

$$SS_A = N_{A_1}(\bar{X}_{A_1} - \bar{X}_{tot})^2 + N_{A_2}(\bar{X}_{A_2} - \bar{X}_{tot})^2$$

where N_{A_1} is the total number of scores in the A_1 cells.

For Factor *B*, you get

$$SS_B = N_{B_1}(\bar{X}_{B_1} - \bar{X}_{tot})^2 + N_{B_2}(\bar{X}_{B_2} - \bar{X}_{tot})^2$$

Computational formulas for the main effects also look like formulas for SS_{bg} in a one-way design:

$$SS_A = \frac{(\Sigma X_{A_1})^2}{N_{A_1}} + \frac{(\Sigma X_{A_2})^2}{N_{A_2}} - \frac{(\Sigma X_{tot})^2}{N_{tot}}$$

[4] Some texts call this the *between-cells sum of squares*.

And for the hunters' contest,

$$SS_{\text{targets}} = \frac{(64)^2}{6} + \frac{(49)^2}{6} - \frac{(113)^2}{12} = 18.7501$$

The computational formula for the *B* main effect simply substitutes *B* for *A* in the previous formula. Thus,

$$SS_B = \frac{(\Sigma X_{B_1})^2}{N_{B_1}} + \frac{(\Sigma X_{B_2})^2}{N_{B_2}} - \frac{(\Sigma X_{\text{tot}})^2}{N_{\text{tot}}}$$

For the hunters' contest,

$$SS_{\text{hunters}} = \frac{(55)^2}{6} + \frac{(58)^2}{6} - \frac{(113)^2}{12} = 0.7501$$

To find the sum of squares for the interaction, use this formula:

$$\begin{aligned}
SS_{AB} = N_g[& (\bar{X}_{A_1B_1} - \bar{X}_{A_1} - \bar{X}_{B_1} + \bar{X}_{\text{tot}})^2 \\
+ & (\bar{X}_{A_2B_1} - \bar{X}_{A_2} - \bar{X}_{B_1} + \bar{X}_{\text{tot}})^2 \\
+ & (\bar{X}_{A_1B_2} - \bar{X}_{A_1} - \bar{X}_{B_2} + \bar{X}_{\text{tot}})^2 \\
+ & (\bar{X}_{A_2B_2} - \bar{X}_{A_2} - \bar{X}_{B_2} + \bar{X}_{\text{tot}})^2]
\end{aligned}$$

For the hunters' contest,

$$\begin{aligned}
SS_{AB} = 3[& (11.00 - 10.6667 - 9.1667 + 9.4167)^2 \\
+ & (7.3333 - 8.1667 - 9.1667 + 9.4167)^2 \\
+ & (10.3333 - 10.6667 - 9.6667 + 9.4167)^2 \\
+ & (9.00 - 8.1667 - 9.6667 + 9.4167)^2] \\
= \; & 4.0836
\end{aligned}$$

Since SS_{bg} contains only the components SS_A, SS_B, and the interaction SS_{AB}, you can also obtain SS_{AB} by subtraction. This subtraction check is the justification for calculating SS_{bg} in a factorial ANOVA.

$$SS_{AB} = SS_{\text{bg}} - SS_A - SS_B$$

For the hunters' contest,

$$SS_{AB} = 23.5833 - 18.7501 - 0.7501 = 4.0831$$

Within-Groups Sum of Squares.   As in the one-way analysis, the within-groups variability is due to the fact that subjects treated alike differ from one another on the dependent variable. Since all were treated the same, this difference must be due to uncontrolled variables and is sometimes called **error variance** or the **error term**. SS_{wg} for a 2×2 design is defined as

$$\begin{aligned}
SS_{\text{wg}} = & \Sigma(X_{A_1B_1} - \bar{X}_{A_1B_1})^2 + \Sigma(X_{A_1B_2} - \bar{X}_{A_1B_2})^2 \\
+ & \Sigma(X_{A_2B_1} - \bar{X}_{A_2B_1})^2 + \Sigma(X_{A_2B_2} - \bar{X}_{A_2B_2})^2
\end{aligned}$$

In words, SS_{wg} is the sum of the sums of squares for each *cell* in the experiment. The computational formula is

$$SS_{wg} = \Sigma\left[\Sigma X_g^2 - \frac{(\Sigma X_g)^2}{N_g}\right]$$

For the hunters' contest,

$$SS_{wg} = \left(365 - \frac{(33)^2}{3}\right) + \left(162 - \frac{(22)^2}{3}\right)$$

$$+ \left(325 - \frac{(31)^2}{3}\right) + \left(245 - \frac{(27)^2}{3}\right) = 9.3333$$

The computational check for the hunters' contest is

$$32.9167 = 23.5833 + 9.3333$$

We will interrupt the analysis of the hunters' contest so that you may practice what you have learned about the sums of squares in a factorial ANOVA.

ERROR DETECTION

> One computational check for a factorial ANOVA is the same as that for the one-way classification: $SS_{tot} = SS_{bg} + SS_{wg}$. A more complete version is $SS_{tot} = SS_A + SS_B + SS_{AB} + SS_{wg}$. As before, this check will not catch errors in ΣX or ΣX^2.

PROBLEMS

8. The dependent variable in this hypothetical study was the number of pounds gained during the 20 weeks following weaning. The subjects were male and female puppies fed three different diets.

Diet
(Factor A)

		A_1	A_2	A_3
	Males (B_1)	7 6 4	4 4 3	9 8 7
Sex (Factor B)	Females (B_2)	6 4 4	7 5 5	5 4 3

a. Identify the design, using $R \times C$ notation.
b. Name the independent variables.
c. Calculate SS_{tot}, SS_{bg}, SS_{diets}, SS_{sex}, SS_{AB}, and SS_{wg}.

9. An educational psychologist was interested in whether different teacher responses affected boys and girls differently. Three classrooms with both boys and girls were used in the experiment. In one classroom, the teacher's response was one of "neglect." The children were not even observed during the time they were working arithmetic problems.

In a second classroom, the teacher's response was "reproof." The children were observed and errors were corrected and criticized. In the third classroom, children were "praised" for correct answers and incorrect answers were ignored. The numbers in the table are the numbers of errors made on a comprehensive examination of arithmetic achievement given later.

a. Identify the design, using $R \times C$ notation.
b. Name the independent variables.
c. Name the dependent variable.
d. Calculate SS_{tot}, SS_{bg}, $SS_{response}$, SS_{gender}, SS_{AB}, and SS_{wg}.

		Teacher response		
		Neglect (A_1)	Reproof (A_2)	Praise (A_3)
Gender	Boys (B_1)	25	20	18
		22	17	17
		20	17	16
		19	16	14
		19	15	14
		18	15	13
		16	13	10
		13	10	7
		10	8	5
		7	5	4
	Girls (B_2)	20	23	16
		18	22	14
		17	20	13
		16	18	13
		16	17	12
		15	16	11
		13	16	9
		12	14	9
		10	11	7
		9	10	5

10. Many experiments have established the concept of *state-dependent memory*. Subjects learn a task in a particular state—say, under the influence of a drug or not under the influence of the drug. They recall what they have learned in the same state or in the other state. The dependent variable is recall score (memory). The data in the table are designed to illustrate the phenomenon of state-dependent memory. Calculate SS_{tot}, SS_{bg}, SS_{learn}, SS_{recall}, SS_{AB}, and SS_{wg}.

			During learning	
			Drug	No drug
During recall	Drug	ΣX	43	20
		ΣX^2	400	100
		N	5	5
	No drug	ΣX	25	57
		ΣX^2	155	750
		N	5	5

11. Examine the following summary statistics and explain why the techniques of this chapter cannot be used for this factorial design.

		A_1	A_2
B_1	ΣX	405	614
	ΣX^2	8503	15,540
	N	20	25
B_2	ΣX	585	246
	ΣX^2	12,060	3325
	N	30	20

Degrees of Freedom, Mean Squares, and *F* Tests

Now that you are skilled at calculating sums of squares, we can proceed with the rest of the analysis of the hunters' contest. Mean squares, as before, are found by dividing the sums of squares by their appropriate degrees of freedom. Degrees of freedom for the sources of variance are

In general:

$$df_{\text{tot}} = N_{\text{tot}} - 1$$
$$df_A = A - 1$$
$$df_B = B - 1$$
$$df_{AB} = (A - 1)(B - 1)$$
$$df_{\text{wg}} = N_{\text{tot}} - (A)(B)$$

For the hunters' contest:

$$df_{\text{tot}} = 12 - 1 = 11$$
$$df_{\text{targets}} = 2 - 1 = 1$$
$$df_{\text{hunters}} = 2 - 1 = 1$$
$$df_{AB} = (1)(1) = 1$$
$$df_{\text{wg}} = 12 - (2)(2) = 8$$

In these equations, *A* and *B* stand for the number of levels of Factor *A* and Factor *B*, respectively.

ERROR DETECTION

$$df_{\text{tot}} = df_A + df_B + df_{AB} + df_{\text{wg}}$$
$$df_{\text{bg}} = df_A + df_B + df_{AB}$$

Mean squares for the hunters' contest are

$$MS_{\text{targets}} = \frac{SS_{\text{targets}}}{df_{\text{targets}}} = \frac{18.7501}{1} = 18.7501$$

$$MS_{\text{hunters}} = \frac{SS_{\text{hunters}}}{df_{\text{hunters}}} = \frac{0.7501}{1} = 0.7501$$

$$MS_{AB} = \frac{SS_{AB}}{df_{AB}} = \frac{4.0831}{1} = 4.0831$$

$$MS_{\text{wg}} = \frac{SS_{\text{wg}}}{df_{\text{wg}}} = \frac{9.3333}{8} = 1.1667$$

F is computed, as usual, by dividing each mean square by MS_{wg}:

$$F_{\text{targets}} = \frac{MS_{\text{targets}}}{MS_{\text{wg}}} = \frac{18.7501}{1.1667} = 16.07$$

$$F_{\text{hunters}} = \frac{MS_{\text{hunters}}}{MS_{\text{wg}}} = \frac{0.7501}{1.1667} = 0.64$$

$$F_{AB} = \frac{MS_{AB}}{MS_{\text{wg}}} = \frac{4.0831}{1.1667} = 3.50$$

Again, you should refer to Table F to determine the significance of these F values. You have 1 degree of freedom in the numerator and 8 degrees of freedom in the denominator for F_{targets}. An F value of 11.26 is required to reject the null hypothesis at the .01 level and an F value of 5.32 to reject at the .05 level. Since 16.07 is larger than 11.26, it is significant beyond the .01 level, and the null hypothesis that $\mu_{\text{still}} = \mu_{\text{moving}}$ is rejected. By examining the marginal means (8.17 and 10.67), you may conclude that the hunters hit significantly fewer moving targets than still targets. F_{hunters} was less than 1, and values of F that are less than 1 are never significant. Thus, there was no significant difference in the mean number of targets hit by the two kinds of hunters. F_{AB} (3.50) is less than 5.32 and is, therefore, not significant. There was no significant interaction between kind of hunter and kind of target. Although squirrel hunters were better on still targets and worse on moving targets, with the quail hunters intermediate, this departure from parallel performance was not great enough to reach significance.

Results of a factorial ANOVA are usually presented in a summary table. **Table 10.8** is a summary table for the example of the hunters' contest.

Table 10.8   ANOVA summary table for the hunters' contest

Source	df	SS	MS	F	p
A (targets)	1	18.7501	18.7501	16.07	<.01
B (hunters)	1	0.7501	0.7501	0.64	>.05
AB	1	4.0831	4.0831	3.50	>.05
Within groups	8	9.3333	1.1667		
Total	11	32.9167			

PROBLEMS

12. For the data in Problem 8, compute *df*, *MS*, and *F* values. Arrange these in a summary table. Plot the cell means. Tell what the analysis shows.
13. For the data in Problem 9, compute *df*, *MS*, and *F* values. Arrange these in a summary table. Plot the cell means. Tell what the analysis shows.
14. For the data in Problem 10, compute *df*, *MS*, and *F* values. Arrange these in a summary table. Plot the cell means. Tell what the analysis shows.

ANALYSIS OF A 3 x 3 DESIGN

This section describes the analysis of a 3×3 design. The procedures are exactly like those for the other designs you have analyzed. This section will emphasize the interpretation of results.

Two experimenters were interested in the Gestalt principle of closure—the drive to have things finished, or closed. An illustration of this drive is the fact that people often see a circle with a gap as closed, even if they are looking for the gap. These experimenters thought that the strength of the closure drive in an anxiety-arousing situation would depend on the subjects' general anxiety level. Thus, the experimenters hypothesized an interaction between the anxiety of a person and the kind of situation he or she is in.

The independent variables for this experiment were (1) anxiety level of the subject [5] and (2) kind of situation the person was in—that is, whether it was anxiety arousing or not. As you probably realized from the title of this section, there were three levels for each of these independent variables. The dependent variable was a measure of the closure drive.

To get subjects, the experimenters administered the Taylor Manifest Anxiety Scale (Taylor, 1953) to a large group of college students. From this large group, they selected the 15 lowest scorers, 15 of the middle scorers, and the 15 highest scorers as participants in the study. The first factor in the experiment, then, was anxiety, with three levels: low (A_1), middle (A_2), and high (A_3).

The second factor was the kind of situation. The three kinds were dim illumination (B_1), normal illumination (B_2), and very bright illumination (B_3). The assumption was that dim and bright illumination would create more anxiety than would normal illumination.

Participants viewed 50 circles projected on a screen. Ten of the circles were closed, ten contained a gap at the top, ten a gap at the bottom, ten a gap on the right, and ten a gap on the left. Participants simply stated whether the gap was at the top, bottom, right, or left, or whether the circle was closed. The experimenters recorded as the dependent variable the number of circles reported as closed by each participant.

The hypothetical data and their analysis are reported in **Table 10.9.** Read the experiment over again and work through the analysis of the data in Table 10.9.

Table 10.10 is the ANOVA summary table. The probabilities in Table 10.10 are from Table F in Appendix C. F_A and F_B have 2 degrees of freedom in the numerator and 36 *df* in the denominator. The critical value of *F* for 2, 36 *df* with $\alpha = .01$ is 5.25. Thus, for the factor anxiety, reject the null hypothesis. It would seem that a person's closure drive *is* related to his or her anxiety level. Also, for the illumination variable, reject the null hypothesis. Again it would seem that the level of illumination has an effect on the number of circles that are seen as closed. For the interaction, the critical value of *F* for 4, 36 *df* with $\alpha = .01$ is 3.89. (The change in critical value results from the increase in *df*.) Thus, reject the hypothesis that the illumination conditions affected high-, medium-, and low-anxious participants in the same way. Conclude that there was an interaction between anxiety level and illumination. As you will soon see, this significant interaction affects the interpretation of the main effects.

The interaction can be seen in **Figure 10.5.** For participants who had high anxiety scores, the dim illumination *and* the bright illumination caused more circles to be seen as closed. As you recall, the experimenter expected both the dim and bright illuminations to be anxiety arousing. Thus, the significant interaction in this case is statistical confirmation of the hypothesis that the closure drive is very great in high-anxious persons placed in an anxiety-arousing situation.

[5] For an experiment that manipulated only this variable, see Calhoun and Johnston (1968).

Table 10.9 ANOVA of hypothetical study of effects of anxiety and illumination on closure drive

		Anxiety (Factor A)						
		Low (A_1)		Middle (A_2)		High (A_3)		Summary values
		X_{A_1}	$X^2_{A_1}$	X_{A_2}	$X^2_{A_2}$	X_{A_3}	$X^2_{A_3}$	
Illumination (Factor B)	Dim (B_1)	17	289	14	196	21	441	$\Sigma X_{B_1} = 206$
		15	225	14	196	19	361	$\Sigma X^2_{B_1} = 3048$
		13	169	12	144	17	289	$\bar{X}_{B_1} = 13.7333$
		10	100	10	100	16	256	
		8	64	7	49	13	169	
	Totals	63	847	57	685	86	1516	
	Means	12.60		11.40		17.20		
	Normal (B_2)	14	196	12	144	12	144	$\Sigma X_{B_2} = 158$
		13	169	11	121	10	100	$\Sigma X^2_{B_2} = 1708$
		11	121	11	121	10	100	$\bar{X}_{B_2} = 10.5333$
		11	121	9	81	9	81	
		9	81	8	64	8	64	
	Totals	58	688	51	531	49	489	
	Means	11.60		10.20		9.80		
	Bright (B_3)	12	144	11	121	18	324	$\Sigma X_{B_3} = 178$
		11	121	11	121	17	289	$\Sigma X^2_{B_3} = 2228$
		10	100	10	100	15	225	$\bar{X}_{B_3} = 11.8667$
		10	100	9	81	14	196	
		9	81	9	81	12	144	
	Totals	52	546	50	504	76	1178	
	Means	10.40		10.00		15.20		
	Summary values	$\Sigma X_{A_1} = 173$		$\Sigma X_{A_2} = 158$		$\Sigma X_{A_3} = 211$		$\Sigma X_{tot} = 542$
		$\Sigma X^2_{A_1} = 2081$		$\Sigma X^2_{A_2} = 1720$		$\Sigma X^2_{A_3} = 3183$		$\Sigma X^2_{tot} = 6984$
		$\bar{X}_{A_1} = 11.5333$		$\bar{X}_{A_2} = 10.5333$		$\bar{X}_{A_3} = 14.0667$		$\bar{X}_{tot} = 12.0444$

Sums of squares

$$SS_{tot} = \Sigma X^2_{tot} - \frac{(\Sigma X_{tot})^2}{N_{tot}} = 6984 - \frac{(542)^2}{45} = 455.9111$$

$$SS_{bg} = \Sigma \left[\frac{(\Sigma X_g)^2}{N_g} \right] - \frac{(\Sigma X_{tot})^2}{N_{tot}} = \frac{(63)^2}{5} + \frac{(57)^2}{5} + \frac{(86)^2}{5} + \frac{(58)^2}{5} + \frac{(51)^2}{5} + \frac{(49)^2}{5}$$

$$+ \frac{(52)^2}{5} + \frac{(50)^2}{5} + \frac{(76)^2}{5} - \frac{(542)^2}{45} = 263.9111$$

$$SS_A = \frac{(\Sigma X_{A_1})^2}{N_{A_1}} + \frac{(\Sigma X_{A_2})^2}{N_{A_2}} + \frac{(\Sigma X_{A_3})^2}{N_{A_3}} - \frac{(\Sigma X_{tot})^2}{N_{tot}} = \frac{(173)^2}{15} + \frac{(158)^2}{15} + \frac{(211)^2}{15} - \frac{(542)^2}{45} = 99.5111$$

$$SS_B = \frac{(\Sigma X_{B_1})^2}{N_{B_1}} + \frac{(\Sigma X_{B_2})^2}{N_{B_2}} + \frac{(\Sigma X_{B_3})^2}{N_{B_3}} - \frac{(\Sigma X_{tot})^2}{N_{tot}} = \frac{(206)^2}{15} + \frac{(158)^2}{15} + \frac{(178)^2}{15} - \frac{(542)^2}{45} = 77.5111$$

$$SS_{AB} = N_g[(\bar{X}_{A_1B_1} - \bar{X}_{A_1} - \bar{X}_{B_1} + \bar{X}_{tot})^2 + (\bar{X}_{A_1B_2} - \bar{X}_{A_1} - \bar{X}_{B_2} + \bar{X}_{tot})^2$$

$$+ (\bar{X}_{A_1B_3} - \bar{X}_{A_1} - \bar{X}_{B_3} + \bar{X}_{tot})^2 + \cdots + (\bar{X}_{A_3B_3} - \bar{X}_{A_3} - \bar{X}_{B_3} + \bar{X}_{tot})^2]$$

$$= 5[(12.60 - 11.5333 - 13.7333 + 12.0444)^2 + (11.60 - 11.5333 - 10.5333 + 12.0444)^2$$

$$+ (10.40 - 11.5333 - 11.8667 + 12.0444)^2 + (11.40 - 10.5333 - 13.7333 + 12.0444)^2$$

$$+ (10.20 - 10.5333 - 10.5333 + 12.0444)^2 + (10.00 - 10.5333 - 11.8667 + 12.0444)^2$$

$$+ (17.20 - 14.0667 - 13.7333 + 12.0444)^2 + (9.80 - 14.0667 - 10.5333 + 12.0444)^2$$

$$+ (15.20 - 14.0667 - 11.8667 + 12.0444)^2] = (5)(17.3778) = 86.8890$$

(continued)

Table 10.9 *(continued)*

Check: $SS_{AB} = SS_{bg} - SS_A - SS_B = 263.9111 - 99.5111 - 77.5111 = 86.8889$

$$SS_{wg} = \Sigma\left[\Sigma X_g^2 - \frac{(\Sigma X_g)^2}{N_g}\right]$$

$$= \left(847 - \frac{(63)^2}{5}\right) + \left(688 - \frac{(58)^2}{5}\right) + \left(546 - \frac{(52)^2}{5}\right) + \left(685 - \frac{(57)^2}{5}\right)$$

$$+ \left(531 - \frac{(51)^2}{5}\right) + \left(504 - \frac{(50)^2}{5}\right) + \left(1516 - \frac{(86)^2}{5}\right)$$

$$+ \left(489 - \frac{(49)^2}{5}\right) + \left(1178 - \frac{(76)^2}{5}\right) = 192$$

Check: $455.9111 = 263.9111 + 192$

Degrees of freedom

$df_A = A - 1 = 3 - 1 = 2$

$df_B = B - 1 = 3 - 1 = 2$

$df_{AB} = (A - 1)(B - 1) = (2)(2) = 4$

$df_{wg} = N_{tot} - (A)(B) = 45 - (3)(3) = 36$

$df_{tot} = N_{tot} - 1 = 45 - 1 = 44$

Mean squares

$MS_A = \dfrac{SS_A}{df_A} = \dfrac{99.51}{2} = 49.76$

$MS_B = \dfrac{SS_B}{df_B} = \dfrac{77.51}{2} = 38.76$

$MS_{AB} = \dfrac{SS_{AB}}{df_{AB}} = \dfrac{86.89}{4} = 21.72$

$MS_{wg} = \dfrac{SS_{wg}}{df_{wg}} = \dfrac{192}{36} = 5.33$

F values

$F_A = \dfrac{MS_A}{MS_{wg}} = \dfrac{49.76}{5.33} = 9.33$

$F_B = \dfrac{MS_B}{MS_{wg}} = \dfrac{38.76}{5.33} = 7.27$

$F_{AB} = \dfrac{MS_{AB}}{MS_{wg}} = \dfrac{21.72}{5.33} = 4.07$

Table 10.10 ANOVA summary table for the closure study

Source	df	SS	MS	F	p
A (anxiety)	2	99.51	49.76	9.33	<.01
B (illumination)	2	77.51	38.76	7.27	<.01
AB	4	86.89	21.72	4.07	<.01
Within groups	36	192.00	5.33		
Total	44	455.91			

$$F_{.01}(2, 36\ df) = 5.25 \qquad F_{.01}(4, 36\ df) = 3.89$$

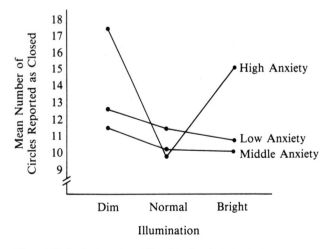

Figure 10.5 Interaction effect for the closure study

Look again at Figure 10.5, the graph of the significant interaction effect from the closure study. The significant F for the anxiety factor indicates that the three anxiety groups differed in closure. However, it would appear that this difference may be due entirely to the performance of the high-anxious group under conditions of dim and bright illumination. Similarly, the significant illumination main effect indicates that the three different amounts of light produced three sets of scores that do not appear to have a common population mean. Figure 10.5 shows that this significant main effect may be due primarily to the high-anxious subjects and not to subjects in general.

When an interaction is significant, the interpretation of the main effects is usually not simple and straightforward. A main effect is a comparison of the average of each level with the grand mean. A significant interaction indicates that the averages may be misleading.

In summary, when an interaction is significant, main effects must be interpreted in the light of the interaction. For a problem like the closure study, summarized in Table 10.10 and Figure 10.5, the experienced researcher would probably test for treatment-contrast interactions, a technique that is beyond the scope of this book.[6] However, when an interaction is significant, you can often correctly interpret the results simply by drawing a graph of cell means and examining it.

PROBLEMS

15. The following three problems are designed to help you learn to *interpret* the results of a factorial experiment. Consider as statistically significant any effect that has a probability less than .05.

   a. A clinical psychologist has been looking at the relationship between humor and aggression. The subject's task in these experiments is to think up as many captions as possible for four cartoons. The test lasts 8 minutes, and the psychologist simply counts the number of captions produced. This clinical psychologist's latest experiment is one in which a subject is either insulted or treated in a neutral way by an experimenter. The subject then responds to the cartoons at the request of either the same experimenter or a different experimenter. The cell means and the summary table are shown in the tables. Identify the independent and dependent variables, fill in the probability values in the summary table, and interpret the results.

		Subject was	
		Insulted	Treated neutrally
Experimenter was	Same	31	18
	Different	19	17

Source	df	MS	F	p
A (between treatments)	1	675.00	9.08	
B (between experimenters)	1	507.00	6.82	
AB	1	363.00	4.88	
Within groups	44	74.31		

[6] For information on treatment-contrast interactions, see Kirk (1982).

b. An educational psychologist was interested in response biases. You exhibit a response bias when you make a decision on the basis of an irrelevant stimulus. For example, a response bias is shown if the grade an English teacher puts on a theme written by a tenth-grader is influenced by the student's name or by the occupation of his father. For this particular experiment, the psychologist picked a "high-prestige" name (David) and a "low-prestige" name (Elmer). In addition, she made up two biographical sketches that differed only in the occupation of the student's father (research chemist or unemployed). The subjects in this experiment were given the task of grading a theme on a scale of 50 to 100. They had been told the name of the writer and had read his biographical sketch. The same theme was given to each subject in the experiment. Identify the independent and dependent variables, fill in the probability values in the summary table, and interpret the results.

		Name	
		David	Elmer
Occupation of father	Research chemist	86	81
	Unemployed	80	88

Source	df	MS	F	p
A (between names)	1	22.50	0.44	
B (between occupations)	1	2.50	0.05	
AB	1	422.50	8.20	
Within groups	36	51.52		

c. A social psychologist was in charge of a large class in community psychology. During the course, 20 outside speakers holding community service jobs each gave a short talk. Ten of these speakers were men, ten women. Five of the men had been at their jobs 2 years or longer, and the other five had been at their jobs less than 6 months. This same job-experience variable divided the women into two groups of five. The social psychologist wanted to see whether these two variables (gender and job experience) were related to the amount of attention the speakers received. She arranged a wide-angle camera to take a picture of the class when the speaker had completed the third minute of his or her talk. The dependent variable was the number of persons who were looking directly at the speaker when the picture was taken. Fortunately, the same number of people were in class on every occasion. The cell means and the summary table are shown here. Identify the independent and dependent variables, fill in the probability values, and interpret the results.

		Gender of speaker	
		Female	Male
Job experience	Less than 6 months	23	28
	Longer than 2 years	31	39

Source	df	MS	F	p
A (between gender)	1	211.25	4.59	
B (between experience)	1	451.25	9.80	
AB	1	11.25	0.24	
Within groups	16	46.03		

COMPARING LEVELS WITHIN A FACTOR

When a factorial ANOVA produces a significant main effect, you usually want to know which groups are responsible. This is exactly the situation you faced with one-way ANOVA. You will be pleased to learn that the solution you learned for one-way ANOVA also applies to factorial ANOVA.

Of course, for some main effects you do not need a test subsequent to ANOVA; the F test on that main effect gives you the answer. This is true for *any main effect that involves only two levels*. Just like a t test, a significant F tells you that the two means are from different populations. Thus, if you are analyzing a 2×2 factorial ANOVA, you do not need a test subsequent to ANOVA for *either* factor.

When there are more than two levels of one factor in a factorial design, Tukey's HSD test can be used to test for significant differences between pairs of levels. There is one restriction, however, which is that Tukey's HSD test is appropriate *only if the interaction is **not** significant*. We will use Tukey's HSD test as our one example of a test subsequent to factorial ANOVA. To keep you from getting bored, we will show you a way to arrange the arithmetic that will save you some time.

In the teacher-response study you analyzed in Problem 9, the interaction was not significant. The "praise" group completed more arithmetic problems than the other groups. Did they complete significantly more? A set of Tukey's HSDs will provide answers. The HSD formula for a factorial ANOVA is the same as that for a one-way ANOVA:

$$HSD = \frac{\bar{X}_1 - \bar{X}_2}{s_{\bar{X}}}$$

where $s_{\bar{X}} = \sqrt{\dfrac{MS_{wg}}{N}}$

If you will think of this as an analysis of teacher responses and ignore the gender variable, you will have no trouble figuring out what means and what N to use. (When one factor is ignored, statisticians say that the design is *collapsed* over the ignored factor.) Thus, the means are the marginal means, not the cell means. N refers to a column N or a row N, not a cell N.

When Tukey's HSDs are calculated on means that are all based on equal N's (as is the case for all factorial ANOVAs in this chapter), there is an arithmetic arrangement that is more efficient than that used in Chapter 9. In this arrangement, $(HSD)(s_{\bar{X}})$ is calculated. HSD comes from Table G and $s_{\bar{X}}$ comes from the data.[7] Any difference between means that is larger than this product is significant. If a mean difference is smaller, it is not significant. (You can play with the formula above and see how this works.) For the neglect-reproof-praise study (where all N's are 20),

$$(HSD_{.05})(s_{\bar{X}}) = (3.44)\sqrt{\frac{19.89}{20}} = 3.43$$

The group that was neglected was not significantly different from the group that received reproof (mean difference = 0.60, which is less than 3.43). The group

[7] Since all comparisons are based on equal N's, the $s_{\bar{X}}$ for every comparison will be the same.

that received praise made significantly fewer errors than the reproof group (mean difference = 3.80) and significantly fewer than the neglect group (mean difference = 4.40). Thus, the conclusion is that praise is better than either neglect or reproof, which do not differ from each other.

PROBLEMS

16. Suppose you were asked to determine whether there is a significant difference between the low-anxiety scores and the high-anxiety scores in the study of the effects of anxiety and illumination on closure drive. Could you use Tukey's HSD test?

17. Two social psychologists asked freshman and senior college students to write an essay supporting recent police action on campus. (The students were known to be against the police action.) The students were given 50 cents, one dollar, five dollars, or ten dollars for their essay. Later each student's attitude toward the police was measured on a scale of 1–20. Perform an ANOVA, fill out a summary table, and write an interpretation of the analysis at this point. (See Brehm and Cohen, 1962, for a similar study of Yale students.) If appropriate, calculate HSDs for all pairwise comparisons. Write an overall interpretation of the results of the study.

			\$10	\$5	\$1	50¢
			Reward			
Students	Freshmen	ΣX	73	83	99	110
		ΣX^2	750	940	1290	1520
		N	8	8	8	8
	Seniors	ΣX	70	81	95	107
		ΣX^2	820	910	1370	1610
		N	8	8	8	8

18. Eckhard Hess used a scientific experiment to show that a person's pupil diameter changes as emotional arousal changes. (Hess reports that his wife first noted this phenomenon while watching her husband leaf through a magazine—a magazine with pictures of birds.) The data in the table illustrate some of Hess's findings. Men and women were shown one picture and their pupil size was recorded (large numbers indicate more arousal). Analyze the data with a factorial ANOVA and, if appropriate, a set of Tukey's HSDs. Write a conclusion.

			Nude man	Nude woman	Infant	Landscape
			Pictures			
Gender	Women	ΣX	22	9	23	15
		ΣX^2	166	29	179	77
		N	3	3	3	3
	Men	ΣX	11	26	9	14
		ΣX^2	41	226	29	70
		N	3	3	3	3

Transition Page

In your study of inferential statistics, you have used two families of curves. The normal curve is appropriate when sampling is random and you know σ (Chapters 5–7). The t and F distributions are appropriate when sampling is random and population scores are normally distributed and have equal variances (Chapters 8–10). In the next two chapters, you will learn about some statistical tests that do not require knowledge of σ, assumptions about the form of the population distribution, or homogeneity of variance. Random sampling, however, will still be required.

In Chapter 11, "The Chi Square Distribution," you will learn to analyze frequency count data. Such data exist when observations are classified into categories and the frequencies in each category are counted. In Chapter 12, "Nonparametric Statistics," you will learn four techniques for analyzing scores that are ranks or can be reduced to ranks.

The techniques in these next two chapters are often described as "less powerful." This means that *if* the population scores satisfy the assumptions of normality and homogeneity of variance, a t or F test is more likely than a chi square test or a nonparametric test to reject H_0 if it should be rejected. To put this same idea another way, t and F tests have a smaller probability of a Type II error if the population scores are normally distributed and have equal variances.

11

THE CHI SQUARE DISTRIBUTION

Objectives for Chapter 11: After studying the text and working the problems in this chapter, you should be able to:

1. Tell when the chi square distribution may be used in hypothesis testing
2. Use the chi square distribution to test how well empirical frequency data fit a theoretical model (goodness of fit)
3. Use the chi square distribution to test the independence of two variables
4. Calculate chi square from 2 x 2 table using the shortcut method

A sociology class was deep into a discussion of methods of population control. As the class ended, the topic was abortion; it was clear that there was strong disagreement about using abortion as a method of population control. Both sides in the argument had declared that "educated people" supported their side. Two empirically minded students left the class determined to get actual data on attitudes toward abortion. They solicited the help of a friendly psychology professor, who encouraged them to distribute a questionnaire to his General Psychology class. The questionnaire asked, "Do you consider abortion to be an acceptable or unacceptable method of population control?" The questionnaire also asked the respondent's gender. The results were presented to the sociology class in the following tabular form:

	Abortion is	
	Acceptable	*Unacceptable*
Females	59	29
Males	15	37
Σ	74	66
	53%	47%

The general conclusion of the class was that college students were pretty evenly divided on this issue, 53 percent to 47 percent.

The attitude data, then, did not seem to indicate that "educated people" were clearly on one side or the other on this issue. The class discussion resumed, with personal opinions dominating, until an observant student said, "Look! Look at that table! The girls are for abortion, and the guys are against it. I wonder whether that is a real difference between males and females, or whether it is just a chance event that occurred in this sample."

Does that question sound familiar? Indeed, it sounds like the kind of question that could be answered by comparing the observed results with a sampling distribution based on chance results. In this case, a proportion of .67 of the females (59 out of 88) considered abortion to be acceptable, but only .29 of the males did so (15 out of 52). Could these two statistics have come from the same population, or should they be considered different and independent?[1]

This is yet another chapter on hypothesis testing; to retain or reject the null hypothesis—that is the question. In this chapter, however, the data are different. In previous chapters, the dependent variables were phenomena such as IQ, racial attitudes, or time—phenomena that are measured on a fairly continuous scale (a quantitative measurement). Thus, the unit of observation (a person, perhaps) would perform a task and be given a score somewhere along the scale.

In this chapter, the unit of observation is observed and assigned to a category. Sometimes the categories are ordered (for example, freshman, sophomore, junior) and sometimes the categories have no inherent order (for example, Australian, Canadian, English). What the statistician analyzes are the frequencies in each category.

Chi square (pronounced "ki" as in *kite,* "square," and symbolized χ^2) is the sampling distribution to use to analyze frequencies. Since frequencies like those in the abortion–gender table are distributed approximately as chi square, the chi square distribution will provide probabilities that help answer questions like those at the end of the story. The chi square statistic is used widely, turning up in psychology, sociology, biology, political science, education, agriculture, and economics.

Karl Pearson (1857–1936), of Pearson product-moment correlation coefficient fame, published his article on chi square in 1900. Pearson wanted a way to measure the "fit" between data generated by a theory and data obtained from observations. Until chi square was dreamed up, theory testers presented theoretical predictions and empirical data side by side, followed by a declaration like "good fit" or "poor fit."

The most popular theories at the end of the 19th century predicted that data would be distributed as a normal curve. Many data gatherers had adopted Quételet's position that measurements of almost any social phenomenon would be normally distributed if the number of cases was large enough. Pearson (and others) thought this was not true and they proposed other curves. By inventing chi square, Pearson provided everyone with a quantitative method of *choosing* the curve of best fit.

Chi square as a test statistic has turned out to be very versatile, being applicable to several problems besides curve-fitting ones. In addition to its use as a test statistic, the chi square *distribution* has come to occupy an important place in the theory of

---

[1] Casting the chi square problems in this chapter into proportions will help you in interpreting the results. However, the test of whether two or more proportions, obtained from separate samples, have a common population proportion is not covered in this book. (See Kirk, 1984, pp. 385–387.)

statistics. As further evidence of the importance of Pearson's chi square, the editors of *Science 84* chose it as one of the 20 most important discoveries of the 20th century (Hacking, 1984).

Pearson, too, turned out to be very versatile, contributing data and theory to both biology and statistics. He is credited with naming the standard deviation, giving it the symbol σ, and coining the word *biometry*. He (and Galton and Weldon) founded the journal *Biometrika* to publicize and promote the marriage between biology and mathematics.[2] In addition to his scientific efforts, Pearson was an advocate of women's rights and the father of an eminent statistician, Egon Pearson.

As a final note, Pearson's overall goal was not to be a biologist or a statistician. His goal was to better life for the human race. An important step in accomplishing this was to "develop a methodology for the exploration of life" (Walker, 1968).[3]

THE CHI SQUARE DISTRIBUTION

The **chi square distribution** is a theoretical distribution, just as t and F are. Like them, as the number of degrees of freedom increases, there is a change in the shape of the distribution. **Figure 11.1** shows four different chi square distributions, one each for 1, 3, 5, and 10 df. As you can see, χ^2, like the F distribution, is a positively skewed curve. **Table E** in Appendix C gives five critical values for χ^2: .10, .05, .02, .01, and .001. We will symbolize these critical values of χ^2 by giving the α level, df, and the χ^2 value.

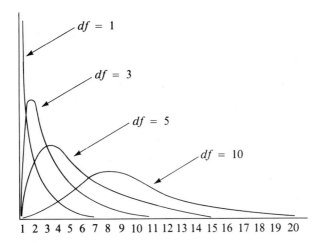

Value of χ^2

Figure 11.1 Chi square distribution for four different degrees of freedom. Adapted with permission from Cornell (1956, p. 198)

[2] In 1901 when *Biometrika* began, The Royal Society, the principal scientific society in Great Britain, accepted articles on biology and articles on mathematics, but they would not permit papers that combined the two. (See Galton, 1901, for a thinly disguised complaint against the establishment.)

[3] For a short biography of Pearson, see Helen M. Walker (1968).

Thus, $\chi^2_{.05}$ (1 *df*) = 3.84. Notice that, with chi square, as degrees of freedom increase (looking down columns), larger and larger values of χ^2 are required to reject the null hypothesis. *This is just the opposite of what occurs in the t and F distributions.* This will make sense if you examine the four curves in Figure 11.1.

When **observed frequencies** are being compared to **theoretical frequencies,** a value for chi square may be computed using the formula

$$\chi^2 = \Sigma\left[\frac{(O - E)^2}{E}\right]$$

where O = observed frequency
E = expected frequency

Obtaining the observed frequency is simple enough—count the events in each category. Finding the expected frequency is a bit more complex. Two methods can be used to determine the expected frequency. One method is used if the problem is one of "goodness of fit" and the other method is used if the problem is a "test of independence." We'll discuss these two tests in some detail.

CHI SQUARE AS A TEST FOR GOODNESS OF FIT

If you have a hypothesis or theory that predicts the expected frequency, χ^2 can be used to test how well the observed frequency conforms to that expected frequency—that is, how good the "fit" is between the actual (observed) frequency and the expected frequency. In testing for **goodness of fit,** *the null hypothesis is that the actual data do fit the expected data.* Thus, rejection of H_0 means that the data *do not* fit the theory; the theory is inadequate. If H_0 is retained, then you must, once again, word your interpretation carefully. *A retained H_0 means that the data are not at odds with the theory.* Such results do not prove the theory (other theories may also predict such a result), but they do support the theory.

The goodness-of-fit application of χ^2 is frequent in population genetics, where ratios such as 3:1, 1:2:1, and 9:3:3:1 are predicted on the basis of Mendelian laws. The theory may predict that a certain hybrid will produce three times as many smooth pea seeds as wrinkled seeds (3:1 ratio). If you perform the crosses, you might get 316 smooth seeds and 84 wrinkled seeds, for a total of 400. How well do these actual frequencies fit the expected frequencies of 300 and 100? Chi square will give you a probability figure that will help you decide if the theory accounts for the observed data.

Although the mechanics of χ^2 require you to manipulate raw frequency counts (like 316 and 84), you can best understand this test by thinking of proportions. Thus, 316 and 84 out of 400 represent proportions of .79 and .21. How likely are such proportions if the population proportions are .75 and .25? Chi square provides you with the probability of obtaining the observed proportions if the theory is true. As a result, you will have a quantitative way to decide "good fit" or "poor fit."

In the genetics example, the expected frequencies were predicted by a theory. Sometimes, in a goodness-of-fit test, the expected frequencies are predicted by chance. In such a case, a rejected H_0 means that something besides chance is at work. When you interpret the analysis you should identify what that "something" is. Here is an example.

Suppose you were so interested in sex stereotypes and job discrimination that you conducted the following experiment (see Mischel, 1974). You made up four one-page resumes and four fictitious names—two female, two male. Names and resumes were randomly combined and each subject was asked to read the resumes and "hire" one of the four "applicants" for a management trainee position. The theory (or model) tested is a "no discrimination" theory. The theory says that gender is *not* being used as a basis for hiring and therefore equal numbers of men and women will be hired. Note that this theory is a statement of the null hypothesis. Thus, if the data cause the null hypothesis to be rejected, the theory will be rejected.

Suppose you had 120 subjects, and they "hired" 75 men and 45 women. Is there statistical evidence that sex stereotypes are leading to discrimination? That is, given the hypothesized result of 60 men and 60 women, how good a fit is the observed data of 75 men and 45 women?

Applying the χ^2 formula, you get

$$\chi^2 = \Sigma\left[\frac{(O-E)^2}{E}\right] = \frac{(75-60)^2}{60} + \frac{(45-60)^2}{60} = 3.75 + 3.75$$
$$= 7.50$$

The number of degrees of freedom for this problem is the number of categories minus one. There are two categories here—hired men and hired women. Thus, $2 - 1 = 1$ *df*. Looking in Table E, you will find in row 1 that, if $\chi^2 \geq 6.64$, the null hypothesis may be rejected at the .01 level. Thus, the model may be rejected. The conclusion from the experiment is that discrimination in favor of men was demonstrated.

The term *degrees of freedom* implies that there are some restrictions. In the case of χ^2, one restriction is always that the sum of the expected events must be equal to the sum of the observed events; that is, $\Sigma E = \Sigma O$. If you manufacture a set of expected frequencies from a model, their sum must be equal to the sum of the observed frequencies. In our sex discrimination example, $\Sigma O = 75 + 45 = 120$, and $\Sigma E = 60 + 60 = 120$.

ERROR DETECTION

> The sum of the expected frequencies must be equal to the sum of the observed frequencies: $\Sigma E = \Sigma O$.

PROBLEMS

1. Using the data on wrinkled and smooth pea seeds on page 236, test how well the data fit a 3:1 hypothesis. Set $\alpha = .05$.

2. Here is a problem that illustrates some work of behaviorist John B. Watson (1879–1958) on the question of whether there are any basic, inherited emotions. Watson (1924) thought there were three: fear, rage, and love. He suspected that the wide variety of emotions experienced by adults was learned over the years. Proving his suspicion would be difficult because "unfortunately there are no facilities in maternity wards for keeping mother and child under close observation for years." So Watson attacked the apparently simpler problem of showing that the emotions of fear, rage, and love could be distinguished in infants. (He recognized the difficulty of proving that these were the *only* basic emotions,

and he made no such claim.) Fear could be elicited by dropping the child onto a soft feather pillow (but not by the dark, dogs, white rats, or a pigeon fluttering its wings in the baby's face). Rage could be elicited by holding the child's arms tightly at its sides, and love by "tickling, shaking, gentle rocking, and patting" among other things.

Here is an experiment reconstructed from the conclusions Watson described in his book. A child was stimulated in a way designed to elicit fear, rage, or love. An observer then looked at the child and judged the emotion the child was experiencing. Each judgment was scored as correct or incorrect—correct meaning that the judgment (say, fear) corresponded to the stimulus (say, dropping). Sixty observers made one judgment each with the results shown in the table. To find the expected frequencies, think about the chance of being correct or incorrect when there are three possible outcomes. Analyze the data with χ^2 and write your conclusion in a sentence.

Correct	Incorrect
32	28

CHI SQUARE AS A TEST OF INDEPENDENCE

Another use of χ^2 is to test the *independence* of two variables on which frequency data are available. **Table 11.1,** the work of empirically minded students, shows the abortion questionnaire data. Such a table is called a *contingency table*; here, the question is whether a person's attitude toward abortion is contingent upon gender. *The null hypothesis is that attitudes and gender in the population are independent—* that knowing a person's gender gives you *no* clue to his or her attitude and vice versa. Rejection of the null hypothesis will support the alternate hypothesis, which is that gender and attitude are not independent but related—that knowing a person's gender does help you to predict his or her attitude and that attitude is contingent upon gender.

Table 11.1 gives the observed frequencies, with the expected frequencies in parentheses. Pay careful attention to the logic behind the calculation of the expected frequencies. We will start with an explanation of the expected frequency in the upper left corner, 46.51, the expected number of females who think abortion is acceptable. Of all the 140 subjects, 88 were female, so if we chose a subject at random, the probability that the person would be female is $88/140 = .6286$. In a similar way, of

Table 11.1   Hypothetical data on attitudes toward abortion (Expected frequencies are in parentheses.)

	Abortion is		
	Acceptable	Unacceptable	Σ
Females	59 (46.51)	29 (41.49)	88
Males	15 (27.49)	37 (24.51)	52
Σ	74	66	140

the 140 subjects, 74 thought that abortion was acceptable. Thus, the probability that a randomly chosen person would think that abortion is acceptable is 74/140 = .5286. If we now ask what the probability is that a person chosen at random is a female who thinks that abortion is acceptable, we can find the answer by multiplying together the probability of the two separate events.[4] Thus, (.6286)(.5286) = .3323.

Finally, we can find the expected frequency of such people by multiplying the probability of such a person by the total number of people. Thus, (.3323)(140) = 46.52. Notice what happens to the arithmetic when these steps are combined:

$$\left(\frac{88}{140}\right)\left(\frac{74}{140}\right)(140) = \frac{(88)(74)}{140} = 46.51$$

The slight difference in the two methods is due to rounding (46.52 vs. 46.51). The second answer is the correct one even though we carried four decimal places in our first answer. (This should serve as a warning for those of you who value precision.)

In a similar way, the expected frequency of males with an attitude that abortion is unacceptable is the probability of a male times the probability of an attitude of unacceptable times the number of subjects. For all the data, here are the calculations of the expected values.

$$\frac{(88)(74)}{140} = 46.51 \qquad \frac{(88)(66)}{140} = 41.49$$

$$\frac{(52)(74)}{140} = 27.49 \qquad \frac{(52)(66)}{140} = 24.51$$

Once the expected values are calculated, a table like **Table 11.2** can be constructed that will lead to a χ^2 value.

To determine the *df* for any $R \times C$ (rows by columns) table like Table 11.1, use the formula $(R - 1)(C - 1)$. In this case, $(2 - 1)(2 - 1) = 1$. To explain this surprising state of affairs ("Hold it! Before, with two categories, there was 1 *df* and now, with four categories, I still only have 1 *df* ???"), we will resort to our "freedom to vary" explanation. For any $R \times C$ table, the *marginal* totals of the expected frequencies are fixed so that they equal the observed frequency.[5] In a 2 × 2 table, this

Table 11.2 Calculation of χ^2 for the data in Table 11.1

O	E	$O - E$	$(O - E)^2$	$\dfrac{(O - E)^2}{E}$
59	46.51	12.49	156.00	3.35
15	27.49	−12.49	156.00	5.67
29	41.49	−12.49	156.00	3.76
37	24.51	12.49	156.00	6.36
				$\chi^2 = 19.14$

[4] This is like determining the chances of obtaining two heads in two tosses of a coin. For each toss, the probability of a head is $\frac{1}{2}$. The probability of two heads in two tosses is obtained by multiplying the two probabilities: $(\frac{1}{2})(\frac{1}{2}) = \frac{1}{4} = .25$. See the section, "Binomial Distribution" in Chapter 5 for a review of this topic.

[5] In Table 11.1, the marginal total of 88 must be the sum of the two expected frequencies, 46.51 and 41.49, as well as the sum of the obtained frequencies, 59 and 29.

leaves only one cell free to vary. Once a value is fixed for one cell, the other cell values are determined if you are to maintain the marginal totals. The general rule again is $df = (R - 1)(C - 1)$, where R and C refer to the number of rows and columns.

To determine the significance of $\chi^2 = 19.14$ with 1 df, look at Table E in Appendix C. In the first row, you will find that, if $\chi^2 = 10.83$, the null hypothesis may be rejected at the .001 level of significance. Since our χ^2 exceeds this, you can conclude that the attitudes toward abortion are definitely influenced by gender; that is, gender and attitudes toward abortion in the population are not independent but related. By examining the proportions, you can conclude that females are more likely to consider abortions to be acceptable.

This test of independence is essentially a test of whether there is an interaction between the two variables—in this case, gender and attitude. The interpretation of a χ^2 test of independence in an $R \times C$ table is the same as the interpretation of an interaction in a two-way ANOVA: the result for one independent variable *depends* on the level of the other independent variable.

SHORTCUT FOR ANY 2 x 2 TABLE

When you have a 2 × 2 table to analyze and a calculator, the following shortcut will save you time. With this shortcut, it is not necessary to calculate the expected frequencies, which reduces calculating time by one-half to three-quarters. For the general case of a 2 × 2 table,

A	B	$A + B$
C	D	$C + D$
$A + C$	$B + D$	N

$$\chi^2 = \frac{N(AD - BC)^2}{(A + B)(C + D)(A + C)(B + D)}$$

The term in the numerator, $AD - BC$, is the difference between the cross-products. The denominator is the product of the four marginal totals.

To illustrate the equivalence of the shortcut method and the method explained in the previous section, we will calculate χ^2 from the data in Table 11.1. Translating cell letters into cell totals, you get $A = 59$, $B = 29$, $C = 15$, $D = 37$, and $N = 140$. Thus,

$$\chi^2 = \frac{(140)[(59)(37) - (29)(15)]^2}{(88)(52)(74)(66)} = 19.14$$

Both methods yield $\chi^2 = 19.14$.

PROBLEMS   3. In the late 1930s and early 1940s an important sociology experiment took place in the Boston area. At that time 650 boys (median age = $10\frac{1}{2}$ years) participated in the

Cambridge-Somerville Youth Study (named for the two economically depressed communities where the boys lived). The participants were *randomly* assigned to either a delinquency-prevention program or a control group. The delinquency-prevention program consisted of a counselor and several years of opportunities for enrichment. At the end of the study, police records were examined for evidence of delinquency among all 650 boys. Analyze the data in the table and write a conclusion.

	Received program	Control
Police record	114	101
No police record	211	224

4. A student of psychology was interested in the effect of group size on the likelihood of joining an informal group. On one part of the campus, he placed a group of *two* people looking intently up into a tree. On a distant part of the campus, a group of *five* stood looking up into a tree. Single passersby were classified as joiners or nonjoiners. If a passerby stopped and looked up for 5 seconds or longer or made some comment to the group, he or she was classified as a joiner. The data in the table were obtained. Use a χ^2 test to determine whether group size had an effect on joining. Use the method you did not choose for Problem 3. Write a conclusion.

	Group size		
	2	5	Sum
Joiners	9	26	35
Nonjoiners	31	34	65
Sum	40	60	100

5. You have probably heard that salmon return to the stream in which they hatched in order to spawn. According to the story, after years of maturing in the ocean, the salmon arrive at the mouth of a river and swim upstream, choosing at each fork the stream that leads to the pool in which they hatched. Arthur D. Hasler (1966) did a number of interesting experiments on this homing instinct. Here are two sets of data that he gathered.
   a. These data help answer the question of whether salmon really do make consistent choices at the fork of a stream in their return upstream. Forty-six salmon were captured from the Issaquah Creek (in Washington State) and another 27 from its East Fork. All salmon were released below the confluence of these two streams. All of the 46 captured in the Issaquah were recaptured there. Of the 27 originally captured in the East Fork, 19 were recaptured from the East Fork and eight from the Issaquah. Use χ^2 to determine if salmon make consistent choices at the confluence of two streams. Write a sentence explanation of the results.
   b. Hasler believed that the salmon were making the sequence of choices at each fork on the basis of olfactory (smell) cues. He thought that young salmon become imprinted on the particular mix of dissolved mineral and vegetable molecules in their home stream. As adults, they simply make choices at forks on the basis of where the smell of home is coming from. To test this, he captured 70 salmon from the two streams, plugged their nasal openings, and released them downstream. The fish were recaptured above the fork in one stream or the other. Compute χ^2 and write a sentence about Hasler's hypothesis.

Recapture site

	Issaquah	East Fork
Issaquah	39	12
East Fork	16	3

Capture site

6. Here are some data that one of your authors worked on for an attorney. Of the 4200 white applicants at a large manufacturing facility, 390 were hired. Of the 850 black applicants, 18 were hired. Analyze the data and write a conclusion. Note: You will have to work on the data some before you set up the table.

CHI SQUARE WITH MORE THAN ONE DEGREE OF FREEDOM

So far the problems we have considered have been analyzed with a χ^2 distribution with 1 df. In this section, you will analyze more complex designs that require χ^2 distributions with more than 1 df.

Suppose you asked the following question of 128 members of a freshman psychology course in the United States, all of whom were born and raised in the United States. "Imagine that you have volunteered to have a foreign roommate next year, and you must tell whether you prefer an African, an Asian, or a European. Which would you choose?" The hypothetical responses could be compiled into the following form:

Preference			
African	Asian	European	Total
40	28	60	128

We may use χ^2 to test the hypothesis that such frequencies could have arisen by chance; that is, we will test these data for goodness of fit to a model in which each of the three geographical areas is preferred equally. In this case, the chance expectation for each cell is one-third of 128, or 42.67. The application of χ^2 is a simple extension of what you have already learned. (See **Table 11.3.**)

What is the df for this χ^2 value? The marginal total of 128 is fixed, and, since ΣE must equal ΣO, only two of the expected cell frequencies are free to vary. Once two

Table 11.3   Calculation of χ^2 for the roommate preference data

Options	O	E	$O - E$	$(O - E)^2$	$\dfrac{(O - E)^2}{E}$
African	40	42.67	−2.67	7.1111	0.1667
Asian	28	42.67	−14.67	215.1111	5.0417
European	60	42.67	17.33	300.4444	7.0417
				$\chi^2 =$	12.2501

are determined, the third is restricted to whatever value will make $\Sigma E = \Sigma O$. Thus, there are $3 - 1 = 2$ *df*. For this design, the number of *df* is the number of categories minus one. From Table E, for $\alpha = .01$, χ^2 with 2 *df* $= 9.21$. Therefore, reject the hypothesis of independence and conclude that the students were responding in a nonchance manner to the questionnaire.

One advantage of χ^2 is its *additive* nature. Each $(O - E)^2/E$ value is a measure of deviation of the data from the model, and the final χ^2 is simply the sum of these measures. Because of this, you can examine the $(O - E)^2/E$ values and see which deviations are contributing the most to χ^2. For the roommate preference data, it is a greater number of European choices and fewer Asian choices that make the final χ^2 significant. The African choices are about what would be predicted by the "equal likelihood" model.

Suppose you showed your roommate preference data to a few of your friends, and one of them (a senior psychology major) said, "That's typical of data in psychology—it's based only on college students! If you do any more studies, why don't you find out what the noncollege population thinks? Try some high school kids; try some business types downtown." Since this attack seems to you to have some validity, you include all three of the suggested groups in your next study, one on family planning. This time the questionnaire reads, "A couple just starting a family this year should plan to have 0 or 1 ( ), 2 or 3 ( ), 4 or 5 ( ) child(ren). Check one." Respondents classified themselves as high school, college, or business types. The compiled data are shown in **Table 11.4.**

For this problem, there is no theoretical expectancy. The question is whether the three groups (high school, college, and business) differ in their responses. What is required is a test of independence; it will tell us whether attitudes toward family size are related to groups or are independent of group affiliation.

To get the expected frequency for each cell, assume independence (H_0) and apply the reasoning we explained earlier about multiplying probabilities.

$$\frac{(92)(104)}{275} = 34.793 \qquad \frac{(92)(130)}{275} = 43.491 \qquad \frac{(92)(41)}{275} = 13.716$$

$$\frac{(117)(104)}{275} = 44.247 \qquad \frac{(117)(130)}{275} = 55.309 \qquad \frac{(117)(41)}{275} = 17.444$$

$$\frac{(66)(104)}{275} = 24.960 \qquad \frac{(66)(130)}{275} = 31.200 \qquad \frac{(66)(41)}{275} = 9.840$$

Table 11.4   Recommended family size by those in high school, college, and business

Subjects	Number of children			
	0–1	2–3	4–5	Σ
High school	26	57	9	92
College	61	38	18	117
Business	17	35	14	66
Σ	104	130	41	275

These expected frequencies are incorporated into **Table 11.5,** which shows the calculation of the χ^2 value.

Table 11.5   Calculation of χ^2 for the recommended family size data

O	E	$O - E$	$(O - E)^2$	$\dfrac{(O - E)^2}{E}$
26	34.793	−8.793	77.317	2.222
57	43.491	13.509	182.493	4.196
9	13.716	−4.716	22.241	1.622
61	44.247	16.753	280.663	6.343
38	55.309	−17.309	299.602	5.417
18	17.444	0.556	0.309	0.018
17	24.960	−7.960	63.362	2.539
35	31.200	3.800	14.440	0.463
14	9.840	4.160	17.306	1.759
				$\chi^2 = \overline{24.579}$

The *df* for a contingency table such as this is obtained from the formula

$$df = (R - 1)(C - 1)$$

Thus, $df = (R - 1)(C - 1) = (3 - 1)(3 - 1) = 4$.

For $df = 4$, $\chi^2 = 18.46$ is required to reject H_0 at the .001 level. The obtained χ^2 exceeds this, so reject H_0 and conclude that attitudes toward family size are related to the groups sampled.[6]

By examining the right-hand column, you can see that 6.343 and 5.417 constitute a large portion of the final χ^2 value. By working backward on those rows, you will discover that college students chose the 0–1 category more often than expected and the 2–3 category less often than expected. The interpretation then is that college students think that families should be smaller than high school students and business people do.

---

7. Professor Stickler always grades "on the curve." "Yes, I see to it that my grades are distributed as 7 percent A's, 24 percent B's, 38 percent C's, 24 percent D's, and 7 percent flunks," he explained to a young colleague. The colleague, convinced that the professor was really a softer touch than he sounded, looked at Professor Stickler's grade distribution for the past year. He found frequencies of 20, 74, 120, 88, and 38, respectively, for the five categories, A through F. Run a χ^2 test for the colleague at the .05 level. What would *you* conclude about Professor Stickler?

8. Is Problem 7 a problem of goodness of fit or of independence?

9. "Snake-eyes," a local gambler, let a group of sophomores know they would be welcome in a dice game. After 3 hours the sophomores went home broke. However, one sharp-

---

[6] The statistical analysis for the problem is correct, but the experimental design makes interpretation of the significant χ^2 difficult. Since the three groups differ in both age and education, we are unsure whether the differences in attitude are related only to age, only to education, or to both of these variables. A better design would have groups that differed only on the age or only on the education variable.

minded, sober lad had recorded the results on each throw. He decided to see if the results of the evening would fit an "unbiased dice" model. Conduct the test and write a sentence summary of your conclusions.

Number of spots	Frequency
6	195
5	200
4	220
3	215
2	190
1	180

10. Identify Problem 9 as a test of independence or goodness of fit.

11. A political science student was struck by the strong differences of opinion among Americans when the Soviet Union invades another country. After thinking and discussing it with friends, she carried out the following experiment. She wrote a one-page description of a hypothetical Soviet invasion, suggesting four responses the United States might take (see the table). She then got 170 people at her university to read the description and give their recommendation for a U.S. response. The 170 respondents were classified as young students (younger than 22), older students (older than 30), or faculty (who varied in age). Analyze the data in the table and write a conclusion.

Response	Young students	Older students	Faculty
Do nothing	29	6	11
Publicly denounce the invasion	31	10	10
Impose economic sanctions	20	13	12
Intervene with military force	7	14	7

12. Is Problem 11 one of independence or goodness of fit?

SMALL EXPECTED FREQUENCIES

The theoretical chi square distribution, which comes from a mathematical formula, is a continuous function that can have any positive numerical value. Chi square test statistics calculated from frequencies do not work this way. They change in discrete steps and, when the expected frequencies are very small, the discrete steps are quite large. It is clear to mathematical statisticians that as the expected frequencies approach zero, the theoretical chi square distribution becomes a less and less reliable way to estimate probabilities. In particular, the fear is that such chi square analyses will reject the null hypothesis more often than warranted.

The question for researchers who hope to analyze their data with chi square has always been, Those expected frequencies—how small is too small? For years the usual answer was less than 5 or, according to some, less than 10. Recent studies, however, have led to a revision of the answer.

When *df* = 1

Several studies have used a computer to draw thousands of random samples from a known population for which the null hypothesis is true. Each sample is analyzed with a chi square, and the proportion of rejections of the null hypothesis (Type I errors) can be calculated. If this proportion is approximately equal to α, the chi square analysis is clearly appropriate. The general trend of these studies has been to show that the theoretical chi square distribution gives accurate conclusions even when the expected frequencies are considerably less than 5.[7] (See Camilli and Hopkins, 1978, for an analysis of a 2 × 2 test of independence.)

When *df* > 1

When $df > 1$, the same uncertainty exists if one or more expected frequencies is small. Is the chi square test still advisable? Bradley, Bradley, McGrath, and Cutcomb (1979) addressed some of the cases in which $df > 1$. Again, they used a computer to draw thousands of samples and analyzed each sample with a chi square test. The proportion of cases in which the null hypothesis was mistakenly rejected was calculated. Nearly all were in the .03 to .06 range but a few fell outside this range, especially when the sample sizes were small ($N = 20$). Once again, the conclusion is that the chi square test gives fairly accurate probabilities, even with sample sizes that are considered small.

An Important Consideration

Now, let's back away from these trees we've been examining and visualize the forest that you are trying to understand. The big question, as always, is, What is the nature of the population? The question is answered by studying a sample, which may lead you to reject or to retain the null hypothesis. Either decision could be in error. The concern of this section has been whether the chi square test is keeping the percentage of mistaken rejections (Type I errors) near an acceptable level (.05).

Since the big question is, What is the nature of the population? you should be reminded of the other error you could make: retaining the null hypothesis when it should be rejected (a Type II error). How likely is a Type II error when expected frequencies are small? Overall (1980) has addressed this question and his answer is "very."

Overall's warning to us is that if the effect of the variable being studied is not just huge, then small expected frequencies have a high probability of dooming us to make the mistake of failing to discover a real difference. You may recall that this issue was discussed in Chapter 7 under the topic of power.

How can this mistake be avoided? Use large *N*'s. The larger the *N*, the more power you have to detect any real differences. And in addition, the larger the *N*, the more confident you are that the actual probability of a Type I error is your adopted α level.

---

[7] **Yates' correction** used to be recommended for 2 × 2 tables that had one expected frequency less than 5. The effect of this was to reduce the size of the obtained chi square. According to current thinking, Yates' correction overcorrected, resulting in α levels that were much smaller than the investigator claimed to be using. Therefore, we do not recommend that Yates' correction be used.

Combining Categories

Combining categories is a technique that allows you to ensure that your chi square analysis will be more accurate. (It reduces the probability of a Type II error and keeps the probability of a Type I error close to α.) We will illustrate this technique with some data from Chapter 5 on the normal distribution.

In Chapter 5, we attempted to convince you that IQ scores are distributed normally by showing you a frequency polygon, Figure 5.7. We asked you simply to look at it and note that this empirical distribution appeared to fit the normal curve. (We adopted the pre-1900 method of presenting evidence.) Now you are in a position to use a Pearson goodness-of-fit test to determine whether a normal curve fits the distribution. For those IQ scores, the mean was 101 and the standard deviation was 13.4. The question this goodness-of-fit test will answer is, Will a theoretical normal curve with a mean of 101 and a standard deviation of 13.4 fit the data graphed in Figure 5.7?

To find the expected frequency of each class interval, we constructed **Table 11.6.** The first three columns show the observed scores arranged into a grouped frequency distribution and the lower limit of each class interval. The z score column shows the z score for the lower limit of the class interval, based on a mean of 101 and a standard deviation of 13.4. The next column was obtained by entering the normal-curve table at the z score and finding the proportion of the curve above the z score. From this proportion we subtracted the proportion of the curve associated with class intervals with higher scores (for example, from the proportion above $z = 1.75$, .0401, we subtracted .0166 to get .0235). Finally, the proportion in the interval was multiplied by 261, the number of students, to get the expected frequency.

Now we have the ingredients necessary for a χ^2 test. The observed frequencies are in the second column. The frequencies expected if the data were distributed as a normal curve with a mean of 101 and a standard deviation of 13.4 are seen in the last

Table 11.6 IQ scores and expected frequencies for 261 fifth-grade students

Class interval	Observed frequency	Lower limit of class interval	z Score	Proportion of normal curve within class interval	Expected frequency (261 × proportion)
130 and up	4	129.5	2.13	.0166	4.3
125–129	5	124.5	1.75	.0235	6.1
120–124	18	119.5	1.38	.0437	11.4
115–119	19	114.5	1.01	.0724	18.9
110–114	25	109.5	0.63	.1081	28.2
105–109	28	104.5	0.26	.1331	34.7
100–104	36	99.5	−0.11	.1464	38.2
95–99	41	94.5	−0.49	.1441	37.6
90–94	31	89.5	−0.86	.1172	30.6
85–89	28	84.5	−1.23	.0856	22.3
80–84	14	79.5	−1.60	.0545	14.2
75–79	7	74.5	−1.98	.0309	8.1
70–74	4	69.5	−2.35	.0145	3.8
65–69	1	64.5	−2.72	.0061	1.6
64 and below	0		below −2.72	.0033	0.9
Sum	261			1.0000	260.9

Table 11.7   A goodness-of-fit test of IQ data to a normal curve

Class interval	O	E	$O - E$	$(O - E)^2$	$\dfrac{(O - E)^2}{E}$
125 and up	9	10.4	−1.4	1.96	0.189
120–124	18	11.4	6.6	43.56	3.821
115–119	19	18.9	0.1	0.01	0.001
110–114	25	28.2	−3.2	10.24	0.363
105–109	28	34.7	−6.7	44.89	1.294
100–104	36	38.2	−2.2	4.84	0.127
95–99	41	37.6	3.4	11.56	0.307
90–94	31	30.6	0.4	0.16	0.005
85–89	28	22.3	5.7	32.49	1.457
80–84	14	14.2	−0.2	0.04	0.003
75–79	7	8.1	−1.1	1.21	0.149
74 and below	5	6.3	−1.3	1.69	0.268
Σ	261	260.9			$\chi^2 = 7.984$

column. We could combine these observed and expected frequencies into the now familiar $(O - E)^2/E$ formula, except that several categories have suspiciously small expected frequencies.

The solution to this problem is to combine some adjacent categories. If we make our top category 125 and up, our expected frequency becomes 10.4 (4.3 + 6.1). In a similar fashion, we can make our bottom category 74 and below and have an expected frequency of 6.3 (0.9 + 1.6 + 3.8).[8] We have made these combinations in **Table 11.7,** a familiar χ^2 table with 12 categories. The χ^2 value is 7.98.

What is the *df* for this problem? This is a case in which *df* is not equal to the number of categories minus one. The reason is that we placed more than our usual one restriction on the expected frequencies. Usually, we simply require that $\Sigma E = \Sigma O$, which costs 1 *df*. This time, however, we also required that the mean and standard deviation of the expected frequency be equal to the observed mean and observed standard deviation. Thus, *df* in this case is the number of categories minus 3, and $12 - 3 = 9$ *df*.

By consulting Table E in Appendix C, you can see that a χ^2 of 16.92 would be required to reject the null hypothesis at the .05 level. The null hypothesis for a goodness-of-fit test is that the data *do* fit the model. Thus, to reject H_0 in this case is to show that the data are not normally distributed. However, since the obtained $\chi^2 = 7.98$ is less than the critical value of 16.92, we can state that these data do not differ significantly from a normal-curve model in which $\mu = 101$ and $\sigma = 13.4$.

There are other situations in which small expected frequencies may be combined. For example, in an opinion survey, the response categories might be "strongly agree," "agree," "no opinion," "disagree," and "strongly disagree." If there are few respondents who check the "strongly" categories, those who do might simply be combined with their "agree" or "disagree" neighbors—reducing the table to

---

[8] We have to confess that the guideline we used for how much combining to do was the old notion about not having expected frequencies of less than 5.

three cells instead of five. The basic rule on combinations is that they must "make sense." It wouldn't make sense to combine the "strongly agree" and the "no opinion" categories in this example.

PEARSON'S χ^2 AND PEARSON'S r

One of the principal joys of a teacher/scholar is showing connections. This section shows a connection between Pearson's correlation coefficient and Pearson's chi square.

We will illustrate this connection using data published by Karl Pearson and Alice Lee in 1903, reprinted in Fisher (1925). **Figure 11.2** shows the heights of fathers and their adult daughters, presented in the kind of scatterplot first used by Francis Galton. The numbers in the cells give the frequency counts for the number of pairs at each intersection. For Figure 11.2 you could calculate r.

Table 11.8 shows the same data collapsed into a 2×2 table. For Table 11.8 you could calculate χ^2. Not surprisingly, both analyses support the interpretation that there is a relationship between the height of fathers and the height of their adult daughters.

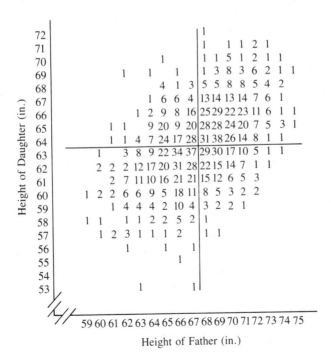

Figure 11.2 Scatterplot of the heights of fathers and their adult daughters (based on Pearson and Lee, 1903)

Table 11.8 Data of Figure 11.2 collapsed into a 2 × 2 table

		Height of father	
		67 inches or less	68 inches or more
Height of daughter	64 inches or more	207	507
	63 inches or less	437	225

WHEN YOU MAY USE CHI SQUARE

1. Chi square, like all the methods of statistical inference that you have studied, requires you to have random (or at least representative) samples from the population to which you wish to generalize.
2. Chi square is appropriate when the data are frequency counts rather than some form of quantitative measurement. Chi square is not appropriate to test a difference between means or a difference between medians.
3. The sampling distribution of χ^2 can be used to test for (a) goodness of fit and (b) independence of two variables. In addition, χ^2 has other applications, which you are likely to encounter if you take a second course in statistics.
4. The procedures presented in this book require that each event be **independent** of the other events. *Independent* here means uncorrelated; *independent* is used in the same sense as it was in the assumption of independent samples for t and F. As one example, if each subject contributed two or more observations to the study, the observations will *not* be independent. In addition, if the outcome of one event depends on the outcome of an already recorded event, the events are not independent. For example, if people are asked for their preference in a foreign roommate in the presence of others who have already responded, their stated preference is not independent of the other preferences.

PROBLEMS

13. A researcher believed that schizophrenia (a form of severe mental illness) has a simple genetic cause. In accordance with the theory, one-fourth (a 1:3 ratio) of the offspring of a certain selected group of parents would be expected to be diagnosed as schizophrenic. After several months of gathering data, the results were as shown. Test these data for goodness of fit to a 1:3 model.

Schizophrenic	*Not schizophrenic*	*Total*
23	107	130

14. What is the difference between the goodness-of-fit test and the test of independence?

15. Suppose a friend of yours decides to gather his own data on sex stereotypes and job discrimination. He makes up ten resumes and ten fictitious names, five male and five female. He asks his 24 subjects each to "hire" five of the candidates rather than just one. The data show that 65 men and 55 women were hired. He asks you for advice on χ^2. What would you say to your friend?

16. As an assignment for a political science class, students were asked to gather data to establish a "voter profile" for backers of each of the two candidates for senator (Hill and Dale). One student stationed herself at a busy intersection and categorized the cars with bumpers stickers that turned left according to whether they signaled the turn or not. After 2 hours, she left with the frequencies shown in the table. Test these data with χ^2 to see whether there is a story to be told about the backers of the two candidates. Use the expected frequency procedure rather than the shortcut. Carefully phrase the student's contribution to the voter profile.

Signaled turn	Sticker	Frequency
No	Hill	11
No	Dale	2
Yes	Hill	57
Yes	Dale	31

17. Another friend, who knows that you are almost finished with this book, comes to you for advice on the statistical analysis of the following data. He wants to know whether there is a significant difference between males and females. What do you tell your friend?

	Males	Females
Mean score	123	206
N	18	23

18. Our political science student from Problem 16 has launched a new experiment, one to determine whether affluence is related to voter choice. This time she looks for yard signs for the two candidates and then classifies the houses as brick or one-story frame. Her reasoning is that the one-story frame homes represent less affluent voters in her city. Houses that do not fit the categories are ignored. After 3 hours of driving and three-fourths of a tank of gas, the frequency counts are as shown in the table. Test the hypothesis that candidate preference and affluence (as measured by style of home) are independent. You select the α level. Carefully write a statement of the results that can be used by the political science class in the voter profile.

Type of house	Yard signs	Frequency
Brick	Hill	17
Brick	Dale	88
Frame	Hill	59
Frame	Dale	51

19. When using χ^2 what is the proper *df* for the following tables?
   a. 1×4 **b.** 4×5 **c.** 2×4 **d.** 6×3

12 NONPARAMETRIC STATISTICS

Objectives for Chapter 12: After studying the text and working the problems in this chapter, you should be able to:

1. Describe the rationale of nonparametric statistical tests
2. Determine when a Mann-Whitney U test is appropriate, perform the test, and interpret the results
3. Determine when a Wilcoxon matched-pairs signed-ranks test is appropriate, perform the test, and interpret the results
4. Determine when a Wilcoxon-Wilcox multiple-comparisons test is appropriate, perform the test, and interpret the results
5. Calculate a Spearman's r_s correlation coefficient and determine the probability that the coefficient came from a population with a correlation of zero

Two child psychologists were talking shop over coffee one morning. (Much research begins with just such sessions. For one example, see Abelson, 1985.) The topic was the effect of intensive early training in athletics. Both psychologists were convinced that such training made the child less sociable as an adult, but one psychologist went even further. "I think that really intensive training of young kids is ultimately detrimental to their performance in the sport. Why, I'll bet that, among the top ten men's singles tennis players, those with intensive early training are not in the highest ranks."

"Well, I certainly wouldn't go that far," said the second psychologist. "I think all that early intensive training is quite helpful."

"Good. In fact, great. We disagree and we may be able to decide who is right. Let's get the ground rules straight. For tennis players, how early is early, and what is intensive?"

"Oh, I'd say early is starting by age 7 and intensive is playing every day for two or more hours."[1]

---

[1] Since the phrase *intensive early training* can mean different things to different people, the first psychologist has provided the second with an operational definition. An **operational definition** is a definition that specifies a concrete meaning for a term. A concrete meaning is one everyone understands. "Seven years old" and "two or more hours of practice every day" are concepts that everyone understands.

"That seems reasonable. Now, let's see, our population is 'excellent tennis players' and these top ten will serve as our representative sample."

"Yes, indeed. What we would have among the top ten players would be two groups to compare. One had intensive early training, and the other didn't. The dependent variable is the player's rank. What we need is some statistical test that will tell us whether the difference in average ranks of the two groups is statistically significant."[2]

"Right. Now, a *t* test won't give us an accurate probability figure because *t* tests assume that the population of dependent variable scores is normally distributed. A distribution of ranks is rectangular with each score having a frequency of one."

"I think there is a nonparametric test that would be proper to use on such data."

Here is a new category of tests that can be used to analyze experiments in which the dependent variable is ranks. This category of tests is called **nonparametric statistics.** As you will see, there are data other than ranks for which nonparametric tests are appropriate. We'll begin with the rationale of these tests.

THE RATIONALE OF NONPARAMETRIC TESTS

Suppose you drew two samples of equal size (for example, $N_1 = N_2 = 10$) from the same population.[3] You then arranged all the scores from both samples into one overall ranking, from 1 to 20. Since the samples are from the same population, the sum of the ranks of one group should be equal to the sum of the ranks of the second group. In this case, the expected sum for each group is 105. (With a little figuring, you can prove this for yourself now, or you can wait for our explanation later in the chapter.) Any difference between the actual sum and 105 would be the result of sampling fluctuation. Clearly, a sampling distribution of such differences could be constructed.

With this rationale in mind, a familiar epistemology follows. Dream up a comparison between two groups and gather the data. Construct a sampling distribution of chance differences based on the assumption that the null hypothesis is true. Find the probability of obtaining the data you gathered by consulting the sampling distribution. If that probability is less than .05, reject chance as an explanation of the obtained results. Does this sound familiar?

Even if the two sample sizes are unequal, the same logic will work. A sampling distribution can be constructed that will show the expected variation in sums of ranks for one of the two groups.

In this chapter, you will learn four new techniques. The first three are examples of hypothesis testing that determine whether samples came from the same population. These three use the rationale described before, although only one of the tests (the Wilcoxon-Wilcox test) uses the arithmetic in exactly the way the rationale suggests. Also, for each sampling distribution, only a few points are given to you in

[2] This is how the experts convert vague questionings into comprehensible ideas that can be communicated to others—they identify the independent and the dependent variables.

[3] As always, drawing two samples from one population is statistically the same as starting with two identical populations and drawing a random sample from each.

Table 12.1   The function of some nonparametric and parametric tests

Nonparametric test	Function	Parametric test
Mann-Whitney U	Tests for a significant difference between two independent samples	Independent-samples t test
Wilcoxon matched-pairs signed-ranks	Tests for a significant difference between two correlated samples	Correlated-samples t test
Wilcoxon-Wilcox multiple-comparisons	Tests for significant differences among all possible pairs of independent samples	One-way ANOVA and Tukey's HSDs
r_s	Describes the degree of correlation between two variables	Pearson product-moment correlation coefficient, r

the tables. Like the tables for t and F, only values that experimenters use as α levels (critical values) are given. The fourth technique in this chapter is a descriptive statistic, a correlation coefficient for ranked data symbolized r_s.

The four nonparametric techniques in this chapter and their functions are listed in **Table 12.1.** In earlier chapters, you studied parametric tests that have similar functions. They are listed on the right side of the table. Study this table carefully now.

COMPARISON OF NONPARAMETRIC AND PARAMETRIC TESTS

In what ways are nonparametric tests similar to parametric tests (t tests and ANOVA), and in what ways are they different? The following are similarities:

1. The tests have the same goal: to determine whether samples came from different populations.
2. Both tests require you to have random samples from the population (or at least to assign subjects randomly to subgroups).
3. The tests are based on the logic of testing the null hypothesis. (If you can show that H_0 is very unlikely, you are left with the alternate hypothesis.) As you will see, though, the null hypotheses are different for the two kinds of tests.

As for differences, the t test and ANOVA assume that the scores in the populations that are sampled are normally distributed and have equal variances, but no such assumptions are necessary if you run a nonparametric test. Also, with parametric tests, the null hypothesis is that the population *means* are the same ($H_0: \mu_1 = \mu_2$). In nonparametric tests, the null hypothesis is that the population *distributions* are the same. Since distributions can differ in form, variability, central value, or all three, the interpretation of a rejection of the null hypothesis may not be quite so clearcut with a nonparametric test.

Recommendations on how to choose between a parametric and a nonparametric test have varied over the years. Two of the issues have been the scale of measurement and the power of the tests.

In the 1950s and after, some texts recommended that nonparametric tests be used if the scale of measurement was nominal or ordinal. After a period of controversy (see Chapter 2 of Kirk, 1972, or Gardner, 1975), this consideration was dropped. It recently resurfaced. (See the titles of the articles by Gaito, 1980, and Townsend and Ashby, 1984, to get the flavor.) Michell (1986), who describes the clash as one of underlying theories, may have an analysis that will lead to resolution.

The other issue, power, is also unresolved. Power, you may recall, comes up when the null hypothesis *should* be rejected. A test's power is a measure of how likely the test is to reject a false null hypothesis. It is clear to mathematical statisticians that if the populations that are being sampled are normally distributed and have equal variances, then parametric tests are more powerful than nonparametric ones. If the populations do not have these characteristics, recommendations are not so clear. In the past, parametric tests were considered robust and able to handle large departures from such population characteristics. Now this robustness is being called into question. For example, Blair, Higgins, and Smitley (1980) show that a nonparametric test (Mann-Whitney *U*) is generally more powerful than its parametric counterpart (*t* test) for the nonnormal distribution they tested. More recently, Blair and Higgins (1985) arrived at a similar conclusion when they compared the Wilcoxon matched-pairs signed-ranks test to the *t* test. We are sorry to leave these issues without giving you more specific advice, but no simple rule of thumb about superiority is possible except for one: if the data are ranks, use a nonparametric test.

Finally, there is not even a satisfactory name for these tests. Besides nonparametric, they are also referred to as **distribution-free statistics.** Although nonparametric and distribution-free mean different things to statisticians, the two words are used almost interchangeably by research workers. Ury (1967) suggested a third term, *assumption-freer* tests, which conveys the fact that these tests have fewer restrictive assumptions than parametric tests. Some texts have adopted Ury's term (for example, Kirk, 1984). We will use the term *nonparametric tests* and examine them in the order given in Table 12.1.

THE MANN-WHITNEY *U* TEST

The **Mann-Whitney *U* test** is used to determine whether two sets of data based on two *independent* samples came from the same population. Thus, it is the appropriate test for the child psychologists to use to test the difference in ranks of tennis players. The Mann-Whitney *U* test is identical to the **Wilcoxon rank-sum test,** which is not covered in this book. Wilcoxon published his test first (1945). However, when Mann and Whitney (1947) independently published a test based on the same logic, they provided tables *and a name* for the statistic (*U*). Currently, the Mann-Whitney *U* test appears to be referred to more often than the Wilcoxon rank-sum test.

The Mann-Whitney *U* test produces a statistic, *U*, that is evaluated by consulting the sampling distribution of *U*. Like all the distributions you have encountered that can be used in the analysis of small samples, the sampling distribution of *U* depends on sample size. Table H in Appendix C, which you will learn to use later, gives you critical values of *U*. Use Table H if neither of your two samples is as large as 20.

When the number of scores in one of the samples is greater than 20, the statistic U is distributed approximately as a normal curve. In this case, a z score is calculated and familiar values like 1.96 and 2.58 are used as critical values for $\alpha = .05$ and $\alpha = .01$.

Mann-Whitney U Test for Small Samples

To give us some data to illustrate the Mann-Whitney U test, we invented some information about the intensive early training of the top ten male singles tennis players (**Table 12.2**).

There are two groups: one had intensive early training ($N_{yes} = N_1 = 4$), and a second did not ($N_{no} = N_2 = 6$).

The sum of the ranks for each group is shown at the bottom of the table. A U value can be calculated for each group and then the smaller of the two is used to enter Table H. For the yes group, the U value is

$$U = (N_1)(N_2) + \frac{N_1(N_1 + 1)}{2} - \Sigma R_1$$

$$= (4)(6) + \frac{(4)(5)}{2} - 27 = 7$$

For the no group, the U value is

$$U = (N_1)(N_2) + \frac{N_2(N_2 + 1)}{2} - \Sigma R_2$$

$$= (4)(6) + \frac{(6)(7)}{2} - 28 = 17$$

A convenient way to check your calculation of U values is to know that the sum of the two U values is equal to $(N_1)(N_2)$. For our example, $7 + 17 = 24 = (4)(6)$.

Now, please examine **Table H.** It appears on two pages. To enter the table use the values for N_1 (to find the correct column) and N_2 (to find the correct row). Each

Table 12.2 Early training of top ten male singles tennis players

Initials	Rank	Intensive early training
Y.O.	1	No
U.E.	2	No
X.P.	3	No
E.C.	4	Yes
T.E.	5	No
D.H.	6	Yes
U.M.	7	No
O.R.	8	Yes
H.E.	9	Yes
R.E.	10	No

$\Sigma R_{yes} = 4 + 6 + 8 + 9 = 27$

$\Sigma R_{no} = 1 + 2 + 3 + 5 + 7 + 10 = 28$

intersection of N_1 and N_2 has two critical values. On the first page of the table the Roman type gives the critical values for α levels of .01 (one-tailed test) and .02 (two-tailed test). The numbers in boldface type are critical values for $\alpha = .005$ (one-tailed test) and $\alpha = .01$ (two-tailed test). In a similar way the second page handles larger α values for both one- and two-tailed tests. Now we are almost ready to enter Table H to find a probability figure for the smaller U value of 7.

From the conversation of the two child psychologists, it is clear that a two-tailed test is appropriate; they would be interested in knowing whether intensive, early training *helps* or *hinders* players. Since an α level was not discussed, we will do what they would do—see if the difference is significant at the .05 level, and if it is, see if it is also significant at some smaller α level. Thus, in Table H we will begin by looking for the critical value of U for a two-tailed test with $\alpha = .05$. This number is on the second page in boldface type at the intersection of $N_1 = 4$, $N_2 = 6$. The critical value is 2. Since our obtained value of U is 7, we must *retain* the null hypothesis and conclude that there is no evidence from our sample that the distribution of players trained early and intensively is significantly different from the distribution of those without such training.

Note that in Table H you reject H_0 when the obtained U value is *smaller than* the tabled critical value.

Although you can easily find a U value using the preceding method and quickly go to Table H and reject or retain the null hypothesis, it would help your understanding of this test to think about small values of U. Under what conditions would you get a small U value? What kind of samples would give you a U value of zero? By examining the formula for U, you can see that $U = 0$ when the members of one sample all rank lower than every member of the other sample. Under such conditions, rejecting the null hypothesis seems reasonable. By playing with numbers in this manner, you can move from the rote memory level to the understanding level.

Assigning Ranks and Tied Scores

Sometimes a researcher chooses a nonparametric test for data that are not already in the form of ranks.[4] In such cases, you will have to rank the scores. Two questions often arise. Is the largest or the smallest score ranked 1, and what should I do about the ranks for scores that are tied?

You will find the answer to the first question to be very satisfactory. It doesn't make any difference whether you call the largest or the smallest score 1.

Ties are handled by giving all tied scores the same rank. This rank is the mean of the ranks the tied scores would have if no ties had occurred. For example, if a distribution of scores was 12, 13, 13, 15, and 18, the corresponding ranks would be 1, 2.5, 2.5, 4, 5. The two scores of 13 would have been 2 and 3 if they had not been tied, and 2.5 is the mean of 2 and 3. As another example, the scores 23, 25, 26, 26, 26, 29 would have ranks of 1, 2, 4, 4, 4, 6. Ranks of 3, 4, and 5 average out to be 4.

Ties do not affect the value of U if they are in the same group. If there are several ties that involve both groups, a correction factor may be advisable.[5]

---

[4] Severe skewness or populations with very unequal variances are often reasons for such a choice.
[5] See Kirk (1984, p. 405) for the correction factor.

Mann-Whitney U Test for Larger Samples

When one sample size is 21 or more, the normal curve should be used to assess probability. The z value is obtained by the formula

$$z = \frac{(U + c) - \mu_U}{\sigma_U}$$

where $c = 0.5$, a correction factor explained below

$$\mu_U = \frac{(N_1)(N_2)}{2}$$

$$\sigma_U = \sqrt{\frac{(N_1)(N_2)(N_1 + N_2 + 1)}{12}}$$

Table 12.3   Number of items recalled by males and females, ranks, and a Mann-Whitney U test

Males ($N = 24$)		Females ($N = 17$)	
Items recalled	*Rank*	*Items recalled*	*Rank*
70	3	85	1
51	6	72	2
40	9	65	4
29	13	52	5
24	15	50	7
21	16.5	43	8
20	18.5	37	10
20	18.5	31	11
17	21	30	12
16	22	27	14
15	23	21	16.5
14	24.5	19	20
13	26.5	14	24.5
13	26.5	12	28.5
11	30.5	12	28.5
11	30.5	10	33
10	33	10	33
9	35.5		$\Sigma R_2 = 258$
9	35.5		
8	37.5		
8	37.5		
7	39		
6	40		
3	41		
	$\Sigma R_1 = 603$		

$$U = (N_1)(N_2) + \frac{N_1(N_1 + 1)}{2} - \Sigma R_1 = (24)(17) + \frac{(24)(25)}{2} - 603 = 105$$

$$\mu_U = \frac{(N_1)(N_2)}{2} = \frac{(24)(17)}{2} = 204$$

$$\sigma_U = \sqrt{\frac{(N_1)(N_2)(N_1 + N_2 + 1)}{12}} = \sqrt{\frac{(24)(17)(42)}{12}} = 37.79$$

$$z = \frac{(U + c) - \mu_U}{\sigma_U} = \frac{105 + 0.5 - 204}{37.79} = -2.61$$

Here c is a correction for continuity. It is used because the normal curve is a continuous function but the values of z obtained in this test are discrete. U, as before, is the smaller of the two possible U values.

Once again, the formula is familiar: the difference between a statistic based on data $(U + c)$ and the expected value of a parameter (μ_U) divided by a measure of variability. After you have obtained z, the decision rules are the ones you have used in the past. For a two-tailed test, reject H_0 if $z \geq 1.96$ ($\alpha = .05$). For a one-tailed test, reject H_0 if $z \geq 1.65$ ($\alpha = .05$). The corresponding values for $\alpha = .01$ are $z \geq 2.58$ and $z \geq 2.33$.

Here is a problem for which the normal curve is necessary. An undergraduate psychology major was devoting a year to the study of memory. The principal independent variable was gender. Among her several experiments was one in which she asked the students in a General Psychology class to write down everything they remembered that was unique to the previous day's class, during which a guest had lectured. Students were encouraged to write down every detail they remembered. This class was routinely videotaped so it was easy to check each recollection for accuracy and uniqueness. Because the samples indicated that the population data were severely skewed, she chose a nonparametric test. (A plot of the scores in Table 12.3 will show this skew.)

The scores, their ranks, and the statistical analysis are presented in **Table 12.3.**

The z score of -2.61 led to rejection of the null hypothesis, so the psychology major returned to the original data in order to interpret the results. Since the mean rank of the females, 15 ($258 \div 17$), is higher than that of the males, 25 ($603 \div 24$), and since higher ranks (those closer to 1) mean more recollections, she concluded that females recalled significantly more items than the males did.

Her conclusion is one that singles out central value for emphasis. On the average, females did better than males. The Mann-Whitney test, however, compares distributions. What our undergraduate has done is what most researchers who use the Mann-Whitney do: she has assumed that the two populations have the same form but differ in central value. Thus, when a significant U value is found, it is common to attribute it to a difference in central value.

ERROR DETECTION

> Here are two checks you can easily make. First, the last rank will be the sum of the two N's. In Table 12.3, $N_1 + N_2 = 41$, which is the lowest rank. Second, when ΣR_1 and ΣR_2 are added together, they will equal $N(N + 1)/2$, where N is the total number of scores. In Table 12.3, $603 + 258 = 861 = (41)(42)/2$.

Now you can see how we figured the expected sum of 105 in the section on the rationale of nonparametric tests. There were 20 scores, so the overall sum of the ranks is

$$\frac{(20)(21)}{2} = 210$$

Half of this total should be found in each group, so the *expected* sum of ranks of each group, both of which came from the same population, is 105.

PROBLEMS

1. In an experiment to determine the effects of estrogen on dominance, seven rats were given injections of that female hormone. The control group ($N = 8$) was injected with sesame oil, which is inert. Each rat was paired once with every other in a narrow runway. The rat that pushed the other one out was considered the more dominant of the two (see Work and Rogers, 1972). The number of bouts each rat won is listed here. Those with asterisks were injected with estrogen. Analyze the results with a Mann-Whitney test and write a conclusion about the effects of estrogen.

   14, 13, 12, 12, *11, 9, 8, *7, *6, 4, 3, *2, *2, *1, *0

2. Grackles, commonly referred to as blackbirds, are hosts for a number of parasites. One variety of parasite is a thin worm that lives in the tissue around the brain. To see if the incidence of this parasite was changing, four and twenty blackbirds were captured one winter from a pine thicket. The number of brain parasites was recorded for each bird. These data were compared with the infestation of 16 birds that had been captured from the same pine thicket 10 years earlier. Analyze the data with a Mann-Whitney test and write a conclusion. Be careful and systematic in assigning ranks. Errors here are quite frustrating.

   PRESENT DAY
   20   16   12   11   51   8   23   68   23   44   0   78
    0   28   53   20   44   20   36   32   64   16   101   0

   TEN YEARS EARLIER
    16   19   43   90   16   72   29   62
   103   39   70   29   110   32   87   57

3. A friend of yours is trying to convince a mutual friend that the ride in an automobile built by (Ford, Chrysler [you choose]) is quieter than the ride in an automobile built by (Chrysler, Ford [no choice this time; you had just one degree of freedom—once the first choice was made, the second was determined]). This friend has arranged to borrow six fairly new cars—three made by each company—and to drive your mutual friend around in the six cars (labeled $A-F$) blindfolded until a stable ranking for quietness is achieved. So convinced is your friend that he insists on adopting $\alpha = .01$, "so that only the recalcitrant will not be convinced."

   Since you are the statistician in the group, you decide to do a Mann-Whitney U test on the results. However, being the type who thinks through the statistics *before* gathering any experimental data, you look at the appropriate subtable in Table H and find that the experiment, as designed, is doomed to retain the null hypothesis. Write an explanation for your friend, explaining why the experiment must be redesigned.

4. Suppose your friend talked three more persons out of their cars for an afternoon, labeled the cars $A-I$, changed α to .05, and conducted his "quiet" test. The results are shown in the table. Perform a Mann-Whitney U test.

Car	Company Y ranks	Car	Company Z ranks
B	1	H	3
I	2	A	5
F	4	G	7
C	6	E	8
		D	9

THE WILCOXON MATCHED-PAIRS SIGNED-RANKS TEST

The **Wilcoxon matched-pairs signed-ranks test** (1945) is appropriate for testing the difference between two correlated samples. In Chapter 8, we discussed three kinds of correlated-samples designs: natural pairs, matched pairs, and repeated measures (before and after). In each of these designs, a score in one group is logically paired with a score in the other group. If you are not sure of your understanding of the difference between a correlated-samples and an independent-samples design, you should review that section in Chapter 8. Being sure of the difference is necessary in order to make a proper choice between a Mann-Whitney U test and a Wilcoxon matched-pairs signed-ranks test.

The result of a Wilcoxon matched-pairs signed-ranks test is a T value,[6] which is interpreted using Table J. We will illustrate the rationale and calculation of T using the four pairs of scores in **Table 12.4.**

First, before any ranking takes place, the difference, D, between each pair of scores is found. The *absolute values* of these differences are then ranked, with the smallest difference given the rank of 1, the next smallest a rank of 2, and so on. The original sign of the difference is then given to the rank, and the positive ranks and the negative ranks are each summed separately. T is the *smaller of the absolute values of the two sums.*[7] For Table 12.4, $T = 4$. Note for this test it is the *differences* that are ranked, and not the scores themselves. Also note that the *smallest difference* is given a rank of 1.

The rationale is that, *if* there is no true difference between the two groups, the absolute value of the negative sum should be equal to the positive sum, with any deviations being due to sampling fluctuations.

Table J shows the critical values by sample size for both one- and two-tailed tests. Reject H_0 when T is equal to or *smaller* than the critical value in the table.[8]

Table 12.4   Illustration of how to calculate T

Pair	Variable 1	Variable 2	D	Rank	Signed rank
A	16	24	−8	3	−3
B	14	17	−3	1	−1
C	23	18	5	2	2
D	23	9	14	4	4

$$\Sigma(\text{positive ranks}) = 6$$
$$\Sigma(\text{negative ranks}) = -4$$
$$T = 4$$

[6] Be alert when you see a capital T in your outside readings; it has uses other than to symbolize the Wilcoxon matched-pairs signed-ranks test. Also note that this T is capitalized, whereas the t in the t test and t distribution is not capitalized except by some computer printers that do not have lowercase letters.

[7] The Wilcoxon test is like the Mann-Whitney test in that you have a choice of two values for your test statistic. For both tests, choose the smaller value.

[8] Like the Mann-Whitney test, your obtained statistic (T) must be smaller than the tabled critical value if you are to reject H_0.

Table 12.5 Data and a Wilcoxon matched-pairs signed-ranks analysis of the effects on perception of others' judgments

Subject	Mean movement (inches)		D	Signed ranks
	Before	*After*		
1	3.7	2.1	1.6	4.5
2	12.0	7.3	4.7	10
3	6.9	5.0	1.9	6
4	2.0	2.6	−0.6	−3
5	17.6	16.0	1.6	4.5
6	9.4	6.3	3.1	8
7	1.1	1.1	0.0	Eliminated
8	15.5	11.4	4.1	9
9	9.7	9.3	0.4	2
10	20.3	11.2	9.1	11
11	7.1	5.0	2.1	7
12	2.2	2.3	−0.1	−1

$$\Sigma \text{ positive} = 62$$
$$\Sigma \text{ negative} = -4$$
$$T = 4$$
$$N = 11$$

Check: $62 + 4 = 66$ and $\dfrac{11(12)}{2} = 66$

We will illustrate the calculation and interpretation of a Wilcoxon matched-pairs signed-ranks test with an experiment based on some early work of Muzafer Sherif (1935). Sherif was interested in whether a person's basic perception could be influenced by others. The basic perception he used was a judgment of the size of the *autokinetic effect*. The autokinetic effect is obtained when a person views a stationary point of light in an otherwise dark room. After a few moments, the light appears to move erratically. Sherif asked his subjects to judge how many inches the light moved. Under such conditions, judgments differ widely between individuals but they are fairly consistent for each individual. After establishing a stable mean for each subject, other observers were brought into the room. These new observers were confederates of the experimenter who always judged the movement of the light to be somewhat less than the subject did. Finally, the confederates left and the subject again made judgments until a stable mean was achieved. The before and after scores and the Wilcoxon matched-pairs signed-ranks test are shown in **Table 12.5**. The *D* column is simply the pretest minus the post-test. These *D* scores are then ranked by absolute size and the sign of the difference attached in the Signed Ranks column. Notice that when $D = 0$, that pair of scores is dropped from further analysis and *N* is reduced by one. The negative ranks have the smaller sum, so $T = 4$.

Since this *T* is smaller than the *T* value of 5 shown in Table J under $\alpha = .01$ (two-tailed test) for $N = 11$, the null hypothesis is rejected. The after scores represent a distribution that is different from the before scores. Now let's interpret this in terms of the experiment.

By examining the *D* column, you can see that all scores except two are positive. This means that, after hearing others give judgments smaller than one's own, the amount of movement seen was less. Thus, you may conclude (as did Sherif) that even basic perceptions tend to conform to perceptions expressed by others.

Tied Scores and *D* = 0

Ties among the *D* scores are handled in the usual fashion of assigning to each tied score the mean of the ranks that would have been assigned if there had been no ties. Ties do not affect the probability of the rank sum unless they are numerous (10 percent or more of the ranks are tied). In the case of numerous ties, the probabilities in Table J associated with a given critical *T* value may be too large. In a situation with numerous ties, the test is described as too conservative because it may fail to ascribe significance to differences that are in fact significant (Wilcoxon and Wilcox, 1964).

As you already know, when *one* of the *D* scores is zero, it is not assigned a rank and *N* is reduced by one. When *two* of the *D* scores are tied at zero, each is given the average rank of 1.5. Each is kept in the computation, with one being assigned a plus sign and the other a minus sign. If *three D* scores are zero, one is dropped, *N* is reduced by one, and the remaining two are given signed ranks of $+1.5$ and -1.5. You can generalize from these three cases to situations with four, five, or more zeros.

Wilcoxon Matched-Pairs Signed-Ranks Test for Large Samples

When the number of pairs exceeds 50, the *T* statistic may be evaluated using the normal curve. The test statistic is

$$z = \frac{(T + c) - \mu_T}{\sigma_T}$$

where T = smaller sum of the signed ranks
   $c = 0.5$
   $\mu_T = \dfrac{N(N + 1)}{4}$
   $\sigma_T = \sqrt{\dfrac{N(N + 1)(2N + 1)}{24}}$
   N = number of pairs

PROBLEMS

5. A private consultant was asked to evaluate a government job-retraining program. As part of the evaluation, she gathered information on 112 individuals' income before retraining and after retraining. She found a *T* value of 4077. Complete the analysis and draw a conclusion. Be especially careful in wording your conclusion.

6. Six industrial workers were chosen for a study of the effects of rest periods on production. Output was measured for 1 week before the new rest periods were instituted and again

during the first week of the new schedule. Perform an appropriate nonparametric statistical test on the results shown in the table.

Worker	Without rests	With rests
1	2240	2421
2	2069	2260
3	2132	2333
4	2095	2314
5	2162	2297
6	2203	2389

7. A political science student was interested in the differences between two sets of Americans in attitudes toward government regulation of business. One set of Americans was made up of Canadians and the other of people from the United States. At a model United Nations session, the student managed to get 30 participants to fill out his questionnaire. High scores on this questionnaire indicated a favorable attitude toward government regulation of business. Analyze the data in the table with the appropriate nonparametric test and write a conclusion.

Canadians	United States	Canadians	United States
12	16	27	26
39	19	33	15
34	6	34	14
29	14	18	25
7	20	31	21
10	13	17	30
17	28	8	3
5	9		

8. On the first day of class, one professor always gave his General Psychology class a beginning-point test. The purpose was to find out the students' beliefs about punishment, reward, mental breakdowns, and so forth—topics that would be covered during the course. Many of the items were phrased so that they represented a common misconception. For example, "mental breakdowns run in families and are usually caused by defective genes" was one item on the test. At the end of the course, the same test was given again. High scores mean lots of misconceptions. Analyze the before and after data and write a conclusion about the effect the General Psychology course had on misconceptions.

Student	Before	After
1	18	4
2	14	14
3	20	10
4	6	9
5	15	10
6	17	5
7	29	16
8	5	4
9	8	8
10	10	4
11	26	15
12	17	9
13	14	10
14	12	12

9. A health specialist conducted an 8-week workshop on weight control during which all 17 of the people who completed the course lost weight. In order to assess the long-term effects of the workshop, she weighed the participants again 10 months later. The weight lost or gained is shown here with a positive sign for those who continued to lose weight and a negative sign for those who gained some back. What can you conclude about the long-term effects of the workshop?

```
 -5    24  0  13    9    6  -7  -3  2
-10  -16  7  12  -19  -4   8  15
```

THE WILCOXON-WILCOX MULTIPLE-COMPARISONS TEST

So far in this chapter on the analysis of ranked data, we have covered both designs for the two-group case (independent and correlated samples). The next step is to analyze results from three or more groups. The method presented here is one that allows you to compare all possible *pairs* of groups, regardless of the number of groups in the experiment. This is the nonparametric equivalent of a one-way ANOVA followed by Tukey's HSD tests. A direct analogue of the overall *F* test is the Kruskal-Wallis one-way ANOVA on ranks, which is explained in many elementary statistics texts.

The **Wilcoxon-Wilcox multiple-comparisons test** (1964) is a method that allows you to compare all possible pairs of treatments. This is like running several Mann-Whitney tests, with a test for each pair of treatments. However, the Wilcoxon-Wilcox multiple-comparisons test keeps your α level at .05 or .01 regardless of how many pairs you have. The test is an extension of the procedures in the Mann-Whitney *U* test and, like it, requires independent samples. (Remember that Wilcoxon devised a test identical to the Mann-Whitney *U* test.)

The Wilcoxon-Wilcox method requires you to order the scores from the *K* samples into one overall ranking. Then the sum of the ranks in each group is computed. The rationale is that these sums should all be equal and that large differences in sums must reflect samples from different populations. Of course, the larger *K* is, the greater the likelihood of large differences by chance alone, and this is taken into account in the table of critical values, Table K.

The Wilcoxon-Wilcox test can be used only when the *N*'s for all groups are equal. A common solution to the problem of unequal *N*'s is to reduce the too-large group(s) by throwing out one or more randomly chosen scores. A better solution is to design the experiment so that you have equal *N*'s.

The data in **Table 12.6** represent the results of an experiment conducted on a solar collector by two designer-entrepreneurs. These two had designed and built a 4-foot by 8-foot solar collector they planned to market, and they wanted to know the optimal rate at which to pump water through the collector. Since the rule of thumb for this is one-half gallon per hour per square foot of collector, they chose values of 14, 15, 16, and 17 gallons per hour for their experiment. Starting with the reservoir full of ice water, the water was pumped for 1 hour through the collector and back to the reservoir. At the end of the hour, the temperature of the water in the reservoir was measured in degrees Celsius. Then the water was replaced with ice water, the

Table 12.6 Experiment on flow rates in a solar collector

				Flow rate (gallons/hour)				
14	Rank	15	Rank	16	Rank	17	Rank
28.1	7	28.9	3	25.1	14	24.7	15
28.8	4	27.7	8.5	25.3	13	23.5	18
29.4	1	27.7	8.5	23.7	17	22.6	19
29.0	2	28.6	5	25.9	11	21.7	20
28.3	6	26.0	10	24.2	16	25.8	12
Σ (ranks)	20		35		71		84

Check: $20 + 35 + 71 + 84 = 210$

$$\frac{N(N + 1)}{2} = \frac{20(21)}{2} = 210$$

flow rate was changed, and the process was repeated. The numbers in the body of Table 12.6 are the temperature measurements (to the nearest tenth of a degree).

There are six ways to make pairs of the four groups. The rate of 14 gallons per hour can be paired with 15, 16, and 17; the rate of 15 can be paired with 16 and 17; and the rate of 16 can be paired with 17. For each pair, a difference in the sum of ranks is found and the absolute value of that difference is compared with the critical value in Table K to see if it is significant.

Table K appears in two parts—one for the .05 level and one for the .01 level. In both cases, critical values are given for a two-tailed test. In the case of the data in Table 12.6, where $K = 4$, $N = 5$, you will find in Table K that rank-sum differences of 48.1 and 58.2 are required to reject H_0 at the .05 and .01 levels, respectively.

A convenient summary table for the Wilcoxon-Wilcox multiple-comparisons test is shown in **Table 12.7.** The table displays, for each pair of independent-variable values, the difference in the sum of the ranks. The next task is simply to see if any of these obtained differences is *larger* than the critical values of 48.1 and 58.2. At the .05 level, rates of 14 and 16 are significantly different from each other, as are 15 and 17. In addition, a rate of 14 is significantly different from a rate of 17 at the .01 level. What does all this mean for our two designer-entrepreneurs? Let's listen to their explanation to their old statistics professor.

Table 12.7 Summary table for differences in the sums of ranks in the flow-rate experiment

		Flow rates		
	ΣR	14 (20)	15 (35)	16 (71)
Flow rates	15(35)	15		
	16(71)	51*	36	
	17(84)	64**	49*	13

*$p < .05$
**$p < .01$

"How did the flow-rate experiment come out, fellows?" inquired the kindly old gentleman.

"OK, but we are going to have to do a followup experiment using different flow rates. We know that 16 and 17 gallons per hour are not as good as 14, but we don't know if 14 is optimal for our design. Fourteen was the best of the rates we tested, though. On our next experiment, we are going to test rates of 12, 13, 14, and 15."

The professor stroked his beard and nodded thoughtfully. "Typical experiment. You know more after it than you did before, . . . but not yet enough."

PROBLEMS

10. Given the summary data in the table, test all possible comparisons. Each group had a sample of eight subjects.

	Sum of ranks for six groups					
	1	2	3	4	5	6
ΣR	196	281	227	214	93	165

11. The effect of three types of leadership on group productivity and satisfaction was investigated.[9] Groups of five children were randomly constituted and assigned an authoritarian, a democratic, or a laissez-faire leader. Nine groups were formed—three with each type of leader. The groups worked for a week on various projects. During this time, four different measures of each child's personal satisfaction were taken. These were pooled to give one score for each child. The data are presented in the table. Test all possible comparisons with a Wilcoxon-Wilcox test.

	Leaders	
Authoritarian	*Democratic*	*Laissez-faire*
120	108	100
102	156	69
141	92	103
90	132	76
130	161	99
153	90	126
77	105	79
97	125	114
135	146	141
121	131	82
100	107	84
147	118	120
137	110	101
128	132	62
86	100	50

12. List the tests presented so far in this chapter and the design for which each is appropriate. Be sure you can do this from memory.

[9] For a summary of a similar investigation, see Lewin (1958).

CORRELATION OF RANKED DATA

We will begin this section with a short review of what you learned about correlation in Chapter 4.

1. Correlation requires a bivariate distribution (a logical pairing of scores).
2. Correlation is a method of describing the degree of relationship between two variables—that is, the degree to which high scores on one variable are associated with low or high scores on the other variable.
3. Correlation coefficients range in value from +1.00 (perfect positive) to −1.00 (perfect negative). A value of .00 indicates that there is no relationship between the two variables.
4. Statements about causal relations may not be made on the basis of a correlation coefficient alone.

In 1901, Charles Spearman (1863–1945) was "inspired by a book by Galton" and began experimenting at a little village school. He wanted to see if there was a relationship between intellect (school grades) and sensory ability (detecting musical discord). He thought there was a relationship and he wanted to determine its *degree*. So he developed a coefficient to express this degree. Later he found that others were already ahead of him in developing a coefficient (Spearman, 1930).

Spearman's name is attached to the coefficient that is used to show the degree of correlation between two sets of *ranked* data. He used the Greek letter ρ (rho) as the symbol for his coefficient. Later statisticians began to reserve Greek letters to indicate parameters, so the modern symbol for Spearman's statistic has become r_s, the *s* honoring Spearman.

The statistic r_s is a special case of the Pearson product-moment correlation coefficient and is most often used when the number of pairs of scores is small (less than 20). Actually, r_s is a *descriptive statistic* that could have been introduced in the first part of this book. We waited until now to introduce it because r_s is a rank-order statistic, and this is a chapter about ranks. We will show you how to calculate this descriptive statistic and then how the statistical significance of r_s can be determined.

Calculation of r_s

The formula for r_s is

$$r_s = 1 - \frac{6\Sigma D^2}{N(N^2 - 1)}$$

where D = difference in ranks of a pair of scores
N = number of pairs of scores

We started this chapter with speculation about male tennis players; we will end it with data about female tennis players. Suppose you were interested in the relationship between age and rank among professional women tennis players. The statistic r_s will give you a numerical index of the degree of the relationship. A high positive r_s would mean that, the older the player, the higher her rank. A high negative r_s would mean that, the older the player, the lower her rank. A zero or near zero r_s would indicate that there is no relationship between age and rank.

Table 12.8 The top ten women tennis players in 1986: their rank in age, and the calculation of Spearman's r_s

Player	Rank in tennis	Rank in age	D	D^2
Navratilova	1	2	-1	1
Lloyd	2	1	1	1
Graf	3	9	-6	36
Mandlikova	4	3	1	1
Sukova	5	7	-2	4
Shriver	6	4	2	4
Sabatini	7	10	-3	9
M. Maleeva	8	8	0	0*
Kohde Kilch	9	5	4	16
McNeil	10	6	4	16
				$\Sigma = 88$

$$r_s = 1 - \frac{6\Sigma D^2}{N(N^2 - 1)} = 1 - \frac{6(88)}{10(99)} = .47$$

*Note that N is *not* reduced for r_s when a difference is zero.

Table 12.8 shows the ten top-ranked women tennis players for 1986, their age as a rank score among the ten, and an $r_s = .47$. Now you can ask how much it means. Can you say with confidence to a friend, "There is a distinct tendency for older women to rank higher in tennis"? Or, perhaps an $r_s = .47$ is not trustworthy and reliable. Phrasing the question in statistical language, Is it likely that such an r_s would come from a population in which the true correlation is zero? You will recall that you answered this kind of question for a Pearson r at the end of Chapter 8.

Testing the Significance of r_s

Table L in Appendix C gives values of r_s that are significant at the .05 and .01 levels when the number of pairs is 30 or less. The tennis data in Table 12.8 produced $r_s = .47$ based on ten pairs. **Table L** shows that a correlation of .65 (either positive or negative) is required for significance at the .05 level. Thus, a correlation of .47 is not statistically significant.

Notice in Table L that rather large correlations are required for significance. As with r, not much confidence can be placed in low or moderate correlation coefficients that are based on only a few pairs of scores.

For samples larger than 30, you may test the significance of r_s by comparing it to the tabled values in Table A, just as you did for a Pearson r. Note, however, that **Table A** requires df, which for r_s, as with r, is the number of pairs minus two ($N - 2$).

Tied Ranks

The formula for r_s that you are working with is not designed to handle ties. With r_s ties are troublesome. A tedious procedure has been devised to overcome ties, but your best solution is to arrange your data collection so there are no ties. Sometimes, however, you are stuck with tied ranks, perhaps as a result of working with someone else's data. Kirk (1984) recommends assigning average ranks to ties, as you did for the three other procedures in this chapter, and then computing a Pearson r on the data.

PROBLEMS

13. Calculate r_s for the hypothetical data in the table. Determine whether it is significant at the .001 level of significance.

Years	Number of marriages ($\times 10,000$)	Total grain crop ($ billions)
1918	131	1.7
1919	142	2.5
1920	125	1.4
1921	129	1.7
1922	145	2.8
1923	151	3.3
1924	142	2.6
1925	160	3.7
1926	157	3.4
1927	163	3.5
1928	141	2.6
1929	138	2.6
1930	173	3.6
1931	166	3.9

14. With $N = 16$ and $\Sigma D^2 = 308$, calculate r_s and test its significance at the .05 level.

15. Two members of the Department of Philosophy (Locke and Kant) had the principal responsibility for hiring a new professor. Each privately ranked from 1 to 10 the ten candidates who met the objective requirements (degree, specialty, and so forth). The rankings are shown in the table. Calculate an r_s. If the correlation is low, the philosophers probably have different sets of criteria of what is important. If so, they should discuss the two different sets. On the other hand, a high correlation probably means they have about the same set of criteria.

Candidates	Professors Locke	Professors Kant
A	7	8
B	10	10
C	3	5
D	9	9
E	1	1
F	8	7
G	5	3
H	2	4
I	6	6
J	4	2

16. You are once again asked to give advice to a friend who comes to you for criticism of an experimental design. This friend has four pairs of scores obtained randomly from a population. Her intention is to calculate a correlation coefficient r_s and decide whether there is any significant correlation in the population. Give advice.

17. For each situation described, tell whether you would use a Mann-Whitney test, a Wilcoxon matched-pairs signed-ranks test, a Wilcoxon-Wilcox multiple-comparisons test, or an r_s.

a. A limnologist (a scientist who studies freshwater streams and lakes) measured algae growth in a lake before and after the construction of a nuclear reactor to see what effect the reactor had.

b. An educational psychologist compared the sociability scores of firstborn children with scores of their next-born brother or sister to see if the two groups differed in sociability.

c. A child psychologist wanted to determine the degree of relationship between IQ scores obtained at age 3 and scores obtained from the same individuals at age 12.

d. A nutritionist randomly and evenly divided boxes of cornflakes cereal into three groups. One group was stored at 40° F, one group at 80° F, and one group alternated from day to day between the two temperatures. After 30 days, the vitamin content of each box was assayed.

e. The effect of STP gas treatment on gasoline mileage was assessed by driving six cars over a 10-mile course, adding STP, and then again driving over the same 10-mile course.

18. Fill in the descriptions for the following table.

	Symbol of statistic (if any)	Appropriate for what design?	Calculated statistic must be (larger, smaller) than the tabled statistic to reject H_0
Mann-Whitney U test			
Wilcoxon matched-pairs signed-ranks test			
Wilcoxon-Wilcox multiple-comparisons test			

OUR FINAL WORD

As we said in Chapter 1, the basic essence of applied statistics is to let numbers stand for things of interest, manipulate the numbers according to the rules of statistics, translate the numbers back into the things, and finally describe the relationship between the things of interest. The question, What things are better understood when translated into numbers? was not raised in this book, but it is important nonetheless. Here is an anecdote by E. F. Schumacher (1979) that helps emphasize that importance.

I will tell you a moment in my life when I almost missed learning something. It was during the war and I was a farm laborer and my task was before breakfast to go to yonder hill and to a field there and count the cattle. I went and I counted the cattle—there were always thirty-two—and then I went back to the bailiff, touched my cap, and said, "Thirty-two, sir," and went and had my breakfast. One day when I arrived at the field an old farmer was standing at the gate, and he said, "Young man, what do you do here every morning?" I said, "Nothing much. I just count the cattle." He shook his head and said, "If you count them every day they won't flourish." I went back, I reported thirty-two, and on the way back I thought, Well, after all, I am a professional statistician, this is only a country yokel, how stupid can he get. One day I went back, I counted and counted again, there were only thirty-one. Well, I didn't want to spend all day there so I went back and reported thirty-one. The bailiff was very angry. He said, "Have your breakfast and then we'll go up there together." And we went together and we searched the place and indeed, under a bush, was a dead beast. I thought to myself, Why have I been counting them all the time? I haven't prevented this beast dying. Perhaps that's what the farmer meant. They won't flourish if you don't look and watch the quality of each individual beast. Look him in the eye. Study the sheen on his coat. Then I might have gone back and said, "Well, I don't know how many I saw but one looks mimsey."

13 TALES OF DISTRIBUTIONS— PAST, PRESENT, AND FUTURE

In the preface we said that we thought you would like this book. We hope you did. Our expectation is that by using this book you have acquired a sound, basic knowledge of statistics. If so, you can use that knowledge to design, analyze, and evaluate experiments, understand statistical arguments, persuade others by using a statistical presentation, or do all of the above.

This chapter is designed with integration in mind. You will review the important things you have learned and read about statistics you are prepared to learn. Most important, there are 34 situations for you to think about and apply your knowledge to.

TALES OF DISTRIBUTIONS—PAST

If you have been reading for integration, the kind of reading in which you are continually asking the question, How does this stuff relate to what I learned before?, you may have clearly in your head a principal thread that ties this book together from Chapter 2 onward. The name of that thread is distributions. It is the importance of distributions that justifies the subtitle of this book, *Tales of Distributions*. We want to summarize explicitly some of what you have learned about distributions to help your integration of this material and to provide the groundwork that will allow us to describe some other statistical techniques that were not covered in this course but that you might encounter in future courses or in future encounters with statistics.

In Chapter 2, you were confronted with a disorganized array of scores on a variable. You learned to organize those scores into a frequency distribution, graph the distribution, and find the central values. In Chapter 3, you learned about the variability of distributions—especially the standard deviation—and how to express scores as z scores. In Chapter 4, where you were introduced to bivariate distributions, you learned to express the relationship between variables with a correlation coefficient and to predict scores on one variable from scores on another. In Chapter 5, you learned about the normal distribution and how to use means and standard deviations to find probabilities associated with distributions. In Chapter 6, you were introduced to the important concept of the sampling distribution. You learned there that distributions of statistics (like the mean) can be used to evaluate the likelihood of parameters. Chapter 7 presented you with the idea that a difference between means (the heart of an experiment) has a sampling distribution and that conclusions about experiments are made on the basis of such sampling distributions. In Chapter 8, you learned that some experiments require a new kind of sampling distribution (*t* distribution) in order to get accurate probability figures. In Chapter 9, you learned about still another sampling distribution (*F* distribution) that is used when experiments with two or more groups are analyzed. The next step (Chapter 10) was an even more complex experiment, one with *two* independent variables for which probability figures could be obtained with the *F* distribution. Chapter 11 taught you how to use the chi square distribution for data that consist of frequencies. In Chapter 12, you learned how to use distributions that do not require the assumptions that the populations of dependent-variable scores be normally distributed and have equal variances.

TALES OF DISTRIBUTIONS—FUTURE

There are several paths that might be taken from here. One more or less obvious next step is the analysis of an experiment with three independent variables. The three main effects and the several interactions are tested with an *F* distribution.

A second possibility is the analysis of experiments with correlated groups. Such experiments are often referred to as repeated-measures designs, and there are repeated-measures analogues for simple ANOVA designs and factorial ANOVA designs. Again, an *F* distribution is the appropriate sampling distribution.

A third possibility is to study techniques for the analysis of experiments that have more than one *dependent* variable. Such techniques are called multivariate statistics. Many of these techniques are analogous to those you have studied in this book, except there are two or more dependent variables instead of just one. For example, Hotelling's T^2 is analogous to the *t* test; one-way multiple analysis of variance (MANOVA) is analogous to one-way ANOVA; and higher-order MANOVA is analogous to factorial ANOVA. [For a good nontechnical introduction to many of the techniques of multivariate statistics, see Harris (1985, chap. 1) or Johnson and Wichern (1988, chap. 1).]

Finally, there is the task of combining the results of many *separate* studies that all deal with the same topic. It is not unusual to have conflicting results among 10 or 15 or 50 studies, and researchers have been at the task of synthesizing since the beginning. Recently the task has been facilitated by a set of statistical procedures called **meta-analysis.** For a short history and a description of meta-analytic techniques, see Bangert-Drowns (1986).

TALES OF DISTRIBUTIONS—PRESENT

An almost universal condition at the end of a first course in statistics is a feeling of "almost understanding." Most students feel that they can analyze data and interpret results of experiments using any of the techniques they have studied, but they lack a "feel" for the whole course. When faced with the description of an experiment, they often have difficulty deciding on the appropriate method of analysis.

Our experience in teaching statistics has taught us that the best way for you to bring together all the bits and pieces you have learned is to reread the whole book, giving particular emphasis to the parts that you recognize as especially important or that you have forgotten, and less emphasis to the parts that are less important or are well remembered. It will take a considerable amount of time for you to do this— probably 8 to 16 hours. However, the benefits are great, according to students who have invested the time. A typical quote is, "Well, during the course I would read a chapter and work the problems without too much trouble; but I didn't relate that to the earlier stuff. When I went back over the whole book, though, I got the big picture. All the pieces do fit together."

As a second exercise in getting the "big picture," we have posed some problems for you; the problems that follow are divided into two sets. The first set requires you to decide what statistical test or descriptive statistic will answer the question the investigator is asking. The second set requires you to interpret the results of an analyzed experiment. You will probably be glad to find out that no number crunching is required for these problems. You can put away your calculator.

Here is a word of warning. For a few of the problems, the statistical techniques necessary for analysis are not covered in this book. Thus, part of your task is to recognize what problems you are prepared to solve and what problems would require some digging into more advanced statistics textbooks.

PROBLEMS **Set A** Determine what descriptive statistic or inferential test is appropriate for each problem in this set.

1. A company had three separate divisions. Based on the capital invested, one year Division A made 10 cents per dollar, Division B made 20 cents per dollar, and Division C made 30 cents per dollar. How can the overall profit for the company be found?

2. Reaction time scores tend to be severely skewed. The majority of the scores are quick responses and there are diminishing numbers of scores in the slower categories. A student wanted to find out the effects of alcohol on reaction time, so he found the reaction time of each subject under both conditions—alcohol and no alcohol.

3. As part of a 2-week treatment for phobias, a therapist measured the general anxiety level of each client four times: before, after 1 week of treatment, at the end of treatment, and 4 months later.

4. "I want a number that will describe the typical score on this test. Most did very well, some scored in the middle, and a very few did quite poorly."

5. The designer for a city park had some benches built that were 16.7 inches high. The mean distance from the bottom of the shoe to the back of the knee is 18.1 inches for women. The standard deviation is 0.70 inch. What proportion of the women who use the bench will have their feet dangle (unless they sit forward on the bench)?

6. A promoter of low-cost, do-it-yourself buildings found the insulation scores (R values) for 12 samples of concrete mixed with clay and for 12 samples of concrete mixed with hay. Each sample was 6 inches thick. He wanted to know if clay or straw was superior as an insulating additive.

7. A large school district wanted to know the range of scores within which they could expect the mean reading achievement of their sixth-grade students to be. A random sample of 50 of these students and their scores on a standardized reading achievement test was available.

8. Based on several large studies, a researcher knew that 40 percent of the public in the United States favored capital punishment, 35 percent were against it, and 25 percent expressed no strong feelings. The researcher polled 150 Canadians on this question with the intention of comparing the two countries in attitudes toward capital punishment.

9. What descriptive statistic could be used to identify the most typical choice of major at your college?

10. An experimentalist wanted to express with a correlation coefficient the strength of the relationship between stimulus intensity and pleasantness. This experimentalist worked with music and she knew that both very low and very high intensities are not pleasant and that the middle range of intensity produces the highest pleasantness ratings. Would you recommend a Pearson r or an r_s?

11. A consumer-testing group compared Boraxo and Tide to determine which got laundry whiter. White towels that had been subjected to a variety of filthy treatments were identified on each end and were cut in half. Each was then washed in either Boraxo or Tide. After washing, each half was tested with a photometer for the amount of light reflected.

12. An investigator wanted to predict a male adult's height from his length at birth. He obtained records of both measures from a sample of male military personnel.

13. An experimenter was interested in the effect of expectations and drugs on alertness. Each subject was given an amphetamine (stimulant) or a placebo (an inert substance). In addition, half the subjects in each group were told that the drug taken was a stimulant, and half that the drug was a depressant. An alertness score for each subject was obtained from a composite of measures, which included a questionnaire and observation of the subjects.

14. A teacher told a class that she found a correlation coefficient of .20 between the alphabetical order of a person's last name and that person's cumulative point total for a general psychology course. One student in the class wondered whether the correlation coefficient was reliable; that is, would a nonzero correlation be likely to be found in other classes. How can a probability figure be found without calculating an r on another class?

15. As part of a test for some advertising copy, a company assembled 120 people who rated four toothpastes. They then read several ads (including the ad being tested). Then they rated all four toothpastes again. The data to be analyzed consisted of the number of people who rated Crest highest before reading the ad and the number who rated it highest after reading the ad.

16. Most light-bulb manufacturing companies claim that their 60-watt bulbs have an average life of 1000 hours. A skeptic with some skill as an electrician wired up some sockets and timers and burned 40 bulbs until they failed. The life of each bulb was recorded automatically. The skeptic wanted to know if the companies' claim was justified.

17. An investigator wanted to know the degree to which a person's education was related to his or her satisfaction in life. The investigator had a way to measure both of these variables.

18. In an effort to find out the effect of a "terrible drug" on reaction time, an investigator administered 0, 25, 50, or 75 milligrams to four groups of volunteer rats. The quickness of their response to a tipping platform was measured in seconds.

19. The experimenters (a male and a female) staged 140 "shoplifting" incidents in a grocery store. In each case, the "shoplifter" (one experimenter) picked up a carton of cigarettes in full view of a customer and then walked out. Half the time the experimenter was

well dressed, and half the time sloppily dressed. Half the incidents involved a male shoplifter and half involved a female. For each incident, the experimenters (with the cooperation of the checkout person) simply noted whether the shoplifter was reported or not.

20. A psychologist wanted to illustrate for a group of seventh-graders the fact that items in the middle of a series are more difficult to learn than those at either end. The students learned an eight-item series. The psychologist found the mean number of repetitions necessary to learn each of the items.

21. A nutritionist, with the help of an anthropologist, gathered data on the incidence of cancer among 21 different cultures. The incidence data were quite skewed, with many cultures having low incidences and the rest scattered among the higher incidences. In eight of the cultures, red meat was a significant portion of the diet. For the other 13 groups, the diet consisted primarily of cereals.

22. A military science teacher wanted to find if there was any relationship between the rank in class of graduates of West Point and their military rank in the service at age 45.

23. The boat dock operators at a lake sponsored a fishing contest. Prizes were to be awarded to the persons who caught the largest bass, the largest crappie, and the largest bream, and to the overall winner. The problem in deciding the overall winner is that the bass are by nature larger than the other two species of fish. Describe an objective way to find the overall winner that will be fair to all three kinds of contestants.

Set B Read each problem, look at the result of the statistical analysis, and write an appropriate conclusion. For each conclusion, use the terms used in the problem; that is, rather than saying "the null hypothesis was rejected at the .05 level," say "those with medium anxiety solved problems more quickly than those with high anxiety."

For some of the problems, the design of the study has flaws. There are uncontrolled extraneous variables that would make it impossible to draw precise conclusions about the experiment. It is a very good thing for you to be able to recognize such design flaws, but what we want you to do in these problems is to interpret the statistics. Thus, don't let design flaws keep you from drawing a conclusion based on the statistic.

For all problems, use a two-tailed test with $\alpha = .05$. If the results are significant at the .01 or .001 level, report that, but treat any difference that has a probability greater than .05 as being due to chance.

24. Four kinds of herbicides were compared for their weed-killing characteristics. Eighty plots were randomly but evenly divided and one herbicide was applied to each. Since the effects of two of the herbicides were quite variable, the dependent variable was reduced to ranks and a Wilcoxon-Wilcox multiple-comparisons test was run. The plot with the fewest weeds remaining was given a rank of 1. The summary table is shown here.

	A (936.5)	B (316.5)	C (1186)
B (316.5)	620		
C (1186)	249.5	869.5	
D (801)	135.5	482.5	385

25. In a particular recycling process, the break-even point for each batch was when 54 kilograms of the raw material was unusable foreign matter. An engineer developed a

screening process that left a mean of 38 kilograms of foreign material. The upper and lower limits of a 95 percent interval were 30 and 46 kilograms.

26. The interference theory of the serial position effect predicts that subjects who are given extended practice will perform more poorly on the initial items of a series than will subjects who are not given extended practice. An experiment was done and the mean number of errors on initial items is shown in the table. A *t* test with 54 *df* produced a value of 3.88.

	Extended practice	*No extended practice*
Mean number of errors	13.7	18.4

27. A developmental psychologist developed a theory that predicted the proportion of children who would, during a period of stress, cling to their mother, attack the mother, or attack a younger sibling. The stress situation was set up and the responses of 50 children recorded. A χ^2 was calculated and found to be 5.30.

28. An experimental psychologist at a Veterans Administration hospital obtained clearance from the ethics committee to conduct a study of the efficacy of Elavil for the treatment of depression. Patients whose principal problem was depression were chosen for the study. The experiment lasted 30 days, during which one group was given placebos, one was given a low dose of Elavil, and one was given a high dose of Elavil. At the end of the experiment, the patients were observed by two outside psychologists who rated several behaviors (eye contact, posture, verbal output, activity, and such) for degree of depression. A composite score was obtained on each patient, with low scores indicating depression. The mean scores were: placebo, 9.86; low dose, 23.00; and high dose, 9.21. An ANOVA summary table and two Tukey HSD tests are shown in the table.

Source	*df*	*F*
Between groups	2	21.60
Within groups	18	
HSD (placebo vs. low) = 13.14		
HSD (placebo vs. high) = 0.65		

29. One sample of 16 third-graders had been taught to read by the "look-say" method. A second sample of 18 had been taught by the phonics method. Both were given a reading achievement test and a Mann-Whitney *U* test was performed on the results. A *U* value of 51 was found.

30. Sociologists sometimes make a profile of the attitudes of a group by asking each individual to rate his or her tolerance of people with particular characteristics. For example, members of a group might be asked to assess, on a scale of 1–7, their tolerance of prostitutes, atheists, former mental patients, intellectuals, college graduates, and so on. The mean score for each category is then calculated and the categories are ranked from low to high. When 18 categories of people were ranked from low to high by both a group of people who owned businesses and a group of college students, an r_s of .753 between the rankings was found.

31. In a large high school, one group of juniors took an English course that included a 9-week unit on poetry. Another group of juniors studied plays during that 9-week period.

Afterward, both groups completed a questionnaire on their attitudes toward poetry. High scores mean favorable attitudes. The means and variances are presented in the table. To compare the means, a t test was run: t (78 df) = .48. To compare the variances, an F test was performed: F (37, 41) = 3.62. (See footnote 7 in Chapter 9.)

	Attitudes toward poetry	
	Studied poetry	Studied plays
Mean	23.7	21.7
Variance	51.3	16.1

32. Here is an example of the results of experiments that asked the question: Can you change the attitudes of an audience by presenting just one side of an argument claiming that it is correct, or should you present both sides and then claim that one side is correct? In the following experiment, a second independent variable is also examined: level of education. The first table contains the mean change in attitude and the second is an ANOVA summary table.

	Presentation	
Level of education	One side	Both sides
Less than a high school diploma	4.3	2.7
At least 1 year of college	2.1	4.5

Source	df	F
Between presentations	1	2.01
Between education	1	1.83
Presentations × education	1	7.93
Within groups	44	

33. A college student was interested in the relationship between handedness and verbal ability. She gave three classes of third-grade students a test of verbal ability that she had devised and then classified each child as right-handed, left-handed, or mixed. The obtained F value was 2.63.

	Handedness		
	Left	Right	Mixed
Mean verbal ability score	21.6	25.9	31.3
N	16	36	12

34. As part of a large-scale study on alcoholism, the alcoholics and the control group were classified according to when they were toilet-trained as children. The three categories were early, normal, or late. A χ^2 of 7.90 was found.

Appendixes

Appendix A
REFERENCES

Abelson, R. P. (1985). A variance explanation paradox: When a little is a lot. *Psychological Bulletin, 97,* 129–133.

Astin, A. W. (1987). *American freshmen: National norms for fall, 1987.* Los Angeles: American Council of Education.

Bangert-Drowns, R. L. (1986). Review of developments in meta-analytic method. *Psychological Bulletin, 99,* 388–399.

Barber, T. X. (1976). Suggested ("hypnotic") behavior: The trance paradigm versus an alternative paradigm. In T. X. Barber (Ed.), *Advances in altered states of consciousness & human potentialities* (Vol. 1, pp. 175–259). New York: Psychological Dimensions.

Barker, R. B., Dembo, T., & Lewin, K. (1941). Frustration and regression: An experiment with young children. *University of Iowa Studies in Child Welfare, 18,* No. 1.

Beirne, P. (1987). Adolphe Quételet and the origins of positivist criminology. *American Journal of Sociology, 92,* 1140–1169.

Berger, R. W., & Hart, T. H. (1986). *Statistical process control.* New York: Marcel Dekker.

Biffen, R. H. (1905). Mendel's laws of inheritance and wheat breeding. *Journal of Agricultural Science, 1,* 4–48.

Birch, H. G., & Rabinowitz, H. S. (1951). The negative effect of previous experience on productive thinking. *Journal of Experimental Psychology, 41,* 121–125.

Blair, R. C., & Higgins, J. J. (1985). Comparison of the power of the paired samples *t* test to that of Wilcoxon's signed-ranks test under various population shapes. *Psychological Bulletin, 97,* 119–128.

Blair, R. C., Higgins, J. J., & Smitley, W. D. S. (1980). On the relative power of the *U* and *t* tests. *British Journal of Mathematical and Statistical Psychology, 33,* 114–120.

Box, J. F. (1981). Gosset, Fisher, and the *t* distribution. *American Statistician, 35,* 61–66.

Bradley, D. R., Bradley, T. D., McGrath, S. G., & Cutcomb, S. D. (1979). Type I error rate of the chi-square test of independence in R × C tables that have small expected frequencies. *Psychological Bulletin, 86,* 1290–1297.

Bradley, J. V. (1978). Robustness? *British Journal of Mathematical Statistical Psychology, 31,* 144–152.

Bransford, J. D., & Franks, J. J. (1971). The abstraction of linguistic ideas. *Cognitive Psychology, 2,* 331–350.

Brehm, J. W., & Cohen, A. R. (1962). *Explorations in cognitive dissonance.* New York: Wiley.

Bureau of Labor Statistics. (December, 1944). Union wages and hours of motor truck drivers and helpers. July 1, 1944, *Monthly Labor Review.*

Calhoun, J. P., & Johnston, J. O. (1968). Manifest anxiety and visual acuity. *Perceptual and Motor Skills, 27,* 1177–1178.

Camilli, G., & Hopkins, K. D. (1978). Applicability of chi-square to 2 × 2 contingency tables with small expected cell frequencies. *Psychological Bulletin, 85,* 163–167.

Campbell, S. K. (1974). *Flaws and fallacies in statistical thinking.* Englewood Cliffs, NJ: Prentice-Hall.

Christensen, L. B., & Stoup, C. M. (1986). *Introduction to statistics for the social and behavioral sciences.* Pacific Grove, CA: Brooks/Cole.

Cohen, I. B. (March, 1984). Florence Nightingale. *Scientific American, 250,* 128–137.

Cornell, F. G. (1956). *The essentials of educational statistics.* New York: Wiley.

David, F. N. (1968). Francis Galton. In D. L. Sills (Ed.), *International encyclopedia of the social sciences* (Vol. 6, pp. 48–53). New York: Macmillan.

Doll, R. (1955). Etiology of lung cancer. *Advances in Cancer Research, 3,* 1–50.

Downie, N. M., & Heath, R. W. (1983). *Basic statistical methods* (5th ed.). New York: Harper & Row.

Dunkling, L., & Gosling, W. (1983). *The facts on file dictionary of first names.* New York: Facts on File.

Durkheim, E. (1951, reprint). *Suicide.* New York: Free Press.

Edwards, A. L. (1972). *Experimental design in psychological research* (4th ed.). New York: Holt, Rinehart & Winston.

Edwards, A. L. (1984). *An introduction to linear regression and correlation* (2nd ed.). San Francisco: Freeman.

Ellis, D. (1938). *A source book of Gestalt psychology.* London: Routledge & Kegan Paul.

Ferguson, G. A. (1981). *Statistical analysis in psychology and education* (5th ed.). New York: McGraw-Hill.

Fisher, R. A. (1925). *Statistical methods for research workers.* London: Oliver and Boyd.

Fisher, R. A., & Yates, F. (1963). *Statistical tables for biological, agricultural, and medical research* (6th ed.). Edinburgh: Oliver and Boyd.

Flynn, J. R. (1987). Massive IQ gains in 14 nations: What IQ tests really measure. *Psychological Bulletin, 101,* 171–191.

Forbs, R., & Meyer, A. B. (1955). *Forestry handbook.* New York: Ronald.

Gaito, J. (1980). Measurement scales and statistics: Resurgence of an old misconception. *Psychological Bulletin, 87,* 564–567.

Galton, F. (1869). *Hereditary genius.* London: Macmillan.

Galton, F. (1889). *Natural inheritance.* London: Macmillan.

Galton, F. (1901). Biometry. *Biometrika, 1,* 7–10.

Gardner, P. L. (1975). Scales and statistics. *Review of Educational Research, 45,* 43–57.

Greene, J. E. (Ed.-in-chief). (1966). *McGraw-Hill modern men of science.* New York: McGraw-Hill.

Guilford, J. P. (1967). *The nature of human intelligence.* New York: McGraw-Hill.

Guilford, J. P., & Fruchter, B. (1978). *Fundamental statistics in psychology and education* (6th ed.). New York: McGraw-Hill.

Hacking, I. (1984). Trial by number. *Science 84, 5,* 69–70.

Harris, R. J. (1985). *A primer of multivariate statistics* (2nd ed.). New York: Academic.

Hasler, A. D. (1966). *Underwater guideposts.* Madison: University of Wisconsin Press.

Hasler, A. D., Scholz, A. T., & Horrall, R. M. (1978). Olfactory imprinting and homing in salmon. *American Scientist, 66,* 347–355.

Hellekson, C. J., Kline, J. A., & Rosenthal, N. E. (1986). Phototherapy for seasonal affective disorder in Alaska. *American Journal of Psychiatry, 143,* 1035–1037.

Howell, D. C. (1987). *Statistical methods for psychology* (2nd ed.). Boston: Duxbury.

Huff, D. (1954). *How to lie with statistics.* New York: Norton.

Irion, A. L. (1976). A survey of the introductory course in psychology. *Teaching of Psychology, 3,* 3–8.

Jaeger, R. M. (1984). *Sampling in education and the social sciences.* New York: Longman.

Jenkins, J. G., & Dallenbach, K. M. (1924). Obliviscence during sleep and waking. *American Journal of Psychology, 35,* 605–612.

Johnson, R. A., & Wichern, D. W. (1988). *Applied multivariate statistical analysis* (2nd ed.). New York: Prentice-Hall.

Johnson, R. C., McClearn, G. E., Yuen, S., Nagoshi, C. T., Ahern, F. M., & Cole, R. E. (1985). Galton's data a century later. *American Psychologist, 40,* 875–892.

Johnston, J. O., & Maertens, N. W. (1972). Effects of parental involvement in arithmetic homework. *School Science and Mathematics, 72,* 117–126.

Johnston, J. O., & Spatz, K. C. (1973). Racial attitudes and self-esteem levels among Southeast Arkansas public school students. *Educational Research in Arkansas, 1971–72.* Little Rock: Arkansas Department of Higher Education.

Keppel, G. (1982). *Design and analysis: A researcher's handbook.* Englewood Cliffs, NJ: Prentice-Hall.

Kirk, R. E. (1972). (Ed.). *Statistical issues: A reader for the behavioral sciences.* Pacific Grove, CA: Brooks/Cole.

Kirk, R. E. (1982). *Experimental design: Procedures for the behavioral sciences* (2nd ed.). Pacific Grove, CA: Brooks/Cole.

Kirk, R. E. (1984). *Introductory statistics* (2nd ed.). Pacific Grove, CA: Brooks/Cole.

Kramer, C. Y. (1956). Extension of multiple range tests to group means with unequal numbers of replications. *Biometrics, 12,* 307–310.

Lewin, K. (1958). Group decision and social change. In E. E. Maccoby, T. M. Newcomb, & E. L. Hartley (Eds.), *Readings in social psychology* (3rd ed.). New York: Holt, Rinehart & Winston.

Loewi, O. (1963). The night prowler. In S. Rapport & H. Wright (Eds.). *Science: Method and meaning.* New York: Washington Square Press.

Loftus, E. F. (1979). *Eyewitness testimony.* Cambridge, MA: Harvard University Press.

Loftus, G. R., & Loftus, E. F. (1988). *Essentials of statistics* (2nd ed.). New York: Random House.

Mann, H. B., & Whitney, D. R. (1947). On a test of whether one or two random variables is stochastically larger than the other. *Annals of Mathematical Statistics, 18,* 50–60.

Maugh, T. H. (1977). Guayule and jojoba: Agriculture in semiarid regions. *Science, 196,* 1189–1190.

Mayo, E. (1946). *The human problems of an industrial civilization.* Boston: Harvard University Press.

McGaugh, J. L., & Petrinovich, L. F. (1965). Effects of drugs on learning and memory. *International review of neurobiology* (Vol. 8). New York: Academic.

McKeown, T., & Gibson, J. R. (1951). Observation on all births (23,970) in Birmingham, 1947: IV. "Premature birth." *British Medical Journal, 2,* 513–517.

McMullen, L., & Pearson, E. S. (1939). William Sealy Gosset, 1876–1937. *Biometrika, 30,* 205–253.

Mendenhall, W., Ott, L., & Scheaffer, R. L. (1971). *Elementary survey sampling.* Boston: Duxbury.

Michell, J. (1986). Measurement scales and statistics: A clash of paradigms. *Psychological Bulletin, 100,* 398–407.

Minium, E. W. (1978). *Statistical reasoning in psychology and education* (2nd ed.). New York: Wiley.

Minium, E. W., & Clarke, R. B. (1982). *Elements of statistical reasoning.* New York: Wiley.

Mischel, H. N. (1974). Sex bias in the evaluation of professional achievements. *Journal of Educational Psychology, 66,* 157–166.

Nelson, N., Rosenthal, R., & Rosnow, R. L. (1986). Interpretation of significance levels and effect sizes by psychological researchers. *American Psychologist, 41,* 1299–1301.

Overall, J. E. (1980). Power of chi-square tests for 2×2 contingency tables with small expected frequencies. *Psychological Bulletin, 87,* 132–135.

Pearson, E. S. (1949). W. S. Gosset. *Dictionary of national biography: 1931–1940.* London: Oxford University Press.

Pearson, E. S., & Hartley, H. O. (Eds.). (1958). *Biometrika tables for statisticians* (2nd ed., Vol. I). New York: Cambridge University Press.

Pearson, K., & Lee, A. (1903). Inheritance of physical characters. *Biometrika, 2,* 357–462.

Porter, J. H., & Hamm, R. J. (1986). *Statistics: Applications for the behavioral sciences.* Pacific Grove, CA: Brooks/Cole.

Powers, E., & Witmer, H. (1951). *An experiment in the prevention of delinquency: The Cambridge-Somerville Youth Study.* New York: Columbia University Press.

Rechtschaffen, A., Gilliland, M. A., Bergmann, B. M., & Winter, J. B. (1983). Physiological correlates of prolonged sleep deprivation in rats. *Science, 221,* 182–184.

Rokeach, M., Homant, R., & Penner, L. (1970). A value analysis of the disputed Federalist papers. *Journal of Personality and Social Psychology, 16,* 245–250.

Runyon, R. P. (1981). *How numbers lie.* Lexington, MA: Lewis.

Schachter, S., & Gross, S. P. (1968). Manipulated time and eating behavior. *Journal of Personality and Social Psychology, 10,* 98–106.

Schumacher, E. F. (1979). *Good work.* New York: Harper & Row.

Sherif, M. (1935). A study of some social factors in perception. *Archives of Psychology,* No. 187.

Sherman, M. (1927). The differentiation of emotional responses in infants: The ability of observers to judge the emotional characteristics of the crying infants, and of the voice of an adult. *Journal of Comparative Psychology, 7,* 335–351.

Smith, E. C. (1969). *American surnames.* Baltimore: Chilton.

Smith, R. A. (1971). The effect of unequal group size on Tukey's HSD procedure. *Psychometrika, 36,* 31–34.

Snedecor, G. W., & Cochran, W. G. (1967). *Statistical methods* (6th ed.). Ames: Iowa State University Press.

Spatz, K. C., & Johnston, J. O. (1973). Internal consistency of the Coopersmith Self-Esteem Inventory. *Educational and Psychological Measurement, 33,* 875–876.

Spearman, C. (1930). Autobiography. In C. Murchison (Ed.). *History of psychology in autobiography.* New York: Russell & Russell.

Statistical abstract of the United States: 1986 (106th ed.). (1985). Washington, DC: U.S. Bureau of the Census.

Stevens, S. S. (1946). On the theory of scales of measurement. *Science, 103,* 677–680.

Stewart, I. (1977). Gauss. *Scientific American, 237,* July, pp. 123–131.

Taylor, J. (1953). A personality scale of manifest anxiety. *Journal of Abnormal and Social Psychology, 48,* 285–290.

Tobias, S. (1978). *Overcoming math anxiety.* Boston: Houghton Mifflin.

Townsend, J. T., & Ashby, F. G. (1984). Measurement scales and statistics: The misconception misconceived. *Psychological Bulletin, 96,* 394–401.

Tufte, E. (1983). *The visual display of quantitative information.* Cheshire, CT: Graphics.

Ury, H. (1967). In response to Noether's letter, "Needed—a new name." *American Statistician, 21*(4), 53.

Wainer, H. (1984). How to display data badly. *American Statistician, 38,* 137–147.

Walker, H. M. (1929). *Studies in the history of statistical method.* Baltimore: Williams & Wilkins.

Walker, H. M. (1940). Degrees of freedom. *Journal of Educational Psychology, 31,* 253–269.

Walker, H. M. (1968). Karl Pearson. In D. L. Sills (Ed.). *International encyclopedia of the social sciences.* New York: Macmillan and Free Press.

Wallace, W. L. (1972). College Board Scholastic Aptitude Test. In O. K. Buros (Ed.). *The seventh mental measurements yearbook.* Highland Park, NJ: Gryphon.

Watson, J. B. (1924). *Psychology from the standpoint of a behaviorist* (2nd ed.). Philadelphia: Lippincott.

Weinberg, G. H., Schumaker, J. A., & Oltman, D. (1981). *Statistics: An intuitive approach* (4th ed.). Pacific Grove, CA: Brooks/Cole.

White, E. B. (1970). *The trumpet of the swan.* New York: Harper & Row.

Wilcoxon, F. (1945). Individual comparisons by ranking methods. *Biometrics, 1,* 80–83.

Wilcoxon, F., & Wilcox, R. A. (1964). *Some rapid approximate statistical procedures* (Rev. ed.). Pearl River, NY: Lederle Laboratories.

Williams, B. (1978). *A sampler on sampling.* New York: Wiley.

Winer, B. F. (1971). *Statistical principles in experimental design* (2nd ed.). New York: McGraw-Hill.

Wood, T. B., & Stratton, F. J. M. (1910). The interpretation of experimental results. *Journal of Agricultural Science, 3,* 417–440.

Woodworth, R. S. (1926). Introduction. In H. E. Garrett (Ed.). *Statistics in psychology and education.* New York: Longmans, Green.

Work, M. S., & Rogers, H. (1972). Effects of estrogen level on food-seeking dominance among male rats. *Journal of Comparative and Physiological Psychology, 79,* 414–418.

Youden, W. J. (1962). *Experimentation and measurement.* Washington, D. C.: National Science Teachers Association.

Zajonc, R. B. (1975). Dumber by the dozen. *Psychology Today, 8,* No. 8, 37–43.

Appendix B
ARITHMETIC AND ALGEBRA REVIEW

Objectives for Appendix B: After studying the text and working the problems, you should be able to:

1. Estimate answers
2. Round numbers
3. Find answers to problems with decimals, fractions, negative numbers, proportions, percents, absolute values, ± signs, exponents, square roots, complex expressions, and simple algebraic expressions

This appendix gives you a review of the basic mathematical skills you will need to work the problems in this course. This math is not complex. The highest level of mathematical sophistication is simple algebra.

Although the mathematical reasoning is not very complex, there is a good bit of arithmetic required. To be good in statistics you need to be good at arithmetic. A wrong answer is wrong, whether the error is arithmetical or logical. Most people are better at arithmetic when they use a calculator with a square root key and a memory. With a calculator you can do your work in less time and with fewer computational errors.

Of course, calculators can do more than simple arithmetic. Answers to many of the computational problems in this book can be found using a calculator hard-wired for means, standard deviations, and correlation coefficients. By using a computer program you can find numerical answers to all the computational problems. The question for you is, At what stage of my statistical education should I use these aids?

Remember that the major goal of this course is for you to understand the logic of statistical reasoning so well that you can tell the story of what the answers mean. Our advice is not to use your calculator or computer to produce final statistics like the mean, standard deviation, correlation coefficient, and *t*-test values until you have gone through the step-by-step, part-by-part calculations several times. In the words of one of the reviewers of this book, "The world already has too many people using computer packages without an adequate 'personal knowledge' of the statistical operations performed (and not performed)." Become a person with "personal knowledge."

This appendix is divided into two parts.

- Part 1 is a self-administered, self-scored test of your skills in arithmetic and algebra. If you do very well on this, we suggest you skip Part 2.

- Part 2 summarizes the rules of arithmetic and algebra and gives you problems and worked-out answers.

The pretest gives you problems that are similar to those you will have to work later in the course. Take the pretest and then check your answers against the answers in the back of the book. If you find that you made any mistakes, work through those sections of Part 2 that explain the problems you missed.

As we stated earlier, to be good at statistics, you need to be good at arithmetic. To be good at arithmetic, you need to know the rules and to be careful in your computations. Part 2 is designed to refresh your memory of the rules. It is up to you to be careful in your computations.

PART 1: PRETEST

Estimating Answers   (Estimate whole-number answers to the problems.)

1. $(4.02)^2$

2. 1.935×7.89

3. $31.219 \div 2.0593$

Rounding Numbers   (Round numbers to the nearest tenth.)

4. 6.06

5. 0.35

6. 10.348

Decimals

Add:

7. 3.12, 6.3, 2.004

8. 12, 8.625, 2.316, 4.2

Multiply:

11. 6.2×8.06

12. 0.35×0.162

Subtract:

9. $28.76 - 8.91$

10. $3.2 - 1.135$

Divide:

13. $64.1 \div 21.25$

14. $0.065 \div 0.0038$

Fractions

Add:

15. $\dfrac{1}{8} + \dfrac{3}{4} + \dfrac{1}{2}$

16. $\dfrac{1}{3} + \dfrac{1}{2}$

Subtract:

17. $\dfrac{11}{16} - \dfrac{1}{2}$

18. $\dfrac{5}{9} - \dfrac{1}{2}$

Multiply:

19. $\dfrac{4}{5} \times \dfrac{3}{4}$

20. $\dfrac{2}{3} \times \dfrac{1}{4} \times \dfrac{3}{5}$

Divide:

21. $\dfrac{3}{8} \div \dfrac{1}{4}$

22. $\dfrac{7}{9} \div \dfrac{1}{4}$

Negative Numbers

Add:

23. -5, 16, -1, -4

24. -11, -2, -12, 3

Subtract:

25. $(-10) - (-3)$

26. $(-8) - (-2)$

Multiply:

27. $(-5) \times (-5)$

28. $(-8) \times (3)$

Divide:

29. $(-10) \div (-3)$

30. $(-21) \div 4$

Percents and Proportions

31. 12 is what percent of 36?

32. Find 27 percent of 84.

33. What proportion of 112 is 21?

34. A proportion .40 of the tagged birds was recovered. 150 were tagged in all. How many were recovered?

Absolute Value

35. $|-5|$

36. $|8 - 12|$

$\pm$ Problems

37. $8 \pm 2 =$

38. $13 \pm 9 =$

Exponents

39. 4^2

40. 2.5^2

41. 0.35^2

Square Roots

42. $\sqrt{9.30}$

43. $\sqrt{0.93}$

44. $\sqrt{0.093}$

Complex Problems.   (Round all answers to two decimal places.)

45. $\dfrac{3 + 4 + 7 + 2 + 5}{5}$

46. $\dfrac{(5 - 3)^2 + (3 - 3)^2 + (1 - 3)^2}{3 - 1}$

47. $\dfrac{(8 - 6.5)^2 + (5 - 6.5)^2}{2 - 1}$

48. $\left(\dfrac{8 + 4}{8 + 4 - 2}\right)\left(\dfrac{1}{8} + \dfrac{1}{4}\right)$

49. $\left(\dfrac{3.6}{1.2}\right)^2 + \left(\dfrac{6.0}{2.4}\right)^2$

50. $\dfrac{(3)(4) + (6)(8) + (5)(6) + (2)(3)}{(4)(4 - 1)}$

51. $\dfrac{190 - (25^2/5)}{5 - 1}$

52. $\dfrac{12(50 - 20)^2}{(8)(10)(12)(11)}$

53. $\dfrac{6[(3 + 5)(4 - 1) + 2]}{5^2 - (6)(2)}$

Simple Algebra   (Solve for x.)

54. $\dfrac{x - 3}{4} = 2.5$

55. $\dfrac{14 - 8.5}{x} = 0.5$

56. $\dfrac{20 - 6}{2} = 4x - 3$

57. $\dfrac{6 - 2}{3} = \dfrac{x + 9}{5}$

Statistical Symbols

Although, as far as we know, there has never been a clinical case of neoiconophobia,[1] some students show a mild form of this behavior. Symbols like $\bar{X}$, σ, μ, and Σ may cause a grimace, a frown, or a droopy eyelid. In more severe cases, the behavior involves avoiding a statistics course entirely. We're sure that you don't have such a severe case, since you have read this far. Even so, if you are a typical beginning student in statistics, symbols like σ, μ, Σ, and $\bar{X}$ are not very meaningful to you, and they may even elicit feelings of uneasiness. We also know from our teaching experience that, by the end of the course, you will know what these symbols mean and be able to approach them with an unruffled psyche—and perhaps even approach them joyously.

Some of the symbols we use will stand for concepts you are already quite familiar with. For example, the capital letter X with a bar over it, $\bar{X}$, stands for the mean or average. ($\bar{X}$ is pronounced "mean" or sometimes "ex-bar.") You already know about means. You just add up the scores and divide by the number of scores. This verbal instruction can be put into symbols: $\bar{X} = \Sigma X/N$. Σ (the Greek uppercase sigma) is the instruction to add and X is the symbol for scores. $\bar{X} = \Sigma X/N$ is something you already know about, even if you have not been using these symbols.

Pay careful attention to symbols. They serve as shorthand notations for the ideas and concepts you are learning. Each time a new symbol is introduced, concentrate on it, learn it, memorize its definition and pronunciation. The more meaning a symbol has for you, the better you understand the concepts it represents and, of course, the easier the course will be.

Sometimes we will need to distinguish between two different σ's or two X's. We will use subscripts, and the results will look like σ_1 and σ_2, or X_1 and X_2. Later, we will use subscripts other than numbers to identify a symbol. You will see σ_X and $\sigma_{\bar{X}}$. The point to learn here is that subscripts are for identification purposes only; they never indicate multiplication. Thus $\sigma_{\bar{X}}$ does not mean $(\sigma)(\bar{X})$.

Two additional comments—to encourage and to caution you. We encourage you to do more in this course than just read the text, work the problems, and pass the tests, however exciting that may be. We encourage you to occasionally get beyond this elementary text and read journal articles or short portions of other statistics textbooks. We will indicate our recommendations with footnotes at appropriate places. The word of caution that goes with this encouragement is that reading statistics texts is like reading a Russian novel—the same characters have different names in different places. For example, the mean of a sample in some texts is symbolized M rather than $\bar{X}$. There is even more variety when it comes to symbolizing the standard deviation. If you expect such differences, it will be less difficult for you to fit the new symbol into your established scheme of understanding.

PART 2: REVIEW OF FUNDAMENTALS

This section gives you a quick review of the pencil and paper rules of arithmetic and simple algebra. We assume that you once knew all these rules but that refresher exercises will be helpful. Thus, we do not include much explanation. To obtain basic explanations, ask a teacher in your school's mathematics department to recommend one of the many "refresher" books that are available.

If your concern about this course comes from your anxiety about math, then we may have another recommendation for you. If you are a good reader, we recommend Sheila Tobias's book, *Overcoming Math Anxiety* (1978). This well-written, 250-page book will give

---

[1] An extreme and unreasonable fear of new symbols.

you some insight into your problem and provide you with exercises that should help you reduce your math anxiety.

Definitions

1. *Sum.*   The answer to an addition problem is called a sum. In Chapter 9, you will calculate a *sum of squares,* a quantity that is obtained by adding together some squared numbers.
2. *Difference.*   The answer to a subtraction problem is called a difference. Much of what you will learn in statistics deals with differences and whether or not they should be attributed to chance. In Chapter 7, you will encounter a statistic called the *standard error of a difference.* Obviously, this statistic involves subtraction.
3. *Product.*   The answer to a multiplication problem is called a product. Chapter 4 is about the *product-moment correlation coefficient,* which requires multiplication. Multiplication problems are indicated either by a $\times$ or by parentheses. Thus, 6×4 and $(6)(4)$ call for the same operation.
4. *Quotient.*   The answer to a division problem is called a quotient. The IQ or *intelligence quotient* is based on the division of two numbers. The two ways to indicate a division problem are $\div$ and $-$. Thus, $9 \div 4$ and $\frac{9}{4}$ (or 9/4) call for the same operation. It is a good idea to think of any common fraction as a division problem. The numerator is to be divided by the denominator.

Estimating Answers

It is a **very good idea** to just look at a problem and make an estimate of the answer before you do any calculating. This is referred to as "eyeballing the data" and Edward Minium (1978) has captured its importance with Minium's First Law of Statistics: "The eyeball is the statistician's most powerful instrument."

Estimating answers should keep you from making gross errors, such as misplacing a decimal point. For example, $\frac{31.5}{5}$ can be estimated as a little more than 6. If you make this estimate before you divide, you are likely to recognize that an answer of 63 or 0.63 is incorrect.

The estimated answer to the problem $(21)(108)$ is 2000, since $(20)(100) = 2000$. The problem $(0.47)(0.20)$ suggests an estimated answer of 0.10, since $(\frac{1}{2})(0.20) = 0.10$. With 0.10 in mind, you are not likely to write 0.94 for the answer, which is 0.094. Estimating answers is also important if you are finding a square root. You can estimate that $\sqrt{95}$ is about 10, since $\sqrt{100} = 10$; $\sqrt{1.034}$ is about 1; $\sqrt{0.093}$ is about 0.1.

To calculate a mean, eyeball the numbers and estimate the mean. If you estimate a mean of 30 for a group of numbers that are primarily in the 20s, 30s, and 40s, a calculated mean of 60 will arouse your suspicion that you have made an error.

Rounding Numbers

There are two parts to the rule for rounding a number. If the first digit of the part that is to be dropped is less than 5, simply drop it. If the first digit of the part to be dropped is 5 or greater, increase the number to the left of it by one. These rules are built into most electronic calculators. Here are some illustrations of these rules.

Rounding to the nearest whole number

$6.2 = 6$	$4.5 = 5$
$12.7 = 13$	$163.5 = 164$
$6.49 = 6$	$9.5 = 10$

Rounding to the nearest hundredth

$$13.614 = 13.61 \qquad 12.065 = 12.07$$
$$0.049 = 0.05 \qquad 4.005 = 4.01$$
$$1.097 = 1.10 \qquad 0.675 = 0.68$$
$$3.6248 = 3.62 \qquad 1.995 = 2.00$$

A reasonable question is, How many decimal places should an answer in statistics have? A good rule of thumb in statistics is to carry all operations to three decimal places and then, for the final answer, round back to two decimal places.

Sometimes this rule of thumb can get you into trouble, though. For example, if halfway through some work you had a division problem of $0.0016 \div 0.0074$, and you dutifully rounded those four decimals to three $(0.002 \div 0.007)$, you would get an answer of 0.2857, which becomes 0.29. However, division without rounding gives you an answer of 0.2162 or 0.22. The difference between 0.22 and 0.29 may be quite substantial. We will often give you cues if more than two decimal places are necessary but you will always need to be alert to the problems of rounding.

Most calculators carry more decimal places in memory than they show in the display. If you have such a calculator, it will protect you from the problem of rounding too much or too soon.

PROBLEMS

1. Define (a) sum, (b) quotient, (c) product, and (d) difference.
2. Estimate answers to the following expressions.
   a. $\sqrt{103.48}$ b. $74.16 \div 9.87$ c. $(11.4)^2$ d. $\sqrt{0.0459}$
   e. 0.41^2 f. 11.92×4.60 g. $\sqrt{0.888}$ h. $\sqrt{0.0098}$
3. Round the following numbers to the nearest whole number.
   a. 13.9   b. 126.4   c. 9.0
   d. 0.4   e. 127.5   f. 12.51
   g. 12.49   h. 12.50   i. 9.46
4. Round the following numbers to the nearest hundredth.
   a. 6.3348   b. 12.997   c. 0.050
   d. 0.965   e. 2.605   f. 0.3445
   g. 0.003   h. 0.015   i. 0.9949

Decimals

1. *Addition and subtraction of decimals.*   There is only one rule about the addition and subtraction of numbers that have decimals: *keep the decimal points in a vertical line.* The decimal point in the answer goes directly below those in the problem. This rule is illustrated in the five problems here.

ADD			SUBTRACT	
	0.004	6.0		
1.26	1.310	18.0	14.032	16.00
10.00	4.039	0.5	8.26	4.32
11.26	5.353	24.5	5.772	11.68

2. *Multiplication of decimals.*   The basic rule for multiplying decimals is that the number of decimal places in the answer is found by adding up the number of decimal places in

the two numbers that are being multiplied. To place the decimal point in the product, count from the right.

$$
\begin{array}{r}
1.3 \\
\times 4.2 \\
\hline
26 \\
52 \\
\hline
5.46
\end{array}
\qquad
\begin{array}{r}
0.21 \\
\times 0.4 \\
\hline
0.084
\end{array}
\qquad
\begin{array}{r}
1.47 \\
\times 3.12 \\
\hline
294 \\
147 \\
441 \\
\hline
4.5864
\end{array}
$$

3. *Division of decimals.*   Two methods have been used to teach division of decimals. The older method required the student to move the decimal in the divisor (the number you are dividing by) enough places to the right to make the divisor a whole number. The decimal in the dividend is then moved to the right the same number of places, and division is carried out in the usual way. The new decimal places are identified with carets, and the decimal place in the quotient is just above the caret in the dividend. For example,

$$
\begin{array}{r}
20. \\
0.016_\wedge \overline{)0.320_\wedge}
\end{array}
\qquad
\begin{array}{r}
38.46 \\
0.39_\wedge \overline{)15.00_\wedge}
\end{array}
$$

Decimal moved three places in
both the divisor and the dividend
    Decimal moved two places in
both the divisor and the dividend

$$
\begin{array}{r}
2.072 \\
6\overline{)12.432}
\end{array}
\qquad
\begin{array}{r}
.004 \\
9.1_\wedge \overline{)0.0_\wedge 369}
\end{array}
$$

Divisor is already a whole
number
    Decimal moved one place in
both the divisor and the dividend

The newer method of teaching the division of decimals is to multiply both the divisor and the dividend by the number that will make both of them whole numbers. (Actually, this is the way the caret method works also.) For example,

$$
\frac{0.32}{0.016} \times \frac{1000}{1000} = \frac{320}{16} = 20.00
$$

$$
\frac{0.0369}{9.1} \times \frac{10{,}000}{10{,}000} = \frac{369}{91{,}000} = 0.004
$$

$$
\frac{15}{0.39} \times \frac{100}{100} = \frac{1500}{39} = 38.46
$$

$$
\frac{12.432}{6} \times \frac{1000}{1000} = \frac{12{,}432}{6000} = 2.072
$$

Both of these methods work. Use the one you are more familiar with.

PROBLEMS   Perform the operations indicated.

5. $0.001 + 10 + 3.652 + 2.5$

7. 0.04×1.26

9. $3.06 \div 0.04$

11. $152.12 - 127.4$

6. $14.2 - 7.31$

8. $143.3 + 16.92 + 2.307 + 8.1$

10. $\dfrac{24}{11.75}$

12. $(0.5)(0.07)$

Fractions

In general, there are two ways to deal with fractions.

1. Convert the fraction to a decimal and perform the operations on the decimals.
2. Work directly with the fractions, using a set of rules for each operation. The rule for addition and subtraction is: convert the fractions to ones with common denominators, add or subtract the numerators, and place the result over the common denominator. The rule for multiplication is: multiply the numerators together to get the numerator of the answer, and multiply the denominators together for the denominator of the answer. The rule for division is: invert the divisor and multiply the fractions.

For statistics problems, it is usually easier to convert the fractions to decimals and then work with the decimals. Therefore, this is the method that we will illustrate. However, if you are a whiz at working directly with fractions, by all means continue with your method. To convert a fraction to a decimal, divide the lower number into the upper one. Thus, $3/4 = 0.75$, and $13/17 = 0.765$.

ADDITION OF FRACTIONS

$$\frac{1}{2} + \frac{1}{4} = 0.50 + 0.25 = 0.75$$

$$\frac{13}{17} + \frac{21}{37} = 0.765 + 0.568 = 1.33$$

$$\frac{2}{3} + \frac{3}{4} = 0.667 + 0.75 = 1.42$$

SUBTRACTION OF FRACTIONS

$$\frac{1}{2} - \frac{1}{4} = 0.50 - 0.25 = 0.25$$

$$\frac{11}{12} - \frac{2}{3} = 0.917 - 0.667 = 0.25$$

$$\frac{41}{53} - \frac{17}{61} = 0.774 - 0.279 = 0.50$$

MULTIPLICATION OF FRACTIONS

$$\frac{1}{2} \times \frac{3}{4} = (0.5)(0.75) = 0.38$$

$$\frac{10}{19} \times \frac{61}{90} = 0.526 \times 0.678 = 0.36$$

$$\frac{1}{11} \times \frac{2}{3} = (0.09)(0.67) = 0.06$$

DIVISION OF FRACTIONS

$$\frac{9}{21} \div \frac{13}{19} = 0.429 \div 0.684 = 0.63$$

$$14 \div \frac{1}{3} = 14 \div 0.33 = 42$$

$$\frac{7}{8} \div \frac{3}{4} = 0.875 \div 0.75 = 1.17$$

PROBLEMS   Perform the operations indicated.

13. $\dfrac{9}{10} + \dfrac{1}{2} + \dfrac{2}{5}$

14. $\dfrac{9}{20} \div \dfrac{19}{20}$

15. $\left(\dfrac{1}{3}\right)\left(\dfrac{5}{6}\right)$

16. $\dfrac{4}{5} - \dfrac{1}{6}$

17. $\dfrac{1}{3} \div \dfrac{5}{6}$

18. $\dfrac{3}{4} \times \dfrac{5}{6}$

19. $18 \div \dfrac{1}{3}$

Negative Numbers

1. *Addition of negative numbers.* (Any number without a sign is understood to be positive.)

   a. To add a series of negative numbers, add the numbers in the usual way and attach a negative sign to the total.

   $$\begin{array}{r} -3 \\ -8 \\ -12 \\ -5 \\ \hline -28 \end{array}$$
   $$(-1) + (-6) + (-3) = -10$$

   b. To add two numbers, one positive and one negative, subtract the smaller number from the larger number and attach the sign of the larger number to the result.

   $$\begin{array}{r} 140 \\ -55 \\ \hline 85 \end{array}$$
   $$\begin{array}{r} -14 \\ 8 \\ \hline -6 \end{array}$$
   $$(28) + (-9) = 19$$
   $$74 + (-96) = -22$$

   c. To add a series of numbers, of which some are positive and some negative, add all the positive numbers together, add all the negative numbers together (see 1a), and then combine the two sums (see 1b).

   $$(-4) + (-6) + (12) + (-5) + (2) + (-9) = 14 + (-24) = -10$$
   $$(-7) + (10) + (4) + (-5) = 14 + (-12) = 2$$

2. *Subtraction of negative numbers.* To subtract a negative number, change it to positive and add it.

   $$\begin{array}{r} (-14) \\ -(-2) \\ \hline -12 \end{array}$$
   $$\begin{array}{r} 5 \\ -(-7) \\ \hline 12 \end{array}$$
   $$\begin{array}{r} 8 \\ -(-3) \\ \hline 11 \end{array}$$
   $$\begin{array}{r} (-7) \\ -(-5) \\ \hline -2 \end{array}$$

3. *Multiplication of negative numbers.* When the two numbers to be multiplied are both negative, the product is positive.

   $$(-3)(-3) = 9 \qquad (-6)(-8) = 48$$

   When one of the numbers is negative and the other is positive, the product is negative.

   $$(-8)(3) = -24 \qquad 14 \times -2 = -28$$

4. *Division of negative numbers.* The rule in division is the same as the rule in multiplication. If the two numbers are both negative, the quotient is positive.

   $$(-10) \div (-2) = 5 \qquad (-4) \div (-20) = 0.20$$

   If one number is negative and the other positive, the quotient is negative.

   $$(-10) \div 2 = -5 \qquad 6 \div (-18) = -0.33$$
   $$14 \div (-7) = -2 \qquad (-12) \div 3 = -4$$

PROBLEMS

20. Add the following numbers.
    a. $-3, 19, -14, 5, -11$ b. $-8, -12, -3$
    c. $-8, 11$ d. $3, -6, -2, 5, -7$

21. $(-8)(5)$

24. $(11)(-3)$

27. $12 - (-3)$

30. $(-10) \div 5$

22. $(-4)(-6)$

25. $(-18) - (-9)$

28. $(-6) - (-7)$

31. $4 \div (-12)$

23. $(4)(12)$

26. $14 \div (-6)$

29. $(-9) \div (-3)$

32. $(-7) - 5$

Proportions and Percents

A **proportion** is a part of a whole and can be expressed as a fraction or as a decimal. Usually proportions are expressed as decimals. If eight students in a class of 44 received A's, we may express 8 as a proportion of the whole (44). Thus, we have 8/44, or .18. The proportion of the class that received A's is .18.

To convert a proportion to a percent (per one hundred), multiply by 100. Thus: $.18 \times 100 = 18$; 18 percent of the students received A's. As you can see, proportions and percents are two ways to express the same idea.

If you know a proportion (or percent) and the size of the original whole, you can find the number that the proportion represents. If .28 of the students were absent due to illness and there are 50 students in all, then .28 of the 50 were absent. $(.28)(50) = 14$ students who were absent. Here are some more examples.

1. 26 out of 31 completed the course. What proportion completed the course?

$$\frac{26}{31} = .83$$

2. What percent completed the course?

$.83 \times 100 = 83$ percent

3. What percent of 19 is 5?

$$\frac{5}{19} = .26 \qquad .26 \times 100 = 26 \text{ percent}$$

4. If 90 percent of the population agreed and the population consisted of 210 members, how many members agreed?

$.90 \times 210 = 189$ members

Absolute Value

The **absolute value** of a number ignores the sign of the number. Thus, the absolute value of -6 is 6. This is expressed with symbols as $|-6| = 6$. It is expressed verbally as "the absolute value of negative six is six." In a similar way, the absolute value of $4 - 7$ is 3; that is, $|4 - 7| = |-3| = 3$.

± Signs

A $\pm$ sign ("plus or minus" sign) means to both add *and* subtract. A $\pm$ problem *always* has two answers.

$10 \pm 4 = 6, 14$

$8 \pm (3)(2) = 8 \pm 6 = 2, 14$

$\pm(4)(3) + 21 = \pm 12 + 21 = 9, 33$

$\pm 4 - 6 = -10, -2$

Exponents

In the expression 5^2 ("five squared"), 2 is the exponent. The 2 means that 5 is to be multiplied by itself. Thus, $5^2 = 5 \times 5 = 25$.

In elementary statistics, the only exponent used is 2, but it will be used frequently. When a number has an exponent of 2, the number is said to be squared. The expression 4^2 ("four squared") means 4×4, and the product is 16.

$$8^2 = 8 \times 8 = 64 \qquad \left(\frac{3}{4}\right)^2 = (0.75)^2 = 0.5625$$

$$1.2^2 = (1.2)(1.2) = 1.44 \qquad 12^2 = 12 \times 12 = 144$$

Square Roots

Statistics problems often require that a square root be found. There are three possible ways to find the square root:

1. A calculator with a square root key.
2. A table of square roots.
3. The paper and pencil method.

Our recommendation is that you use a calculator with a square root key. If this is not possible, find a book with a table of square roots. The third solution, the paper and pencil method, is so tedious and error-prone that we do not recommend it.

PROBLEMS

33. Six is what proportion of 13?
34. Six is what percent of 13?
35. What proportion of 25 is 18?
36. What percent of 115 is 85?
37. Find 72 percent of 36.
38. $|-31|$
39. $|21 - 25|$
40. $12 \pm (2)(5)$
41. $\pm(5)(6) + 10$
42. $\pm(2)(2) - 6$
43. $(2.5)^2$
44. 9^2
45. $(0.3)^2$
46. $\left(\frac{1}{4}\right)^2$

47. Find the square root of the following numbers to two decimal places (or more if appropriate).
   a. 625
   b. 6.25
   c. 0.625
   d. 0.0625
   e. 16.85
   f. 0.003
   g. 181,476
   h. 0.25
   i. 22.51

Complex Expressions

Two rules will suffice for the kinds of complex expressions encountered in statistics.

1. Perform the operations within the parentheses first. If there are brackets in the expression, perform the operations within the parentheses and then the operations within the brackets.
2. Perform the operations in the numerator separately from those in the denominator and, finally, carry out the division.

$$\frac{(8-6)^2 + (7-6)^2 + (3-6)^2}{3-1} = \frac{2^2 + 1^2 + (-3)^2}{2} = \frac{4+1+9}{2} = \frac{14}{2} = 7.00$$

$$\left(\frac{10+12}{4+3-2}\right)\left(\frac{1}{4}+\frac{1}{3}\right) = \left(\frac{22}{5}\right)(0.25 + 0.333) = (4.40)(0.583) = 2.57$$

$$\left(\frac{8.2}{4.1}\right)^2 + \left(\frac{4.2}{1.2}\right)^2 = (2)^2 + (3.5)^2 = 4 + 12.25 = 16.25$$

$$\frac{6[(13-10)^2 - 5]}{6(6-1)} = \frac{6(3^2 - 5)}{6(5)} = \frac{6(9-5)}{30} = \frac{6(4)}{30} = \frac{24}{30} = 0.80$$

$$\frac{18 - \frac{6^2}{4}}{4-1} = \frac{18 - \frac{36}{4}}{3} = \frac{18-9}{3} = \frac{9}{3} = 3.00$$

Simple Algebra

To solve a simple algebra problem, isolate the unknown (x) on one side of the equal sign and combine the numbers on the other side. To do this, remember that you can multiply or divide both sides of the equation by the same number without affecting the value of the unknown. For example,

$$\frac{x}{6} = 9 \qquad 11x = 30 \qquad \frac{3}{x} = 14$$

$$(6)\left(\frac{x}{6}\right) = (6)(9) \qquad \frac{11x}{11} = \frac{30}{11} \qquad x\left(\frac{3}{x}\right) = (14)(x)$$

$$x = 54 \qquad x = 2.73 \qquad 3 = 14x$$

$$\frac{3}{14} = \frac{14x}{14}$$

$$0.21 = x$$

In a similar way, the same number can be *added to* or *subtracted from* both sides of the equation without affecting the value of the unknown. For example,

$$x - 5 = 12 \qquad x + 7 = 9$$

$$x - 5 + 5 = 12 + 5 \qquad x + 7 - 7 = 9 - 7$$

$$x = 17 \qquad x = 2$$

We will combine some of these steps in the problems we will work for you. Be sure you see what operation is being performed on both sides in each step.

$$\frac{x - 2.5}{1.3} = 1.96 \qquad \frac{21.6 - 15}{x} = 0.04$$

$$x - 2.5 = (1.3)(1.96) \qquad 6.6 = 0.04x$$

$$x = 2.548 + 2.5 \qquad \frac{6.6}{0.04} = x$$

$$x = 5.048 \qquad 165 = x$$

$$4x - 9 = 5^2$$
$$4x = 25 + 9$$
$$x = \frac{34}{4}$$
$$x = 8.50$$

$$\frac{14 - x}{6} = 1.9$$
$$14 - x = (1.9)(6)$$
$$14 = 11.4 + x$$
$$2.6 = x$$

PROBLEMS   Reduce these complex expressions to a single number rounded to two decimal places.

48. $\dfrac{(4 - 2)^2 + (0 - 2)^2}{6}$

49. $\dfrac{(12 - 8)^2 + (8 - 8)^2 + (5 - 8)^2 + (7 - 8)^2}{4 - 1}$

50. $\left(\dfrac{5 + 6}{3 + 2 - 2}\right)\left(\dfrac{1}{3} + \dfrac{1}{2}\right)$

51. $\left(\dfrac{13 + 18}{6 + 8 - 2}\right)\left(\dfrac{1}{6} + \dfrac{1}{8}\right)$

52. $\dfrac{8[(6 - 2)^2 - 5]}{(3)(2)(4)}$

53. $\dfrac{[(8 - 2)(5 - 1)]^2}{5(10 - 7)}$

54. $\dfrac{6}{1/2} + \dfrac{8}{1/3}$

55. $\left(\dfrac{9}{2/3}\right)^2 + \left(\dfrac{8}{3/4}\right)^2$

56. $\dfrac{10 - (6^2/9)}{8}$

57. $\dfrac{104 - (12^2/6)}{5}$

Find the value of x in the problems.

58. $\dfrac{x - 4}{2} = 2.58$

59. $\dfrac{x - 21}{6.1} = 1.04$

60. $x = \dfrac{14 - 11}{2.5}$

61. $x = \dfrac{36 - 41}{8.2}$

Appendix C
TABLES

Table A Critical values for Pearson product-moment correlation coefficients, r*

df		α Levels *(two-tailed test)*			
	.1	.05	.02	.01	.001
(df = N − 2)		α Levels *(one-tailed test)*			
	.05	.025	.01	.005	.0005
1	.98769	.99692	.999507	.999877	.9999988
2	.90000	.95000	.98000	.990000	.99900
3	.8054	.8783	.93433	.95873	.99116
4	.7293	.8114	.8822	.91720	.97406
5	.6694	.7545	.8329	.8745	.95074
6	.6215	.7067	.7887	.8343	.92493
7	.5822	.6664	.7498	.7977	.8982
8	.5494	.6319	.7155	.7646	.8721
9	.5214	.6021	.6851	.7348	.8371
10	.4973	.5760	.6581	.7079	.8233
11	.4762	.5529	.6339	.6835	.8010
12	.4575	.5324	.6120	.6614	.7800
13	.4409	.5139	.5923	.6411	.7603
14	.4259	.4973	.5742	.6226	.7420
15	.4124	.4821	.5577	.6055	.7246
16	.4000	.4683	.5425	.5897	.7084
17	.3887	.4555	.5285	.5751	.6932
18	.3783	.4438	.5155	.5614	.6787
19	.3687	.4329	.5034	.5487	.6652
20	.3598	.4227	.4921	.5368	.6524
25	.3233	.3809	.4451	.4869	.5974
30	.2960	.3494	.4093	.4487	.5541
35	.2746	.3246	.3810	.4182	.5189
40	.2573	.3044	.3578	.3932	.4896
45	.2428	.2875	.3384	.3721	.4648
50	.2306	.2732	.3218	.3541	.4433
60	.2108	.2500	.2948	.3248	.4078
70	.1954	.2319	.2737	.3017	.3799
80	.1829	.2172	.2565	.2830	.3568
90	.1726	.2050	.2422	.2673	.3375
100	.1638	.1946	.2301	.2540	.3211

*To be significant the r obtained from the data must be equal to or larger than the value shown in the table.

SOURCE: Table VII of Fisher and Yates (1963), *Statistical Tables for Biological, Agricultural and Medical Research,* published by Longman Group Ltd., London (previously published by Oliver & Boyd, Edinburgh), and by permission of the authors and publishers.

Table B Random digits

	00–04	05–09	10–14	15–19	20–24	25–29	30–34	35–39	40–44	45–49
00	54463	22662	65905	70639	79365	67382	29085	69831	47058	08186
01	15389	85205	18850	39226	42249	90669	96325	23248	60933	26927
02	85941	40756	82414	02015	13858	78030	16269	65978	01385	15345
03	61149	69440	11286	88218	58925	03638	52862	62733	33451	77455
04	05219	81619	10651	67079	92511	59888	84502	72095	83463	75577
05	41417	98326	87719	92294	46614	50948	64886	20002	97365	30976
06	28357	94070	20652	35774	16249	75019	21145	05217	47286	76305
07	17783	00015	10806	83091	91530	36466	39981	62481	49177	75779
08	40950	84820	29881	85966	62800	70326	84740	62660	77379	90279
09	82995	64157	66164	41180	10089	41757	78258	96488	88629	37231
10	96754	17676	55659	44105	47361	34833	86679	23930	53249	27083
11	34357	88040	53364	71726	45690	66334	60332	22554	90600	71113
12	06318	37403	49927	57715	50423	67372	63116	48888	21505	80182
13	62111	52820	07243	79931	89292	84767	85693	73947	22278	11551
14	47534	09243	67879	00544	23410	12740	02540	54440	32949	13491
15	98614	75993	84460	62846	59844	14922	48730	73443	48167	34770
16	24856	03648	44898	09351	98795	18644	39765	71058	90368	44104
17	96887	12479	80621	66223	86085	78285	02432	53342	42846	94771
18	90801	21472	42815	77408	37390	76766	52615	32141	30268	18106
19	55165	77312	83666	36028	28420	70219	81369	41943	47366	41067
20	75884	12952	84318	95108	72305	64620	91318	89872	45375	85436
21	16777	37116	58550	42958	21460	43910	01175	87894	81378	10620
22	46230	43877	80207	88877	89380	32992	91380	03164	98656	59337
23	42902	66892	46134	01432	94710	23474	20423	60137	60609	13119
24	81007	00333	39693	28039	10154	95425	39220	19774	31782	49037
25	68089	01122	51111	72373	06902	74373	96199	97017	41273	21546
26	20411	67081	89950	16944	93054	87687	96693	87236	77054	33848
27	58212	13160	06468	15718	82627	76999	05999	58680	96739	63700
28	70577	42866	24969	61210	76046	67699	42054	12696	93758	03283
29	94522	74358	71659	62038	79643	79169	44741	05437	39038	13163
30	42626	86819	85651	88678	17401	03252	99547	32404	17918	62880
31	16051	33763	57194	16752	54450	19031	58580	47629	54132	60631
32	08244	27647	33851	44705	94211	46716	11738	55784	95374	72655
33	59497	04392	09419	89964	51211	04894	72882	17805	21896	83864
34	97155	13428	40293	09985	58434	01412	69124	82171	59058	82859
35	98409	66162	95763	47420	20792	61527	20441	39435	11859	41567
36	45476	84882	65109	96597	25930	66790	65706	61203	53634	22557
37	89300	69700	50741	30329	11658	23166	05400	66669	48708	03887
38	50051	95137	91631	66315	91428	12275	24816	68091	71710	33258
39	31753	85178	31310	89642	98364	02306	24617	09609	83942	22716
40	79152	53829	77250	20190	56535	18760	69942	77448	33278	48805
41	44560	38750	83635	56540	64900	42912	13953	79149	18710	68618
42	68328	83378	63369	71381	39564	05615	42451	64559	97501	65747
43	46939	38689	58625	08342	30459	85863	20781	09284	26333	91777
44	83544	86141	15707	96256	23068	13782	08467	89469	93842	55349
45	91621	00881	04900	54224	46177	55309	17852	27491	89415	23466
46	91896	67126	04151	03795	59077	11848	12630	98375	52068	60142
47	55751	62515	21108	80830	02263	29303	37204	96926	30506	09808
48	85156	87689	95493	88842	00664	55017	55539	17771	69448	87530
49	07521	56898	12236	60277	39102	62315	12239	07105	11844	01117

(continued)

SOURCE: *Statistical Methods* (6th ed.), by G. W. Snedecor and W. G. Cochran (1967). Copyright © by Iowa State University Press, Ames, Iowa. Reprinted by permission.

	50-54	55-59	60-64	65-69	70-74	75-79	80-84	85-89	90-94	95-99
00	59391	58030	52098	82718	87024	82848	04190	96574	90464	29065
01	99567	76364	77204	04615	27062	96621	43918	01896	83991	51141
02	10363	97518	51400	25670	98342	61891	27101	37855	06235	33316
03	86859	19558	64432	16706	99612	59798	32803	67708	15297	28612
04	11258	24591	36863	55368	31721	94335	34936	02566	80972	08188
05	95068	88628	35911	14530	33020	80428	39936	31855	34334	64865
06	54463	47237	73800	91017	36239	71824	83671	39892	60518	37092
07	16874	62677	57412	13215	31389	62233	80827	73917	82802	84420
08	92494	63157	76593	91316	03505	72389	96363	52887	01087	66091
09	15669	56689	35682	40844	53256	81872	35213	09840	34471	74441
10	99116	75486	84989	23476	52967	67104	39495	39100	17217	74073
11	15696	10703	65178	90637	63110	17622	53988	71087	84148	11670
12	97720	15369	51269	69620	03388	13699	33423	67453	43269	56720
13	11666	13841	71681	98000	35979	39719	81899	07449	47985	46967
14	71628	73130	78783	75691	41632	09847	61547	18707	85489	69944
15	40501	51089	99943	91843	41995	88931	73631	69361	05375	15417
16	22518	55576	98215	82068	10798	86211	36584	67466	69373	40054
17	75112	30485	62173	02132	14878	92879	22281	16783	86352	00077
18	80327	02671	98191	84342	90813	49268	95441	15496	20168	09271
19	60251	45548	02146	05597	48228	81366	34598	72856	66762	17002
20	57430	82270	10421	05540	43648	75888	66049	21511	47676	33444
21	73528	39559	34434	88596	54086	71693	43132	14414	79949	85193
22	25991	65959	70769	64721	86413	33475	42740	06175	82758	66248
23	78388	16638	09134	59880	63806	48472	39318	35434	24057	74739
24	12477	09965	96657	57994	59439	76330	24596	77515	09577	91871
25	83266	32883	42451	15579	38155	29793	40914	65990	16255	17777
26	76970	80876	10237	39515	79152	74798	39357	09054	73579	92359
27	37074	65198	44785	68624	98336	84481	97610	78735	46703	98265
28	83712	06514	30101	78295	54656	85417	43189	60048	72781	72606
29	20287	56862	69727	94443	64936	08366	27227	05158	50326	59566
30	74261	32592	86538	27041	65172	85532	07571	80609	39285	65340
31	64081	49863	08478	96001	18888	14810	70545	89755	59064	07210
32	05617	75818	47750	67814	29575	10526	66192	44464	27058	40467
33	26793	74951	95466	74307	13330	42664	85515	20632	05497	33625
34	65988	72850	48737	54719	52056	01596	03845	35067	03134	70322
35	27366	42271	44300	73399	21105	03280	73457	43093	05192	48657
36	56760	10909	98147	34736	33863	95256	12731	66598	50771	83665
37	72880	43338	93643	58904	59543	23943	11231	83268	65938	81581
38	77888	38100	03062	58103	47961	83841	25878	23746	55903	44115
39	28440	07819	21580	51459	47971	29882	13990	29226	23608	15873
40	63525	94441	77033	12147	51054	49955	58312	76923	96071	05813
41	47606	93410	16359	89033	89696	47231	64498	31776	05383	39902
42	52669	45030	96279	14709	52372	87832	02735	50803	72744	88208
43	16738	60159	07425	62369	07515	82721	37875	71153	21315	00132
44	59348	11695	45751	15865	74739	05572	32688	20271	65128	14551
45	12900	71775	29845	60774	94924	21810	38636	33717	67598	82521
46	75086	23537	49939	33595	13484	97588	28617	17979	70749	35234
47	99495	51434	29181	09993	38190	42553	68922	52125	91077	40197
48	26075	31671	45386	36583	93159	48599	52022	41330	60651	91321
49	13636	93596	23377	51133	95126	61496	42474	45141	46660	42338

(continued)

Table B (*continued*)

	00–04	05–09	10–14	15–19	20–24	25–29	30–34	35–39	40–44	45–49
50	64249	63664	39652	40646	97306	31741	07294	84149	46797	82487
51	26538	44249	04050	48174	65570	44072	40192	51153	11397	58212
52	05845	00512	78630	55328	18116	69296	91705	86224	29503	57071
53	74897	68373	67359	51014	33510	83048	17056	72506	82949	54600
54	20872	54570	35017	88132	25730	22626	86723	91691	13191	77212
55	31432	96156	89177	75541	81355	24480	77243	76690	42507	84362
56	66890	61505	01240	00660	05873	13568	76082	79172	57913	93448
57	48194	57790	79970	33106	86904	48119	52503	24130	72824	21627
58	11303	87118	81471	52936	08555	28420	49416	44448	04269	27029
59	54374	57325	16947	45356	78371	10563	97191	53798	12693	27928
60	64852	34421	61046	90849	13966	39810	42699	21753	76192	10508
61	16309	20384	09491	91588	97720	89846	30376	76970	23063	35894
62	42587	37065	24526	72602	57589	98131	37292	05967	26002	51945
63	40177	98590	97161	41682	84533	67588	62036	49967	01990	72308
64	82309	76128	93965	26743	24141	04838	40254	26065	07938	76236
65	79788	68243	59732	04257	27084	14743	17520	95401	55811	76099
66	40538	79000	89559	25026	42274	23489	34502	75508	06059	86682
67	64016	73598	18609	73150	62463	33102	45205	87440	96767	67042
68	49767	12691	17903	93871	99721	79109	09425	26904	07419	76013
69	76974	55108	29795	08404	82684	00497	51126	79935	57450	55671
70	23854	08480	85983	96025	50117	64610	99425	62291	86943	21541
71	68973	70551	25098	78033	98573	79848	31778	29555	61446	23037
72	36444	93600	65350	14971	25325	00427	52073	64280	18847	24768
73	03003	87800	07391	11594	21196	00781	32550	57158	58887	73041
74	17540	26188	36647	78386	04558	61463	57842	90382	77019	24210
75	38916	55809	47982	41968	69760	79422	80154	91486	19180	15100
76	64288	19843	69122	42502	48508	28820	59933	72998	99942	10515
77	86809	51564	38040	39418	49915	19000	58050	16899	79952	57849
78	99800	99566	14742	05028	30033	94889	53381	23656	75787	59223
79	92345	31890	95712	08279	91794	94068	49337	88674	35355	12267
80	90363	65162	32245	82279	79256	80834	06088	99462	56705	06118
81	64437	32242	48431	04835	39070	59702	31508	60935	22390	52246
82	91714	53662	28373	34333	55791	74758	51144	18827	10704	76803
83	20902	17646	31391	31459	33315	03444	55743	74701	58851	27427
84	12217	86007	70371	52281	14510	76094	96579	54863	78339	20839
85	45177	02863	42307	53571	22532	74921	17735	42201	80540	54721
86	28325	90814	08804	52746	47913	54577	47525	77705	95330	21866
87	29019	28776	56116	54791	64604	08815	46049	71186	34650	14994
88	84979	81353	56219	67062	26146	82567	33122	14124	46240	92973
89	50371	26347	48513	63915	11158	25563	91915	18431	92978	11591
90	53422	06825	69711	67950	64716	18003	49581	45378	99878	61130
91	67453	35651	89316	41620	32048	70225	47597	33137	31443	51445
92	07294	85353	74819	23445	68237	07202	99515	62282	53809	26685
93	79544	00302	45338	16015	66613	88968	14595	63836	77716	79596
94	64144	85442	82060	46471	24162	39500	87351	36637	42833	71875
95	90919	11883	58318	00042	52402	28210	34075	33272	00840	73268
96	06670	57353	86275	92276	77591	46924	60839	55437	03183	13191
97	36634	93976	52062	83678	41256	60948	18685	48992	19462	96062
98	75101	72891	85745	67106	26010	62107	60885	37503	55461	71213
99	05112	71222	72654	51583	05228	62056	57390	42746	39272	96659

(*continued*)

Table B (*continued*)

	50–54	55–59	60–64	65–69	70–74	75–79	80–84	85–89	90–94	95–99
50	32847	31282	03345	89593	69214	70381	78285	20054	91018	16742
51	16916	00041	30236	55023	14253	76582	12092	86533	92426	37655
52	66176	34047	21005	27137	03191	48970	64625	22394	39622	79085
53	46299	13335	12180	16861	38043	59292	62675	63631	37020	78195
54	22847	47839	45385	23289	47526	54098	45683	55849	51575	64689
55	41851	54160	92320	69936	34803	92479	33399	71160	64777	83378
56	28444	59497	91586	95917	68553	28639	06455	34174	11130	91994
57	47520	62378	98855	83174	13088	16561	68559	26679	06238	51254
58	34978	63271	13142	82681	05271	08822	06490	44984	49307	62717
59	37404	80416	69035	92980	49486	74378	75610	74976	70056	15478
60	32400	65482	52099	53676	74648	94148	65095	69597	52771	71551
61	89262	86332	51718	70663	11623	29834	79820	73002	84886	03591
62	86866	09127	98021	03871	27789	58444	44832	36505	40672	30180
63	90814	14833	08759	74645	05046	94056	99094	65091	32663	73040
64	19192	82756	20553	58446	55376	88914	75096	26119	83898	43816
65	77585	52593	56612	95766	10019	29531	73064	20953	53523	58136
66	23757	16364	05096	03192	62386	45389	85332	18877	55710	96459
67	45989	96257	23850	26216	23309	21526	07425	50254	19455	29315
68	92970	94243	07316	41467	64837	52406	25225	51553	31220	14032
69	74346	59596	40088	98176	17896	86900	20249	77753	19099	48885
70	87646	41309	27636	45153	29988	94770	07255	70908	05340	99751
71	50099	71038	45146	06146	55211	99429	43169	66259	97786	59180
72	10127	46900	64984	75348	04115	33624	68774	60013	35515	62556
73	67995	81977	18984	64091	02785	27762	42529	97144	80407	64524
74	26304	80217	84934	82657	69291	35397	98714	35104	08187	48109
75	81994	41070	56642	64091	31229	02595	13513	45148	78722	30144
76	59537	34662	79631	89403	65212	09975	06118	86197	58208	16162
77	51228	10937	62396	81460	47331	91403	95007	06047	16846	64809
78	31089	37995	29577	07828	42272	54016	21950	86192	99046	84864
79	38207	97938	93459	75174	79460	55436	57206	87644	21296	43395
80	88666	31142	09474	89712	63153	62333	42212	06140	42594	43671
81	53365	56134	67582	92557	89520	33452	05134	70628	27612	33738
82	89807	74530	38004	90102	11693	90257	05500	79920	62700	43325
83	18682	81038	85662	90915	91631	22223	91588	80774	07716	12548
84	63571	32579	63942	25371	09234	94592	98475	76884	37635	33608
85	68927	56492	67799	95398	77642	54913	91853	08424	81450	76229
86	56401	63186	39389	88798	31356	89235	97036	32341	33292	73757
87	24333	95603	02359	72942	46287	95382	08452	62862	97869	71775
88	17025	84202	95199	62272	06366	16175	97577	99304	41587	03686
89	02804	08253	52133	20224	68034	50865	57868	22343	55111	03607
90	08298	03879	20995	19850	73090	13191	18963	82244	78479	99121
91	59883	01785	82403	96062	03785	03488	12970	64896	38336	30030
92	46982	06682	62864	91837	74021	89094	39952	64158	79614	78235
93	31121	47266	07661	02051	67599	24471	69843	83696	71402	76287
94	97867	56641	63416	17577	30161	87320	37752	73276	48969	41915
95	57364	86746	08415	14621	49430	22311	15836	72492	49372	44103
96	09559	26263	69511	28064	75999	44540	13337	10918	79846	54809
97	53873	55571	00608	42661	91332	63956	74087	59008	47493	99581
98	35531	19162	86406	05299	77511	24311	57257	22826	77555	05941
99	28229	88629	25695	94932	30721	16197	78742	34974	97528	45447

Table C Area under the normal curve between μ and z and beyond z

A z	B Area between mean and z	C Area beyond z	A z	B Area between mean and z	C Area beyond z	A z	B Area between mean and z	C Area beyond z
0.00	.0000	.5000	0.55	.2088	.2912	1.10	.3643	.1357
0.01	.0040	.4960	0.56	.2123	.2877	1.11	.3665	.1335
0.02	.0080	.4920	0.57	.2157	.2843	1.12	.3686	.1314
0.03	.0120	.4880	0.58	.2190	.2810	1.13	.3708	.1292
0.04	.0160	.4840	0.59	.2224	.2776	1.14	.3729	.1271
0.05	.0199	.4801	0.60	.2257	.2743	1.15	.3749	.1251
0.06	.0239	.4761	0.61	.2291	.2709	1.16	.3770	.1230
0.07	.0279	.4721	0.62	.2324	.2676	1.17	.3790	.1210
0.08	.0319	.4681	0.63	.2357	.2643	1.18	.3810	.1190
0.09	.0359	.4641	0.64	.2389	.2611	1.19	.3830	.1170
0.10	.0398	.4602	0.65	.2422	.2578	1.20	.3849	.1151
0.11	.0438	.4562	0.66	.2454	.2546	1.21	.3869	.1131
0.12	.0478	.4522	0.67	.2486	.2514	1.22	.3888	.1112
0.13	.0517	.4483	0.68	.2517	.2483	1.23	.3907	.1093
0.14	.0557	.4443	0.69	.2549	.2451	1.24	.3925	.1075
0.15	.0596	.4404	0.70	.2580	.2420	1.25	.3944	.1056
0.16	.0636	.4364	0.71	.2611	.2389	1.26	.3962	.1038
0.17	.0675	.4325	0.72	.2642	.2358	1.27	.3980	.1020
0.18	.0714	.4286	0.73	.2673	.2327	1.28	.3997	.1003
0.19	.0753	.4247	0.74	.2704	.2296	1.29	.4015	.0985
0.20	.0793	.4207	0.75	.2734	.2266	1.30	.4032	.0968
0.21	.0832	.4168	0.76	.2764	.2236	1.31	.4049	.0951
0.22	.0871	.4129	0.77	.2794	.2206	1.32	.4066	.0934
0.23	.0910	.4090	0.78	.2823	.2177	1.33	.4082	.0918
0.24	.0948	.4052	0.79	.2852	.2148	1.34	.4099	.0901
0.25	.0987	.4013	0.80	.2881	.2119	1.35	.4115	.0885
0.26	.1026	.3974	0.81	.2910	.2090	1.36	.4131	.0869
0.27	.1064	.3936	0.82	.2939	.2061	1.37	.4147	.0853
0.28	.1103	.3897	0.83	.2967	.2033	1.38	.4162	.0838
0.29	.1141	.3859	0.84	.2995	.2005	1.39	.4177	.0823
0.30	.1179	.3821	0.85	.3023	.1977	1.40	.4192	.0808
0.31	.1217	.3783	0.86	.3051	.1949	1.41	.4207	.0793
0.32	.1255	.3745	0.87	.3078	.1922	1.42	.4222	.0778
0.33	.1293	.3707	0.88	.3106	.1894	1.43	.4236	.0764
0.34	.1331	.3669	0.89	.3133	.1867	1.44	.4251	.0749
0.35	.1368	.3632	0.90	.3159	.1841	1.45	.4265	.0735
0.36	.1406	.3594	0.91	.3186	.1814	1.46	.4279	.0721
0.37	.1443	.3557	0.92	.3212	.1788	1.47	.4292	.0708
0.38	.1480	.3520	0.93	.3238	.1762	1.48	.4306	.0694
0.39	.1517	.3483	0.94	.3264	.1736	1.49	.4319	.0681
0.40	.1554	.3446	0.95	.3289	.1711	1.50	.4332	.0668
0.41	.1591	.3409	0.96	.3315	.1685	1.51	.4345	.0655
0.42	.1628	.3372	0.97	.3340	.1660	1.52	.4357	.0643
0.43	.1664	.3336	0.98	.3365	.1635	1.53	.4370	.0630
0.44	.1700	.3300	0.99	.3389	.1611	1.54	.4382	.0618
0.45	.1736	.3264	1.00	.3413	.1587	1.55	.4394	.0606
0.46	.1772	.3228	1.01	.3438	.1562	1.56	.4406	.0594
0.47	.1808	.3192	1.02	.3461	.1539	1.57	.4418	.0582
0.48	.1844	.3156	1.03	.3485	.1515	1.58	.4429	.0571
0.49	.1879	.3121	1.04	.3508	.1492	1.59	.4441	.0559
0.50	.1915	.3085	1.05	.3531	.1469	1.60	.4452	.0548
0.51	.1950	.3050	1.06	.3554	.1446	1.61	.4463	.0537
0.52	.1985	.3015	1.07	.3577	.1423	1.62	.4474	.0526
0.53	.2019	.2981	1.08	.3599	.1401	1.63	.4484	.0516
0.54	.2054	.2946	1.09	.3621	.1379	1.64	.4495	.0505

(continued)

Table C (*continued*)

A z	B Area between mean and z	C Area beyond z	A z	B Area between mean and z	C Area beyond z	A z	B Area between mean and z	C Area beyond z
1.65	.4505	.0495	2.22	.4868	.0132	2.79	.4974	.0026
1.66	.4515	.0485	2.23	.4871	.0129	2.80	.4974	.0026
1.67	.4525	.0475	2.24	.4875	.0125	2.81	.4975	.0025
1.68	.4535	.0465	2.25	.4878	.0122	2.82	.4976	.0024
1.69	.4545	.0455	2.26	.4881	.0119	2.83	.4977	.0023
1.70	.4554	.0446	2.27	.4884	.0116	2.84	.4977	.0023
1.71	.4564	.0436	2.28	.4887	.0113	2.85	.4978	.0022
1.72	.4573	.0427	2.29	.4890	.0110	2.86	.4979	.0021
1.73	.4582	.0418	2.30	.4893	.0107	2.87	.4979	.0021
1.74	.4591	.0409	2.31	.4896	.0104	2.88	.4980	.0020
1.75	.4599	.0401	2.32	.4898	.0102	2.89	.4981	.0019
1.76	.4608	.0392	2.33	.4901	.0099	2.90	.4981	.0019
1.77	.4616	.0384	2.34	.4904	.0096	2.91	.4982	.0018
1.78	.4625	.0375	2.35	.4906	.0094	2.92	.4982	.0018
1.79	.4633	.0367	2.36	.4909	.0091	2.93	.4983	.0017
1.80	.4641	.0359	2.37	.4911	.0089	2.94	.4984	.0016
1.81	.4649	.0351	2.38	.4913	.0087	2.95	.4984	.0016
1.82	.4656	.0344	2.39	.4916	.0084	2.96	.4985	.0015
1.83	.4664	.0336	2.40	.4918	.0082	2.97	.4985	.0015
1.84	.4671	.0329	2.41	.4920	.0080	2.98	.4986	.0014
1.85	.4678	.0322	2.42	.4922	.0078	2.99	.4986	.0014
1.86	.4686	.0314	2.43	.4925	.0075	3.00	.4987	.0013
1.87	.4693	.0307	2.44	.4927	.0073	3.01	.4987	.0013
1.88	.4699	.0301	2.45	.4929	.0071	3.02	.4987	.0013
1.89	.4706	.0294	2.46	.4931	.0069	3.03	.4988	.0012
1.90	.4713	.0287	2.47	.4932	.0068	3.04	.4988	.0012
1.91	.4719	.0281	2.48	.4934	.0066	3.05	.4989	.0011
1.92	.4726	.0274	2.49	.4936	.0064	3.06	.4989	.0011
1.93	.4732	.0268	2.50	.4938	.0062	3.07	.4989	.0011
1.94	.4738	.0262	2.51	.4940	.0060	3.08	.4990	.0010
1.95	.4744	.0256	2.52	.4941	.0059	3.09	.4990	.0010
1.96	.4750	.0250	2.53	.4943	.0057	3.10	.4990	.0010
1.97	.4756	.0244	2.54	.4945	.0055	3.11	.4991	.0009
1.98	.4761	.0239	2.55	.4946	.0054	3.12	.4991	.0009
1.99	.4767	.0233	2.56	.4948	.0052	3.13	.4991	.0009
2.00	.4772	.0228	2.57	.4949	.0051	3.14	.4992	.0008
2.01	.4778	.0222	2.58	.4951	.0049	3.15	.4992	.0008
2.02	.4783	.0217	2.59	.4952	.0048	3.16	.4992	.0008
2.03	.4788	.0212	2.60	.4953	.0047	3.17	.4992	.0008
2.04	.4793	.0207	2.61	.4955	.0045	3.18	.4993	.0007
2.05	.4798	.0202	2.62	.4956	.0044	3.19	.4993	.0007
2.06	.4803	.0197	2.63	.4957	.0043	3.20	.4993	.0007
2.07	.4808	.0192	2.64	.4959	.0041	3.21	.4993	.0007
2.08	.4812	.0188	2.65	.4960	.0040	3.22	.4994	.0006
2.09	.4817	.0183	2.66	.4961	.0039	3.23	.4994	.0006
2.10	.4821	.0179	2.67	.4962	.0038	3.24	.4994	.0006
2.11	.4826	.0174	2.68	.4963	.0037	3.25	.4994	.0006
2.12	.4830	.0170	2.69	.4964	.0036	3.30	.4995	.0005
2.13	.4834	.0166	2.70	.4965	.0035	3.35	.4996	.0004
2.14	.4838	.0162	2.71	.4966	.0034	3.40	.4997	.0003
2.15	.4842	.0158	2.72	.4967	.0033	3.45	.4997	.0003
2.16	.4846	.0154	2.73	.4968	.0032	3.50	.4998	.0002
2.17	.4850	.0150	2.74	.4969	.0031	3.60	.4998	.0002
2.18	.4854	.0146	2.75	.4970	.0030	3.70	.4999	.0001
2.19	.4857	.0143	2.76	.4971	.0029	3.80	.4999	.0001
2.20	.4861	.0139	2.77	.4972	.0028	3.90	.49995	.00005
2.21	.4864	.0136	2.78	.4973	.0027	4.00	.49997	.00003

Table D   The *t* distribution*

df	\u03b1 Levels for two-tailed test					
	.2	.1	.05	.02	.01	.001
	\u03b1 Levels for one-tailed test					
	.10	.05	.025	.01	.005	.0005
1	3.078	6.314	12.706	31.821	63.657	636.619
2	1.886	2.920	4.303	6.965	9.925	31.598
3	1.638	2.353	3.182	4.541	5.841	12.924
4	1.533	2.132	2.776	3.747	4.604	8.610
5	1.476	2.015	2.571	3.365	4.032	6.869
6	1.440	1.943	2.447	3.143	3.707	5.959
7	1.415	1.895	2.365	2.998	3.499	5.408
8	1.397	1.860	2.306	2.896	3.355	5.041
9	1.383	1.833	2.262	2.821	3.250	4.781
10	1.372	1.812	2.228	2.764	3.169	4.587
11	1.363	1.796	2.201	2.718	3.106	4.437
12	1.356	1.782	2.179	2.681	3.055	4.318
13	1.350	1.771	2.160	2.650	3.012	4.221
14	1.345	1.761	2.145	2.624	2.977	4.140
15	1.341	1.753	2.131	2.602	2.947	4.073
16	1.337	1.746	2.120	2.583	2.921	4.015
17	1.333	1.740	2.110	2.567	2.898	3.965
18	1.330	1.734	2.101	2.552	2.878	3.922
19	1.328	1.729	2.093	2.539	2.861	3.883
20	1.325	1.725	2.086	2.528	2.845	3.850
21	1.323	1.721	2.080	2.518	2.831	3.819
22	1.321	1.717	2.074	2.508	2.819	3.792
23	1.319	1.714	2.069	2.500	2.807	3.767
24	1.318	1.711	2.064	2.492	2.797	3.745
25	1.316	1.708	2.060	2.485	2.787	3.725
26	1.315	1.706	2.056	2.479	2.779	3.707
27	1.314	1.703	2.052	2.473	2.771	3.690
28	1.313	1.701	2.048	2.467	2.763	3.674
29	1.311	1.699	2.045	2.462	2.756	3.659
30	1.310	1.697	2.042	2.457	2.750	3.646
40	1.303	1.684	2.021	2.423	2.704	3.551
60	1.296	1.671	2.000	2.390	2.660	3.460
120	1.289	1.658	1.980	2.358	2.617	3.373
∞	1.282	1.645	1.960	2.326	2.576	3.291

*To be significant the *t* obtained from the data must be equal to or larger than the value shown in the table.

SOURCE: Table III of Fisher and Yates (1963), *Statistical Tables for Biological, Agricultural and Medical Research,* published by Longman Group Ltd., London (previously published by Oliver & Boyd, Edinburgh), and by permission of the authors and publishers.

Table E   Chi square distribution*

df	.10	.05	.02	.01	.001
			α levels		
1	2.71	3.84	5.41	6.64	10.83
2	4.60	5.99	7.82	9.21	13.82
3	6.25	7.82	9.84	11.34	16.27
4	7.78	9.49	11.67	13.28	18.46
5	9.24	11.07	13.39	15.09	20.52
6	10.64	12.59	15.03	16.81	22.46
7	12.02	14.07	16.62	18.48	24.32
8	13.36	15.51	18.17	20.09	26.12
9	14.68	16.92	19.68	21.67	27.88
10	15.99	18.31	21.16	23.21	29.59
11	17.28	19.68	22.62	24.72	31.26
12	18.55	21.03	24.05	26.22	32.91
13	19.81	22.36	25.47	27.69	34.53
14	21.06	23.68	26.87	29.14	36.12
15	22.31	25.00	28.26	30.58	37.70
16	23.54	26.30	29.63	32.00	39.25
17	24.77	27.59	31.00	33.41	40.79
18	25.99	28.87	32.35	34.80	42.31
19	27.20	30.14	33.69	36.19	43.82
20	28.41	31.41	35.02	37.57	45.32
21	29.62	32.67	36.34	38.93	46.80
22	30.81	33.92	37.66	40.29	48.27
23	32.01	35.17	38.97	41.64	49.73
24	33.20	36.42	40.27	42.98	51.18
25	34.38	37.65	41.57	44.31	52.62
26	35.56	38.88	42.86	45.64	54.05
27	36.74	40.11	44.14	46.96	55.48
28	37.92	41.34	45.42	48.28	56.89
29	39.09	42.56	46.69	49.59	58.30
30	40.26	43.77	47.96	50.89	59.70

*To be significant the χ^2 obtained from the data must be equal to or larger than the value shown in the table.

SOURCE: Table IV of Fisher and Yates (1963), *Statistical Tables for Biological, Agricultural and Medical Research,* published by Longman Group Ltd., London (previously published by the Oliver & Boyd, Edinburgh), and by permission of the authors and publishers.

Table F   The F distribution*

.05 (roman) and .01 (boldface) α levels for the distribution of F

Degrees of freedom (for the numerator) — df $\frac{b}{W}$

Degrees of freedom (for the denominator) — df $\frac{W}{n}$

df (denom)	1	2	3	4	5	6	7	8	9	10	11	12	14	16	20	24	30	40	50	75	100	200	500	∞
1	161	200	216	225	230	234	237	239	241	242	243	244	245	246	248	249	250	251	252	253	253	254	254	254
	4,052	**4,999**	**5,403**	**5,625**	**5,764**	**5,859**	**5,928**	**5,981**	**6,022**	**6,056**	**6,082**	**6,106**	**6,142**	**6,169**	**6,208**	**6,234**	**6,258**	**6,286**	**6,302**	**6,323**	**6,334**	**6,352**	**6,361**	**6,366**
2	18.51	19.00	19.16	19.25	19.30	19.33	19.36	19.37	19.38	19.39	19.40	19.41	19.42	19.43	19.44	19.45	19.46	19.47	19.47	19.48	19.49	19.49	19.50	19.50
	98.49	**99.00**	**99.17**	**99.25**	**99.30**	**99.33**	**99.34**	**99.36**	**99.38**	**99.40**	**99.41**	**99.42**	**99.43**	**99.44**	**99.45**	**99.46**	**99.47**	**99.48**	**99.48**	**99.49**	**99.49**	**99.49**	**99.50**	**99.50**
3	10.13	9.55	9.28	9.12	9.01	8.94	8.88	8.84	8.81	8.78	8.76	8.74	8.71	8.69	8.66	8.64	8.62	8.60	8.58	8.57	8.56	8.54	8.54	8.53
	34.12	**30.82**	**29.46**	**28.71**	**28.24**	**27.91**	**27.67**	**27.49**	**27.34**	**27.23**	**27.13**	**27.05**	**26.92**	**26.83**	**26.69**	**26.60**	**26.50**	**26.41**	**26.35**	**26.27**	**26.23**	**26.18**	**26.14**	**26.12**
4	7.71	6.94	6.59	6.39	6.26	6.16	6.09	6.04	6.00	5.96	5.93	5.91	5.87	5.84	5.80	5.77	5.74	5.71	5.70	5.68	5.66	5.65	5.64	5.63
	21.20	**18.00**	**16.69**	**15.98**	**15.52**	**15.21**	**14.98**	**14.80**	**14.66**	**14.54**	**14.45**	**14.37**	**14.24**	**14.15**	**14.02**	**13.93**	**13.83**	**13.74**	**13.69**	**13.61**	**13.57**	**13.52**	**13.48**	**13.46**
5	6.61	5.79	5.41	5.19	5.05	4.95	4.88	4.82	4.78	4.74	4.70	4.68	4.64	4.60	4.56	4.53	4.50	4.46	4.44	4.42	4.40	4.38	4.37	4.36
	16.26	**13.27**	**12.06**	**11.39**	**10.97**	**10.67**	**10.45**	**10.27**	**10.15**	**10.05**	**9.96**	**9.89**	**9.77**	**9.68**	**9.55**	**9.47**	**9.38**	**9.29**	**9.24**	**9.17**	**9.13**	**9.07**	**9.04**	**9.02**
6	5.99	5.14	4.76	4.53	4.39	4.28	4.21	4.15	4.10	4.06	4.03	4.00	3.96	3.92	3.87	3.84	3.81	3.77	3.75	3.72	3.71	3.69	3.68	3.67
	13.74	**10.92**	**9.78**	**9.15**	**8.75**	**8.47**	**8.26**	**8.10**	**7.98**	**7.87**	**7.79**	**7.72**	**7.60**	**7.52**	**7.39**	**7.31**	**7.23**	**7.14**	**7.09**	**7.02**	**6.99**	**6.94**	**6.90**	**6.88**
7	5.59	4.74	4.35	4.12	3.97	3.87	3.79	3.73	3.68	3.63	3.60	3.57	3.52	3.49	3.44	3.41	3.38	3.34	3.32	3.29	3.28	3.25	3.24	3.23
	12.25	**9.55**	**8.45**	**7.85**	**7.46**	**7.19**	**7.00**	**6.84**	**6.71**	**6.62**	**6.54**	**6.47**	**6.35**	**6.27**	**6.15**	**6.07**	**5.98**	**5.90**	**5.85**	**5.78**	**5.75**	**5.70**	**5.67**	**5.65**
8	5.32	4.46	4.07	3.84	3.69	3.58	3.50	3.44	3.39	3.34	3.31	3.28	3.23	3.20	3.15	3.12	3.08	3.05	3.03	3.00	2.98	2.96	2.94	2.93
	11.26	**8.65**	**7.59**	**7.01**	**6.63**	**6.37**	**6.19**	**6.03**	**5.91**	**5.82**	**5.74**	**5.67**	**5.56**	**5.48**	**5.36**	**5.28**	**5.20**	**5.11**	**5.06**	**5.00**	**4.96**	**4.91**	**4.88**	**4.86**
9	5.12	4.26	3.86	3.63	3.48	3.37	3.29	3.23	3.18	3.13	3.10	3.07	3.02	2.98	2.93	2.90	2.86	2.82	2.80	2.77	2.76	2.73	2.72	2.71
	10.56	**8.02**	**6.99**	**6.42**	**6.06**	**5.80**	**5.62**	**5.47**	**5.35**	**5.26**	**5.18**	**5.11**	**5.00**	**4.92**	**4.80**	**4.73**	**4.64**	**4.56**	**4.51**	**4.45**	**4.41**	**4.36**	**4.33**	**4.31**
10	4.96	4.10	3.71	3.48	3.33	3.22	3.14	3.07	3.02	2.97	2.94	2.91	2.86	2.82	2.77	2.74	2.70	2.67	2.64	2.61	2.59	2.56	2.55	2.54
	10.04	**7.56**	**6.55**	**5.99**	**5.64**	**5.39**	**5.21**	**5.06**	**4.95**	**4.85**	**4.78**	**4.71**	**4.60**	**4.52**	**4.41**	**4.33**	**4.25**	**4.17**	**4.12**	**4.05**	**4.01**	**3.96**	**3.93**	**3.91**
11	4.84	3.98	3.59	3.36	3.20	3.09	3.01	2.95	2.90	2.86	2.82	2.79	2.74	2.70	2.65	2.61	2.57	2.53	2.50	2.47	2.45	2.42	2.41	2.40
	9.65	**7.20**	**6.22**	**5.67**	**5.32**	**5.07**	**4.88**	**4.74**	**4.63**	**4.54**	**4.46**	**4.40**	**4.29**	**4.21**	**4.10**	**4.02**	**3.94**	**3.86**	**3.80**	**3.74**	**3.70**	**3.66**	**3.62**	**3.60**
12	4.75	3.88	3.49	3.26	3.11	3.00	2.92	2.85	2.80	2.76	2.72	2.69	2.64	2.60	2.54	2.50	2.46	2.42	2.40	2.36	2.35	2.32	2.31	2.30
	9.33	**6.93**	**5.95**	**5.41**	**5.06**	**4.82**	**4.65**	**4.50**	**4.39**	**4.30**	**4.22**	**4.16**	**4.05**	**3.98**	**3.86**	**3.78**	**3.70**	**3.61**	**3.56**	**3.49**	**3.46**	**3.41**	**3.38**	**3.36**
13	4.67	3.80	3.41	3.18	3.02	2.92	2.84	2.77	2.72	2.67	2.63	2.60	2.55	2.51	2.46	2.42	2.38	2.34	2.32	2.28	2.26	2.24	2.22	2.21
	9.07	**6.70**	**5.74**	**5.20**	**4.86**	**4.62**	**4.44**	**4.30**	**4.19**	**4.10**	**4.02**	**3.96**	**3.85**	**3.78**	**3.67**	**3.59**	**3.51**	**3.42**	**3.37**	**3.30**	**3.27**	**3.21**	**3.18**	**3.16**

(continued)

*To be significant the F obtained from the data must be equal to or larger than the value shown in the table.

SOURCE: *Statistical Methods* (6th ed.), by G. W. Snedecor and W. G. Cochran. Copyright © 1967 by Iowa State University Press, Ames, Iowa. Reprinted by permission.

Table F *(continued)*

Degrees of freedom (for the numerator)

denom	1	2	3	4	5	6	7	8	9	10	11	12	14	16	20	24	30	40	50	75	100	200	500	∞
14	4.60 **8.86**	3.74 **6.51**	3.34 **5.56**	3.11 **5.03**	2.96 **4.69**	2.85 **4.46**	2.77 **4.28**	2.70 **4.14**	2.65 **4.03**	2.60 **3.94**	2.56 **3.86**	2.53 **3.80**	2.48 **3.70**	2.44 **3.62**	2.39 **3.51**	2.35 **3.43**	2.31 **3.34**	2.27 **3.26**	2.24 **3.21**	2.21 **3.14**	2.19 **3.11**	2.16 **3.06**	2.14 **3.02**	2.13 **3.00**
15	4.54 **8.68**	3.68 **6.36**	3.29 **5.42**	3.06 **4.89**	2.90 **4.56**	2.79 **4.32**	2.70 **4.14**	2.64 **4.00**	2.59 **3.89**	2.55 **3.80**	2.51 **3.73**	2.48 **3.67**	2.43 **3.56**	2.39 **3.48**	2.33 **3.36**	2.29 **3.29**	2.25 **3.20**	2.21 **3.12**	2.18 **3.07**	2.15 **3.00**	2.12 **2.97**	2.10 **2.92**	2.08 **2.89**	2.07 **2.87**
16	4.49 **8.53**	3.63 **6.23**	3.24 **5.29**	3.01 **4.77**	2.85 **4.44**	2.74 **4.20**	2.66 **4.03**	2.59 **3.89**	2.54 **3.78**	2.49 **3.69**	2.45 **3.61**	2.42 **3.55**	2.37 **3.45**	2.33 **3.37**	2.28 **3.25**	2.24 **3.18**	2.20 **3.10**	2.16 **3.01**	2.13 **2.96**	2.09 **2.89**	2.07 **2.86**	2.04 **2.80**	2.02 **2.77**	2.01 **2.75**
17	4.45 **8.40**	3.59 **6.11**	3.20 **5.18**	2.96 **4.67**	2.81 **4.34**	2.70 **4.10**	2.62 **3.93**	2.55 **3.79**	2.50 **3.68**	2.45 **3.59**	2.41 **3.52**	2.38 **3.45**	2.33 **3.35**	2.29 **3.27**	2.23 **3.16**	2.19 **3.08**	2.15 **3.00**	2.11 **2.92**	2.08 **2.86**	2.04 **2.79**	2.02 **2.76**	1.99 **2.70**	1.97 **2.67**	1.96 **2.65**
18	4.41 **8.28**	3.55 **6.01**	3.16 **5.09**	2.93 **4.58**	2.77 **4.25**	2.66 **4.01**	2.58 **3.85**	2.51 **3.71**	2.46 **3.60**	2.41 **3.51**	2.37 **3.44**	2.34 **3.37**	2.29 **3.27**	2.25 **3.19**	2.19 **3.07**	2.15 **3.00**	2.11 **2.91**	2.07 **2.83**	2.04 **2.78**	2.00 **2.71**	1.89 **2.68**	1.95 **2.62**	1.93 **2.59**	1.92 **2.57**
19	4.38 **8.18**	3.52 **5.93**	3.13 **5.01**	2.90 **4.50**	2.74 **4.17**	2.63 **3.94**	2.55 **3.77**	2.48 **3.63**	2.43 **3.52**	2.38 **3.43**	2.34 **3.36**	2.31 **3.30**	2.26 **3.19**	2.21 **3.12**	2.15 **3.00**	2.11 **2.92**	2.07 **2.84**	2.02 **2.76**	2.00 **2.70**	1.96 **2.63**	1.94 **2.60**	1.91 **2.54**	1.90 **2.51**	1.88 **2.49**
20	4.35 **8.10**	3.49 **5.85**	3.10 **4.94**	2.87 **4.43**	2.71 **4.10**	2.60 **3.87**	2.52 **3.71**	2.45 **3.56**	2.40 **3.45**	2.35 **3.37**	2.31 **3.30**	2.28 **3.23**	2.23 **3.13**	2.18 **3.05**	2.12 **2.94**	2.08 **2.86**	2.04 **2.77**	1.99 **2.69**	1.96 **2.63**	1.92 **2.56**	1.90 **2.53**	1.87 **2.47**	1.85 **2.44**	1.84 **2.42**
21	4.32 **8.02**	3.47 **5.78**	3.07 **4.87**	2.84 **4.37**	2.68 **4.04**	2.57 **3.81**	2.49 **3.65**	2.42 **3.51**	2.37 **3.40**	2.32 **3.31**	2.28 **3.24**	2.25 **3.17**	2.20 **3.07**	2.15 **2.99**	2.09 **2.88**	2.05 **2.80**	2.00 **2.72**	1.96 **2.63**	1.93 **2.58**	1.89 **2.51**	1.87 **2.47**	1.84 **2.42**	1.82 **2.38**	1.81 **2.36**
22	4.30 **7.94**	3.44 **5.72**	3.05 **4.82**	2.82 **4.31**	2.66 **3.99**	2.55 **3.76**	2.47 **3.59**	2.40 **3.45**	2.35 **3.35**	2.30 **3.26**	2.26 **3.18**	2.23 **3.12**	2.18 **3.02**	2.13 **2.94**	2.07 **2.83**	2.03 **2.75**	1.98 **2.67**	1.93 **2.58**	1.91 **2.53**	1.87 **2.46**	1.84 **2.42**	1.81 **2.37**	1.80 **2.33**	1.78 **2.31**
23	4.28 **7.88**	3.42 **5.66**	3.03 **4.76**	2.80 **4.26**	2.64 **3.94**	2.53 **3.71**	2.45 **3.54**	2.38 **3.41**	2.32 **3.30**	2.28 **3.21**	2.24 **3.14**	2.20 **3.07**	2.14 **2.97**	2.10 **2.89**	2.04 **2.78**	2.00 **2.70**	1.96 **2.62**	1.91 **2.53**	1.88 **2.48**	1.84 **2.41**	1.82 **2.37**	1.79 **2.32**	1.77 **2.28**	1.76 **2.26**
24	4.26 **7.82**	3.40 **5.61**	3.01 **4.72**	2.78 **4.22**	2.62 **3.90**	2.51 **3.67**	2.43 **3.50**	2.36 **3.36**	2.30 **3.25**	2.26 **3.17**	2.22 **3.09**	2.18 **3.03**	2.13 **2.93**	2.09 **2.85**	2.02 **2.74**	1.98 **2.66**	1.94 **2.58**	1.89 **2.49**	1.86 **2.44**	1.82 **2.36**	1.80 **2.33**	1.76 **2.27**	1.74 **2.23**	1.73 **2.21**
25	4.24 **7.77**	3.38 **5.57**	2.99 **4.68**	2.76 **4.18**	2.60 **3.86**	2.49 **3.63**	2.41 **3.46**	2.34 **3.32**	2.28 **3.21**	2.24 **3.13**	2.20 **3.05**	2.16 **2.99**	2.11 **2.89**	2.06 **2.81**	2.00 **2.70**	1.96 **2.62**	1.92 **2.54**	1.87 **2.45**	1.84 **2.40**	1.80 **2.32**	1.77 **2.29**	1.74 **2.23**	1.72 **2.19**	1.71 **2.17**
26	4.22 **7.72**	3.37 **5.53**	2.98 **4.64**	2.74 **4.14**	2.59 **3.82**	2.47 **3.59**	2.39 **3.42**	2.32 **3.29**	2.27 **3.17**	2.22 **3.09**	2.18 **3.02**	2.15 **2.96**	2.10 **2.86**	2.05 **2.77**	1.99 **2.66**	1.95 **2.58**	1.90 **2.50**	1.85 **2.41**	1.82 **2.36**	1.78 **2.28**	1.76 **2.25**	1.72 **2.19**	1.70 **2.15**	1.69 **2.13**

Degrees of freedom (for the denominator)

(continued)

Table F (continued)

	Degrees of freedom (for the numerator)																							
	1	2	3	4	5	6	7	8	9	10	11	12	14	16	20	24	30	40	50	75	100	200	500	∞
27	4.21 **7.68**	3.35 **5.49**	2.96 **4.60**	2.73 **4.11**	2.57 **3.79**	2.46 **3.56**	2.37 **3.39**	2.30 **3.26**	2.25 **3.14**	2.20 **3.06**	2.16 **2.98**	2.13 **2.93**	2.08 **2.83**	2.03 **2.74**	1.97 **2.63**	1.93 **2.55**	1.88 **2.47**	1.84 **2.38**	1.80 **2.33**	1.76 **2.25**	1.74 **2.21**	1.71 **2.16**	1.68 **2.12**	1.67 **2.10**
28	4.20 **7.64**	3.34 **5.45**	2.95 **4.57**	2.71 **4.07**	2.56 **3.76**	2.44 **3.53**	2.36 **3.36**	2.29 **3.23**	2.24 **3.11**	2.19 **3.03**	2.15 **2.95**	2.12 **2.90**	2.06 **2.80**	2.02 **2.71**	1.96 **2.60**	1.91 **2.52**	1.87 **2.44**	1.81 **2.35**	1.78 **2.30**	1.75 **2.22**	1.72 **2.18**	1.69 **2.13**	1.67 **2.09**	1.65 **2.06**
29	4.18 **7.60**	3.33 **5.42**	2.93 **4.54**	2.70 **4.04**	2.54 **3.73**	2.43 **3.50**	2.35 **3.33**	2.28 **3.20**	2.22 **3.08**	2.18 **3.00**	2.14 **2.92**	2.10 **2.87**	2.05 **2.77**	2.00 **2.68**	1.94 **2.57**	1.90 **2.49**	1.85 **2.41**	1.80 **2.32**	1.77 **2.27**	1.73 **2.19**	1.71 **2.15**	1.68 **2.10**	1.65 **2.06**	1.64 **2.03**
30	4.17 **7.56**	3.32 **5.39**	2.92 **4.51**	2.69 **4.02**	2.53 **3.70**	2.42 **3.47**	2.34 **3.30**	2.27 **3.17**	2.21 **3.06**	2.16 **2.98**	2.12 **2.90**	2.09 **2.84**	2.04 **2.74**	1.99 **2.66**	1.93 **2.55**	1.89 **2.47**	1.84 **2.38**	1.79 **2.29**	1.76 **2.24**	1.72 **2.16**	1.69 **2.13**	1.66 **2.07**	1.64 **2.03**	1.62 **2.01**
32	4.15 **7.50**	3.30 **5.34**	2.90 **4.46**	2.67 **3.97**	2.51 **3.66**	2.40 **3.42**	2.32 **3.25**	2.25 **3.12**	2.19 **3.01**	2.14 **2.94**	2.10 **2.86**	2.07 **2.80**	2.02 **2.70**	1.97 **2.62**	1.91 **2.51**	1.86 **2.42**	1.82 **2.34**	1.76 **2.25**	1.74 **2.20**	1.69 **2.12**	1.67 **2.08**	1.64 **2.02**	1.61 **1.98**	1.59 **1.96**
34	4.13 **7.44**	3.28 **5.29**	2.88 **4.42**	2.65 **3.93**	2.49 **3.61**	2.38 **3.38**	2.30 **3.21**	2.23 **3.08**	2.17 **2.97**	2.12 **2.89**	2.08 **2.82**	2.05 **2.76**	2.00 **2.66**	1.95 **2.58**	1.89 **2.47**	1.84 **2.38**	1.80 **2.30**	1.74 **2.21**	1.71 **2.15**	1.67 **2.08**	1.64 **2.04**	1.61 **1.98**	1.59 **1.94**	1.57 **1.91**
36	4.11 **7.39**	3.26 **5.25**	2.86 **4.38**	2.63 **3.89**	2.48 **3.58**	2.36 **3.35**	2.28 **3.18**	2.21 **3.04**	2.15 **2.94**	2.10 **2.86**	2.06 **2.78**	2.03 **2.72**	1.98 **2.62**	1.93 **2.54**	1.87 **2.43**	1.82 **2.35**	1.78 **2.26**	1.72 **2.17**	1.69 **2.12**	1.65 **2.04**	1.62 **2.00**	1.59 **1.94**	1.56 **1.90**	1.55 **1.87**
38	4.10 **7.35**	3.25 **5.21**	2.85 **4.34**	2.62 **3.86**	2.46 **3.54**	2.35 **3.32**	2.26 **3.15**	2.19 **3.02**	2.14 **2.91**	2.09 **2.82**	2.05 **2.75**	2.02 **2.69**	1.96 **2.59**	1.92 **2.51**	1.85 **2.40**	1.80 **2.32**	1.76 **2.22**	1.71 **2.14**	1.67 **2.08**	1.63 **2.00**	1.60 **1.97**	1.57 **1.90**	1.54 **1.86**	1.53 **1.84**
40	4.08 **7.31**	3.23 **5.18**	2.84 **4.31**	2.61 **3.83**	2.45 **3.51**	2.34 **3.29**	2.25 **3.12**	2.18 **2.99**	2.12 **2.88**	2.07 **2.80**	2.04 **2.73**	2.00 **2.66**	1.95 **2.56**	1.90 **2.49**	1.84 **2.37**	1.79 **2.29**	1.74 **2.20**	1.69 **2.11**	1.66 **2.05**	1.61 **1.97**	1.59 **1.94**	1.55 **1.88**	1.53 **1.84**	1.51 **1.81**
42	4.07 **7.27**	3.22 **5.15**	2.83 **4.29**	2.59 **3.80**	2.44 **3.49**	2.32 **3.26**	2.24 **3.10**	2.17 **2.96**	2.11 **2.86**	2.06 **2.77**	2.02 **2.70**	1.99 **2.64**	1.94 **2.54**	1.89 **2.46**	1.82 **2.35**	1.78 **2.26**	1.73 **2.17**	1.68 **2.08**	1.64 **2.02**	1.60 **1.94**	1.57 **1.91**	1.54 **1.85**	1.51 **1.80**	1.49 **1.78**
44	4.06 **7.24**	3.21 **5.12**	2.82 **4.26**	2.58 **3.78**	2.43 **3.46**	2.31 **3.24**	2.23 **3.07**	2.16 **2.94**	2.10 **2.84**	2.05 **2.75**	2.01 **2.68**	1.98 **2.62**	1.92 **2.52**	1.88 **2.44**	1.81 **2.32**	1.76 **2.24**	1.72 **2.15**	1.66 **2.06**	1.63 **2.00**	1.58 **1.92**	1.56 **1.88**	1.52 **1.82**	1.50 **1.78**	1.48 **1.75**
46	4.05 **7.21**	3.20 **5.10**	2.81 **4.24**	2.57 **3.76**	2.42 **3.44**	2.30 **3.22**	2.22 **3.05**	2.14 **2.92**	2.09 **2.82**	2.04 **2.73**	2.00 **2.66**	1.97 **2.60**	1.91 **2.50**	1.87 **2.42**	1.80 **2.30**	1.75 **2.22**	1.71 **2.13**	1.65 **2.04**	1.62 **1.98**	1.57 **1.90**	1.54 **1.86**	1.51 **1.80**	1.48 **1.76**	1.46 **1.72**
48	4.04 **7.19**	3.19 **5.08**	2.80 **4.22**	2.56 **3.74**	2.41 **3.42**	2.30 **3.20**	2.21 **3.04**	2.14 **2.90**	2.08 **2.80**	2.03 **2.71**	1.99 **2.64**	1.96 **2.58**	1.90 **2.48**	1.86 **2.40**	1.79 **2.28**	1.74 **2.20**	1.70 **2.11**	1.64 **2.02**	1.61 **1.96**	1.56 **1.88**	1.53 **1.84**	1.50 **1.78**	1.47 **1.73**	1.45 **1.70**

Degrees of freedom (for the denominator)

(continued)

Degrees of freedom (for the numerator)

	1	2	3	4	5	6	7	8	9	10	11	12	14	16	20	24	30	40	50	75	100	200	500	∞
50	4.03 / 7.17	3.18 / 5.06	2.79 / 4.20	2.56 / 3.72	2.40 / 3.41	2.29 / 3.18	2.20 / 3.02	2.13 / 2.88	2.07 / 2.78	2.02 / 2.70	1.98 / 2.62	1.95 / 2.56	1.90 / 2.46	1.85 / 2.39	1.78 / 2.26	1.74 / 2.18	1.69 / 2.10	1.63 / 2.00	1.60 / 1.94	1.55 / 1.86	1.52 / 1.82	1.48 / 1.76	1.46 / 1.71	1.44 / 1.68
55	4.02 / 7.12	3.17 / 5.01	2.78 / 4.16	2.54 / 3.68	2.38 / 3.37	2.27 / 3.15	2.18 / 2.98	2.11 / 2.85	2.05 / 2.75	2.00 / 2.66	1.97 / 2.59	1.93 / 2.53	1.88 / 2.43	1.83 / 2.35	1.76 / 2.23	1.72 / 2.15	1.67 / 2.06	1.61 / 1.96	1.58 / 1.90	1.52 / 1.82	1.50 / 1.78	1.46 / 1.71	1.43 / 1.66	1.41 / 1.64
60	4.00 / 7.08	3.15 / 4.98	2.76 / 4.13	2.52 / 3.65	2.37 / 3.34	2.25 / 3.12	2.17 / 2.95	2.10 / 2.82	2.04 / 2.72	1.99 / 2.63	1.95 / 2.56	1.92 / 2.50	1.86 / 2.40	1.81 / 2.32	1.75 / 2.20	1.70 / 2.12	1.65 / 2.03	1.59 / 1.93	1.56 / 1.87	1.50 / 1.79	1.48 / 1.74	1.44 / 1.68	1.41 / 1.63	1.39 / 1.60
65	3.99 / 7.04	3.14 / 4.95	2.75 / 4.10	2.51 / 3.62	2.36 / 3.31	2.24 / 3.09	2.15 / 2.93	2.08 / 2.79	2.02 / 2.70	1.98 / 2.61	1.94 / 2.54	1.90 / 2.47	1.85 / 2.37	1.80 / 2.30	1.73 / 2.18	1.68 / 2.09	1.63 / 2.00	1.57 / 1.90	1.54 / 1.84	1.49 / 1.76	1.46 / 1.71	1.42 / 1.64	1.39 / 1.60	1.37 / 1.56
70	3.98 / 7.01	3.13 / 4.92	2.74 / 4.08	2.50 / 3.60	2.35 / 3.29	2.23 / 3.07	2.14 / 2.91	2.07 / 2.77	2.01 / 2.67	1.97 / 2.59	1.93 / 2.51	1.89 / 2.45	1.84 / 2.35	1.79 / 2.28	1.72 / 2.15	1.67 / 2.07	1.62 / 1.98	1.56 / 1.88	1.53 / 1.82	1.47 / 1.74	1.45 / 1.69	1.40 / 1.62	1.37 / 1.56	1.35 / 1.53
80	3.96 / 6.96	3.11 / 4.88	2.72 / 4.04	2.48 / 3.56	2.33 / 3.25	2.21 / 3.04	2.12 / 2.87	2.05 / 2.74	1.99 / 2.64	1.95 / 2.55	1.91 / 2.48	1.88 / 2.41	1.82 / 2.32	1.77 / 2.24	1.70 / 2.11	1.65 / 2.03	1.60 / 1.94	1.54 / 1.84	1.51 / 1.78	1.45 / 1.70	1.42 / 1.65	1.38 / 1.57	1.35 / 1.52	1.32 / 1.49
100	3.94 / 6.90	3.09 / 4.82	2.70 / 3.98	2.46 / 3.51	2.30 / 3.20	2.19 / 2.99	2.10 / 2.82	2.03 / 2.69	1.97 / 2.59	1.92 / 2.51	1.88 / 2.43	1.85 / 2.36	1.79 / 2.26	1.75 / 2.19	1.68 / 2.06	1.63 / 1.98	1.57 / 1.89	1.51 / 1.79	1.48 / 1.73	1.42 / 1.64	1.39 / 1.59	1.34 / 1.51	1.30 / 1.46	1.28 / 1.43
125	3.92 / 6.84	3.07 / 4.78	2.68 / 3.94	2.44 / 3.47	2.29 / 3.17	2.17 / 2.95	2.08 / 2.79	2.01 / 2.65	1.95 / 2.56	1.90 / 2.47	1.86 / 2.40	1.83 / 2.33	1.77 / 2.23	1.72 / 2.15	1.65 / 2.03	1.60 / 1.94	1.55 / 1.85	1.49 / 1.75	1.45 / 1.68	1.39 / 1.59	1.36 / 1.54	1.31 / 1.46	1.27 / 1.40	1.25 / 1.37
150	3.91 / 6.81	3.06 / 4.75	2.67 / 3.91	2.43 / 3.44	2.27 / 3.14	2.16 / 2.92	2.07 / 2.76	2.00 / 2.62	1.94 / 2.53	1.89 / 2.44	1.85 / 2.37	1.82 / 2.30	1.76 / 2.20	1.71 / 2.12	1.64 / 2.00	1.59 / 1.91	1.54 / 1.83	1.47 / 1.72	1.44 / 1.66	1.37 / 1.56	1.34 / 1.51	1.29 / 1.43	1.25 / 1.37	1.22 / 1.33
200	3.89 / 6.76	3.04 / 4.71	2.65 / 3.88	2.41 / 3.41	2.26 / 3.11	2.14 / 2.90	2.05 / 2.73	1.98 / 2.60	1.92 / 2.50	1.87 / 2.41	1.83 / 2.34	1.80 / 2.28	1.74 / 2.17	1.69 / 2.09	1.62 / 1.97	1.57 / 1.88	1.52 / 1.79	1.45 / 1.69	1.42 / 1.62	1.35 / 1.53	1.32 / 1.48	1.26 / 1.39	1.22 / 1.33	1.19 / 1.28
400	3.86 / 6.70	3.02 / 4.66	2.62 / 3.83	2.39 / 3.36	2.23 / 3.06	2.12 / 2.85	2.03 / 2.69	1.96 / 2.55	1.90 / 2.46	1.85 / 2.37	1.81 / 2.29	1.78 / 2.23	1.72 / 2.12	1.67 / 2.04	1.60 / 1.92	1.54 / 1.84	1.49 / 1.74	1.42 / 1.64	1.38 / 1.57	1.32 / 1.47	1.28 / 1.42	1.22 / 1.32	1.16 / 1.24	1.13 / 1.19
1000	3.85 / 6.66	3.00 / 4.62	2.61 / 3.80	2.38 / 3.34	2.22 / 3.04	2.10 / 2.82	2.02 / 2.66	1.95 / 2.53	1.89 / 2.43	1.84 / 2.34	1.80 / 2.26	1.76 / 2.20	1.70 / 2.09	1.65 / 2.01	1.58 / 1.89	1.53 / 1.81	1.47 / 1.71	1.41 / 1.61	1.36 / 1.54	1.30 / 1.44	1.26 / 1.38	1.19 / 1.28	1.13 / 1.19	1.08 / 1.11
∞	3.84 / 6.64	2.99 / 4.60	2.60 / 3.78	2.37 / 3.32	2.21 / 3.02	2.09 / 2.80	2.01 / 2.64	1.94 / 2.51	1.88 / 2.41	1.83 / 2.32	1.79 / 2.24	1.75 / 2.18	1.69 / 2.07	1.64 / 1.99	1.57 / 1.87	1.52 / 1.79	1.46 / 1.69	1.40 / 1.59	1.35 / 1.52	1.28 / 1.41	1.24 / 1.36	1.17 / 1.25	1.11 / 1.15	1.00 / 1.00

Degrees of freedom (for the denominator)

Table G   Critical values of the studentized range statistic

$\alpha = .05$

df_{wg}	\multicolumn{14}{c	}{Number of means}
	2	3	4	5	6	7	8	9	10	11	12	13	14	15
1	17.97	26.98	32.82	37.08	40.41	43.12	45.40	47.36	49.07	50.59	51.96	53.20	54.33	55.36
2	6.08	8.33	9.80	10.88	11.74	12.44	13.03	13.54	13.99	14.39	14.75	15.08	15.38	15.65
3	4.50	5.91	6.82	7.50	8.04	8.48	8.85	9.18	9.46	9.72	9.95	10.15	10.35	10.53
4	3.93	5.04	5.76	6.29	6.71	7.05	7.35	7.60	7.83	8.03	8.21	8.37	8.52	8.66
5	3.64	4.60	5.22	5.67	6.03	6.33	6.58	6.80	7.00	7.17	7.32	7.47	7.60	7.72
6	3.46	4.34	4.90	5.31	5.63	5.90	6.12	6.32	6.49	6.65	6.79	6.92	7.03	7.14
7	3.34	4.16	4.68	5.06	5.36	5.61	5.82	6.00	6.16	6.30	6.43	6.55	6.66	6.76
8	3.26	4.04	4.53	4.89	5.17	5.40	5.60	5.77	5.92	6.05	6.18	6.29	6.39	6.48
9	3.20	3.95	4.42	4.76	5.02	5.24	5.43	5.60	5.74	5.87	5.98	6.09	6.19	6.28
10	3.15	3.88	4.33	4.65	4.91	5.12	5.30	5.46	5.60	5.72	5.83	5.94	6.03	6.11
11	3.11	3.82	4.26	4.57	4.82	5.03	5.20	5.35	5.49	5.60	5.71	5.81	5.90	5.98
12	3.08	3.77	4.20	4.51	4.75	4.95	5.12	5.26	5.40	5.51	5.62	5.71	5.79	5.88
13	3.06	3.74	4.15	4.45	4.69	4.88	5.05	5.19	5.32	5.43	5.53	5.63	5.71	5.79
14	3.03	3.70	4.11	4.41	4.64	4.83	4.99	5.13	5.25	5.36	5.46	5.55	5.64	5.71
15	3.01	3.67	4.08	4.37	4.60	4.78	4.94	5.08	5.20	5.31	5.40	5.49	5.57	5.65
16	3.00	3.65	4.05	4.33	4.56	4.74	4.90	5.03	5.15	5.26	5.35	5.44	5.52	5.59
17	2.98	3.63	4.02	4.30	4.52	4.70	4.86	4.99	5.11	5.21	5.31	5.39	5.47	5.54
18	2.97	3.61	4.00	4.28	4.50	4.67	4.82	4.96	5.07	5.17	5.27	5.35	5.43	5.50
19	2.96	3.59	3.98	4.25	4.47	4.64	4.79	4.92	5.04	5.14	5.23	5.32	5.39	5.46
20	2.95	3.58	3.96	4.23	4.44	4.62	4.77	4.90	5.01	5.11	5.20	5.28	5.36	5.43
24	2.92	3.53	3.90	4.17	4.37	4.54	4.68	4.81	4.92	5.01	5.10	5.18	5.25	5.32
30	2.89	3.49	3.84	4.10	4.30	4.46	4.60	4.72	4.82	4.92	5.00	5.08	5.15	5.21
40	2.86	3.44	3.79	4.04	4.23	4.39	4.52	4.64	4.74	4.82	4.90	4.98	5.04	5.11
60	2.83	3.40	3.74	3.98	4.16	4.31	4.44	4.55	4.65	4.73	4.81	4.88	4.94	5.00
120	2.80	3.36	3.69	3.92	4.10	4.24	4.36	4.47	4.56	4.64	4.71	4.78	4.84	4.90
∞	2.77	3.31	3.63	3.86	4.03	4.17	4.29	4.39	4.47	4.55	4.62	4.68	4.74	4.80

(continued)

SOURCE: This table is abridged from Harter, H. L. (1960). Tables of range and studentized range. *Annals of Mathematical Statistics, 31*, 1122–1147, with permission of the author and the publisher.

Table G *(continued)*

$\alpha = .01$

df_{wg}	Number of means													
	2	3	4	5	6	7	8	9	10	11	12	13	14	15
1	90.03	135.0	164.3	185.6	202.2	215.8	227.2	237.0	245.6	253.2	260.0	266.2	271.8	277.0
2	14.04	19.02	22.29	24.72	26.63	28.20	29.53	30.68	31.69	32.59	33.40	34.13	34.81	35.43
3	8.26	10.62	12.17	13.33	14.24	15.00	15.64	12.60	16.69	17.13	17.53	17.89	18.22	18.52
4	6.51	8.12	9.17	9.96	10.58	11.10	11.55	11.93	12.27	12.57	12.84	13.09	13.32	13.53
5	5.70	6.98	7.80	8.42	8.91	9.32	9.67	9.97	10.24	10.48	10.70	10.89	11.08	11.24
6	5.24	6.33	7.03	7.56	7.97	8.32	8.62	8.87	9.10	9.30	9.48	9.65	9.81	9.95
7	4.95	5.92	6.54	7.00	7.37	7.68	7.94	8.17	8.37	8.55	8.71	8.86	9.00	9.12
8	4.75	5.64	6.20	6.62	6.96	7.24	7.47	7.68	7.86	8.03	8.18	8.31	8.44	8.55
9	4.60	5.43	5.96	6.35	6.66	6.92	7.13	7.32	7.50	7.65	7.78	7.91	8.02	8.13
10	4.48	5.27	5.77	6.14	6.43	6.67	6.88	7.06	7.21	7.36	7.48	7.60	7.71	7.81
11	4.39	5.15	5.62	5.97	6.25	6.48	6.67	6.84	6.99	7.13	7.25	7.36	7.46	7.56
12	4.32	5.05	5.50	5.84	6.10	6.32	6.51	6.67	6.81	6.94	7.06	7.17	7.26	7.36
13	4.26	4.96	5.40	5.73	5.98	6.19	6.37	6.53	6.67	6.79	6.90	7.01	7.10	7.19
14	4.21	4.90	5.32	5.63	5.88	6.08	6.26	6.41	6.54	6.66	6.77	6.87	6.96	7.05
15	4.17	4.84	5.25	5.56	5.80	5.99	6.16	6.31	6.44	6.56	6.66	6.76	6.84	6.93
16	4.13	4.79	5.19	5.49	5.72	5.92	6.08	6.22	6.35	6.46	6.56	6.66	6.74	6.82
17	4.10	4.74	5.14	5.43	5.66	5.85	6.01	6.15	6.27	6.38	6.48	6.57	6.66	6.73
18	4.07	4.70	5.09	5.38	5.60	5.79	5.94	6.08	6.20	6.31	6.41	6.50	6.58	6.66
19	4.05	4.67	5.05	5.33	5.55	5.74	5.89	6.02	6.14	6.25	6.34	6.43	6.51	6.58
20	4.02	4.64	5.02	5.29	5.51	5.69	5.84	5.97	6.09	6.19	6.28	6.37	6.45	6.52
24	3.96	4.55	4.91	5.17	5.37	5.54	5.69	5.81	5.92	6.02	6.11	6.19	6.26	6.33
30	3.89	4.46	4.80	5.05	5.24	5.40	5.54	5.65	5.76	5.85	5.93	6.01	6.08	6.14
40	3.82	4.37	4.70	4.93	5.11	5.26	5.39	5.50	5.60	5.69	5.76	5.84	5.90	5.96
60	3.76	4.28	4.60	4.82	4.99	5.13	5.25	5.36	5.45	5.53	5.60	5.67	5.73	5.78
120	3.70	4.20	4.50	4.71	4.87	5.0	5.12	5.21	5.30	5.38	5.44	5.51	5.56	5.61
∞	3.64	4.12	4.40	4.60	4.76	4.88	4.99	5.08	5.16	5.23	5.29	5.35	5.40	5.45

Table H   Critical values for the Mann-Whitney U test*

One-tailed test
$\alpha = .01$ (roman)
$\alpha = .005$ (boldface)

Two-tailed test
$\alpha = .02$ (roman)
$\alpha = .01$ (boldface)

N_2 \ N_1	1	2	3	4	5	6	7	8	9	10	11	12	13	14	15	16	17	18	19	20
1	—	—	—	—	—	—	—	—	—	—	—	—	—	—	—	—	—	—	—	—
2	—	—	—	—	—	—	—	—	—	—	—	—	0	0	0	0	0	0	1	1
													—	—	—	—	—	—	**0**	**0**
3	—	—	—	—	—	—	0	0	1	1	1	2	2	2	3	3	4	4	4	5
							—	—	**0**	**0**	**0**	**1**	**1**	**1**	**2**	**2**	**2**	**2**	**3**	**3**
4	—	—	—	—	0	1	1	2	3	3	4	5	5	6	7	7	8	9	9	10
					—	**0**	**0**	**1**	**1**	**2**	**2**	**3**	**3**	**4**	**5**	**5**	**6**	**6**	**7**	**8**
5	—	—	—	0	1	2	3	4	5	6	7	8	9	10	11	12	13	14	15	16
				—	**0**	**1**	**1**	**2**	**3**	**4**	**5**	**6**	**7**	**7**	**8**	**9**	**10**	**11**	**12**	**13**
6	—	—	—	1	2	3	4	6	7	8	9	11	12	13	15	16	18	19	20	22
				0	**1**	**2**	**3**	**4**	**5**	**6**	**7**	**9**	**10**	**11**	**12**	**13**	**15**	**16**	**17**	**18**
7	—	—	0	1	3	4	6	7	9	11	12	14	16	17	19	21	23	24	26	28
			—	**0**	**1**	**3**	**4**	**6**	**7**	**9**	**10**	**12**	**13**	**15**	**16**	**18**	**19**	**21**	**22**	**24**
8	—	—	0	2	4	6	7	9	11	13	15	17	20	22	24	26	28	30	32	34
			—	**1**	**2**	**4**	**6**	**7**	**9**	**11**	**13**	**15**	**17**	**18**	**20**	**22**	**24**	**26**	**28**	**30**
9	—	—	1	3	5	7	9	11	14	16	18	21	23	26	28	31	33	36	38	40
			0	**1**	**3**	**5**	**7**	**9**	**11**	**13**	**16**	**18**	**20**	**22**	**24**	**27**	**29**	**31**	**33**	**36**
10	—	—	1	3	6	8	11	13	16	19	22	24	27	30	33	36	38	41	44	47
			0	**2**	**4**	**6**	**9**	**11**	**13**	**16**	**18**	**21**	**24**	**26**	**29**	**31**	**34**	**37**	**39**	**42**
11	—	—	1	4	7	9	12	15	18	22	25	28	31	34	37	41	44	47	50	53
			0	**2**	**5**	**7**	**10**	**13**	**16**	**18**	**21**	**24**	**27**	**30**	**33**	**36**	**39**	**42**	**45**	**48**
12	—	—	2	5	8	11	14	17	21	24	28	31	35	38	42	46	49	53	56	60
			1	**3**	**6**	**9**	**12**	**15**	**18**	**21**	**24**	**27**	**31**	**34**	**37**	**41**	**44**	**47**	**51**	**54**
13	—	0	2	5	9	12	16	20	23	27	31	35	39	43	47	51	55	59	63	67
		—	**1**	**3**	**7**	**10**	**13**	**17**	**20**	**24**	**27**	**31**	**34**	**38**	**42**	**45**	**49**	**53**	**56**	**60**
14	—	0	2	6	10	13	17	22	26	30	34	38	43	47	51	56	60	65	69	73
		—	**1**	**4**	**7**	**11**	**15**	**18**	**22**	**26**	**30**	**34**	**38**	**42**	**46**	**50**	**54**	**58**	**63**	**67**
15	—	0	3	7	11	15	19	24	28	33	37	42	47	51	56	61	66	70	75	80
		—	**2**	**5**	**8**	**12**	**16**	**20**	**24**	**29**	**33**	**37**	**42**	**46**	**51**	**55**	**60**	**64**	**69**	**73**
16	—	0	3	7	12	16	21	26	31	36	41	46	51	56	61	66	71	76	82	87
		—	**2**	**5**	**9**	**13**	**18**	**22**	**27**	**31**	**36**	**41**	**45**	**50**	**55**	**60**	**65**	**70**	**74**	**79**
17	—	0	4	8	13	18	23	28	33	38	44	49	55	60	66	71	77	82	88	93
		—	**2**	**6**	**10**	**15**	**19**	**24**	**29**	**34**	**39**	**44**	**49**	**54**	**60**	**65**	**70**	**75**	**81**	**86**
18	—	0	4	9	14	19	24	30	36	41	47	53	59	65	70	76	82	88	94	100
		—	**2**	**6**	**11**	**16**	**21**	**26**	**31**	**37**	**42**	**47**	**53**	**58**	**64**	**70**	**75**	**81**	**87**	**92**
19	—	1	4	9	15	20	26	32	38	44	50	56	63	69	75	82	88	94	101	107
		0	**3**	**7**	**12**	**17**	**22**	**28**	**33**	**39**	**45**	**51**	**56**	**63**	**69**	**74**	**81**	**87**	**93**	**99**
20	—	1	5	10	16	22	28	34	40	47	53	60	67	73	80	87	93	100	107	114
		0	**3**	**8**	**13**	**18**	**24**	**30**	**36**	**42**	**48**	**54**	**60**	**67**	**73**	**79**	**86**	**92**	**99**	**105**

(continued)

*To be significant the U obtained from the data must be *equal to* or *less than* the value shown in the table. Dashes in the body of the table indicate that no decision is possible at the stated level of significance.

SOURCE: *Introductory Statistics*, Second Edition, by Roger E. Kirk. Copyright © 1984, 1978, by Wadsworth Publishing Company, Inc. Reprinted by permission of the publisher, Brooks/Cole Publishing Company, Pacific Grove, Calif.

Table H   (*continued*)

One-tailed test
$\alpha = .05$ (roman)
$\alpha = 0.025$ (**boldface**)

Two-tailed test
$\alpha = .10$ (roman)
$\alpha = .05$ (**boldface**)

In each cell the upper value is roman ($\alpha = .05$ one-tailed / $\alpha = .10$ two-tailed) and the lower value is **boldface** ($\alpha = .025$ one-tailed / $\alpha = .05$ two-tailed).

N_2	1	2	3	4	5	6	7	8	9	10	11	12	13	14	15	16	17	18	19	20
1	—	—	—	—	—	—	—	—	—	—	—	—	—	—	—	—	—	—	0	0
2	—	—	—	—	0	0	0	1	1	1	1	2	2	2	3	3	3	4	4	4
	—	—	—	—	—	—	—	**0**	**0**	**0**	**0**	**1**	**1**	**1**	**1**	**1**	**2**	**2**	**2**	**2**
3	—	—	0	0	1	2	2	3	3	4	5	5	6	7	7	8	9	9	10	11
	—	—	—	—	**0**	**1**	**1**	**2**	**2**	**3**	**3**	**4**	**4**	**5**	**5**	**6**	**6**	**7**	**7**	**8**
4	—	—	0	1	2	3	4	5	6	7	8	9	10	11	12	14	15	16	17	18
	—	—	**0**	**0**	**1**	**2**	**3**	**4**	**4**	**5**	**6**	**7**	**8**	**9**	**10**	**11**	**11**	**12**	**13**	**13**
5	—	0	1	2	4	5	6	8	9	11	12	13	15	16	18	19	20	22	23	25
	—	—	**0**	**1**	**2**	**3**	**5**	**6**	**7**	**8**	**9**	**11**	**12**	**13**	**14**	**15**	**17**	**18**	**19**	**20**
6	—	0	2	3	5	7	8	10	12	14	16	17	19	21	23	25	26	28	30	32
	—	—	**1**	**2**	**3**	**5**	**6**	**8**	**10**	**11**	**13**	**14**	**16**	**17**	**19**	**21**	**22**	**24**	**25**	**27**
7	—	0	2	4	6	8	11	13	15	17	19	21	24	26	28	30	33	35	37	39
	—	—	**1**	**3**	**5**	**6**	**8**	**10**	**12**	**14**	**16**	**18**	**20**	**22**	**24**	**26**	**28**	**30**	**32**	**34**
8	—	1	3	5	8	10	13	15	18	20	23	26	28	31	33	36	39	41	44	47
	—	**0**	**2**	**4**	**6**	**8**	**10**	**13**	**15**	**17**	**19**	**22**	**24**	**26**	**29**	**31**	**34**	**36**	**38**	**41**
9	—	1	3	6	9	12	15	18	21	24	27	30	33	36	39	42	45	48	51	54
	—	**0**	**2**	**4**	**7**	**10**	**12**	**15**	**17**	**20**	**23**	**26**	**28**	**31**	**34**	**37**	**39**	**42**	**45**	**48**
10	—	1	4	7	11	14	17	20	24	27	31	34	37	41	44	48	51	55	58	62
	—	**0**	**3**	**5**	**8**	**11**	**14**	**17**	**20**	**23**	**26**	**29**	**33**	**36**	**39**	**42**	**45**	**48**	**52**	**55**
11	—	1	5	8	12	16	19	23	27	31	34	38	42	46	50	54	57	61	65	69
	—	**0**	**3**	**6**	**9**	**13**	**16**	**19**	**23**	**26**	**30**	**33**	**37**	**40**	**44**	**47**	**51**	**55**	**58**	**62**
12	—	2	5	9	13	17	21	26	30	34	38	42	47	51	55	60	64	68	72	77
	—	**1**	**4**	**7**	**11**	**14**	**18**	**22**	**26**	**29**	**33**	**37**	**41**	**45**	**49**	**53**	**57**	**61**	**65**	**69**
13	—	2	6	10	15	19	24	28	33	37	42	47	51	56	61	65	70	75	80	84
	—	**1**	**4**	**8**	**12**	**16**	**20**	**24**	**28**	**33**	**37**	**41**	**45**	**50**	**54**	**59**	**63**	**67**	**72**	**76**
14	—	2	7	11	16	21	26	31	36	41	46	51	56	61	66	71	77	82	87	92
	—	**1**	**5**	**9**	**13**	**17**	**22**	**26**	**31**	**36**	**40**	**45**	**50**	**55**	**59**	**64**	**69**	**74**	**78**	**83**
15	—	3	7	12	18	23	28	33	39	44	50	55	61	66	72	77	83	88	94	100
	—	**1**	**5**	**10**	**14**	**19**	**24**	**29**	**34**	**39**	**44**	**49**	**54**	**59**	**64**	**70**	**75**	**80**	**85**	**90**
16	—	3	8	14	19	25	30	36	42	48	54	60	65	71	77	83	89	95	101	107
	—	**1**	**6**	**11**	**15**	**21**	**26**	**31**	**37**	**42**	**47**	**53**	**59**	**64**	**70**	**75**	**81**	**86**	**92**	**98**
17	—	3	9	15	20	26	33	39	45	51	57	64	70	77	83	89	96	102	109	115
	—	**2**	**6**	**11**	**17**	**22**	**28**	**34**	**39**	**45**	**51**	**57**	**63**	**69**	**75**	**81**	**87**	**93**	**99**	**105**
18	—	4	9	16	22	28	35	41	48	55	61	68	75	82	88	95	102	109	116	123
	—	**2**	**7**	**12**	**18**	**24**	**30**	**36**	**42**	**48**	**55**	**61**	**67**	**74**	**80**	**86**	**93**	**99**	**106**	**112**
19	0	4	10	17	23	30	37	44	51	58	65	72	80	87	94	101	109	116	123	130
	—	**2**	**7**	**13**	**19**	**25**	**32**	**38**	**45**	**52**	**58**	**65**	**72**	**78**	**85**	**92**	**98**	**105**	**112**	**119**
20	0	4	11	18	25	32	39	47	54	62	69	77	84	92	100	107	115	123	130	138
	—	**2**	**8**	**13**	**20**	**27**	**34**	**41**	**48**	**55**	**62**	**69**	**76**	**83**	**90**	**98**	**105**	**112**	**119**	**127**

Table J   Critical values for the Wilcoxon matched-pairs signed-ranks *T* test*

	Level of significance for a one-tailed test					Level of significance for a one-tailed test		
	.05	.025	.01	.005		.05	.025	.01	.005
	Level of significance for a two-tailed test					Level of significance for a two-tailed test		
N	.10	.05	.02	.01	N	.10	.05	.02	.01
5	0	—	—	—	28	130	116	101	91
6	2	0	—	—	29	140	126	110	100
7	3	2	0	—	30	151	137	120	109
8	5	3	1	0	31	163	147	130	118
9	8	5	3	1	32	175	159	140	128
10	10	8	5	3	33	187	170	151	138
11	13	10	7	5	34	200	182	162	148
12	17	13	9	7	35	213	195	173	159
13	21	17	12	9	36	227	208	185	171
14	25	21	15	12	37	241	221	198	182
15	30	25	19	15	38	256	235	211	194
16	35	29	23	19	39	271	249	224	207
17	41	34	27	23	40	286	264	238	220
18	47	40	32	27	41	302	279	252	233
19	53	46	37	32	42	319	294	266	247
20	60	52	43	37	43	336	310	281	261
21	67	58	49	42	44	353	327	296	276
22	75	65	55	48	45	371	343	312	291
23	83	73	62	54	46	389	361	328	307
24	91	81	69	61	47	407	378	345	322
25	100	89	76	68	48	426	396	362	339
26	110	98	84	75	49	446	415	379	355
27	119	107	92	83	50	466	434	397	373

* To be significant the *T* obtained from the data must be *equal to* or *less than* the value shown in the table.
SOURCE: *Introductory Statistics,* Second Edition, by Roger E. Kirk. Copyright © 1984, 1978 by Wadsworth Publishing Company, Inc. Reprinted by permission of the publisher, Brooks/Cole Publishing Company, Pacific Grove, Calif.

Table K Critical differences for the Wilcoxon-Wilcox multiple-comparisons test*

N				$\alpha = .01$ (*two-tailed*)				
	$K = 3$	$K = 4$	$K = 5$	$K = 6$	$K = 7$	$K = 8$	$K = 9$	$K = 10$
1	4.1	5.7	7.3	8.9	10.5	12.2	13.9	15.6
2	10.9	15.3	19.7	24.3	28.9	33.6	38.3	43.1
3	19.5	27.5	35.7	44.0	52.5	61.1	69.8	78.6
4	29.7	41.9	54.5	67.3	80.3	93.6	107.0	120.6
5	41.2	58.2	75.8	93.6	111.9	130.4	149.1	168.1
6	53.9	76.3	99.3	122.8	146.7	171.0	195.7	220.6
7	67.6	95.8	124.8	154.4	184.6	215.2	246.3	277.7
8	82.4	116.8	152.2	188.4	225.2	262.6	300.6	339.0
9	98.1	139.2	181.4	224.5	268.5	313.1	358.4	404.2
10	114.7	162.8	212.2	262.7	314.2	366.5	419.5	473.1
11	132.1	187.6	244.6	302.9	362.2	422.6	483.7	545.6
12	150.4	213.5	278.5	344.9	412.5	481.2	551.0	621.4
13	169.4	240.6	313.8	388.7	464.9	542.4	621.0	700.5
14	189.1	268.7	350.5	434.2	519.4	606.0	693.8	782.6
15	209.6	297.8	388.5	481.3	575.8	671.9	769.3	867.7
16	230.7	327.9	427.9	530.1	634.2	740.0	847.3	955.7
17	252.5	359.0	468.4	580.3	694.4	810.2	927.8	1046.5
18	275.0	391.0	510.2	632.1	756.4	882.6	1010.6	1140.0
19	298.1	423.8	553.1	685.4	820.1	957.0	1095.8	1236.2
20	321.8	457.6	597.2	740.0	885.5	1033.3	1183.3	1334.9
21	346.1	492.2	642.4	796.0	952.6	1111.6	1273.0	1436.0
22	371.0	527.6	688.7	853.4	1021.3	1191.8	1364.8	1539.7
23	396.4	563.8	736.0	912.1	1091.5	1273.8	1458.8	1645.7
24	422.4	600.9	784.4	972.1	1163.4	1357.6	1554.8	1754.0
25	449.0	638.7	833.8	1033.3	1236.7	1443.2	1652.8	1864.6

N				$\alpha = .05$ (*two-tailed*)				
	$K = 3$	$K = 4$	$K = 5$	$K = 6$	$K = 7$	$K = 8$	$K = 9$	$K = 10$
1	3.3	4.7	6.1	7.5	9.0	10.5	12.0	13.5
2	8.8	12.6	16.5	20.5	24.7	28.9	33.1	37.4
3	15.7	22.7	29.9	37.3	44.8	52.5	60.3	68.2
4	23.9	34.6	45.6	57.0	68.6	80.4	92.4	104.6
5	33.1	48.1	63.5	79.3	95.5	112.0	128.8	145.8
6	43.3	62.9	83.2	104.0	125.3	147.0	169.1	191.4
7	54.4	79.1	104.6	130.8	157.6	184.9	212.8	240.9
8	66.3	96.4	127.6	159.6	192.4	225.7	259.7	294.1
9	78.9	114.8	152.0	190.2	229.3	269.1	309.6	350.6
10	92.3	134.3	177.8	222.6	268.4	315.0	362.4	410.5
11	106.3	154.8	205.0	256.6	309.4	363.2	417.9	473.3
12	120.9	176.2	233.4	292.2	352.4	413.6	476.0	539.1
13	136.2	198.5	263.0	329.3	397.1	466.2	536.5	607.7
14	152.1	221.7	293.8	367.8	443.6	520.8	599.4	679.0
15	168.6	245.7	325.7	407.8	491.9	577.4	664.6	752.8
16	185.6	270.6	358.6	449.1	541.7	635.9	732.0	829.2
17	203.1	296.2	392.6	491.7	593.1	696.3	801.5	907.9
18	221.2	322.6	427.6	535.5	646.1	758.5	873.1	989.0
19	239.8	349.7	463.6	580.6	700.5	822.4	946.7	1072.4
20	258.8	377.6	500.5	626.9	756.4	888.1	1022.3	1158.1
21	278.4	406.1	538.4	674.4	813.7	955.4	1099.8	1245.9
22	298.4	435.3	577.2	723.0	872.3	1024.3	1179.1	1335.7
23	318.9	465.2	616.9	772.7	932.4	1094.8	1260.3	1427.7
24	339.8	495.8	657.4	823.5	993.7	1166.8	1343.2	1521.7
25	361.1	527.0	698.8	875.4	1056.3	1240.4	1427.9	1617.6

*To be significant the difference obtained from the data must be equal to or larger than the tabled value.
SOURCE: *Some Rapid Approximate Statistical Procedures*, by F. Wilcoxon and R. Wilcox (1964). Reprinted by permission.

Table L   Critical values for Spearman's r_s*

N (number of pairs)	Level of significance (α)				
	.20	.10	.05	.02	.01
4		1.00			
5	.80	.90		1.00	
6	.66	.83	.89	.94	1.00
7	.57	.71	.79	.89	.93
8	.52	.64	.74	.83	.88
9	.48	.60	.68	.78	.83
10	.45	.56	.65	.73	.79
11	.41	.52	.61	.71	.77
12	.39	.50	.59	.68	.75
13	.37	.47	.56	.65	.71
14	.36	.46	.54	.63	.69
15	.34	.44	.52	.60	.66
16	.33	.42	.51	.58	.64
17	.32	.41	.49	.57	.62
18	.31	.40	.48	.55	.61
19	.30	.39	.46	.54	.60
20	.29	.38	.45	.53	.58
21	.29	.37	.44	.51	.56
22	.28	.36	.43	.50	.55
23	.27	.35	.42	.49	.54
24	.27	.34	.41	.48	.53
25	.26	.34	.40	.47	.52
26	.26	.33	.39	.46	.51
27	.25	.32	.38	.45	.50
28	.25	.32	.38	.44	.49
29	.24	.31	.37	.44	.48
30	.24	.31	.36	.43	.47

*To be significant, the r_s obtained from the data must be equal to or larger than the value shown in the table.

SOURCE: Reproduced by permission from E. G. Olds, Distribution of sums of squares of rank differences, and the 5% significance levels for sums of squares of rank differences and a correction. *Annals of Mathematical Statistics, 9*: 133–148, 1938 and *20*: 117–118, 1949.

Appendix D
GLOSSARY OF WORDS

Abscissa   The horizontal or X axis of a graph.

Absolute Value   The value of a number without consideration of its algebraic sign.

Alternative Hypothesis   The hypothesis that the means of the populations are not equal; symbolized H_1.

Analysis of Variance (ANOVA)   A statistical method for determining the statistical significance of differences among a set of two or more means.

A Priori Tests   A category of tests designed to test for significant differences among means after an analysis of variance. The number of comparisons permissible is limited and must be planned before examining the data.

Asymptotic   A line that continually approaches but never reaches a specified level.

Bar Graph   A frequency distribution for nominal or qualitative data.

Biased Sample   A sample selected in such a way that all possible samples from the population do not have an equal chance of being chosen.

Bimodal Distribution   A distribution with two modes.

Binomial Distribution   A distribution of events that has only two possible outcomes.

Bivariate Distribution   A joint distribution of two variables, the individual scores of which are paired in some logical way.

Cell   The portion of an ANOVA table containing the scores of subjects that are treated alike.

Central Limit Theorem   The theorem in mathematical statistics that the sampling distribution of the mean approaches a normal curve as N gets larger. The standard deviation of this sampling distribution is equal to $\sigma/\sqrt{N}$.

Central Value   The mean, median, or mode; a statistic that describes the typical score in a distribution.

Chi Square Distribution   A theoretical sampling distribution of chi square values. The shape of the chi square distribution varies with the degrees of freedom.

Class Interval   A range of scores grouped together in a grouped frequency distribution.

Coefficient of Determination   A squared correlation coefficient; an estimate of common variance.

Common Variance   Variance held in common by two variables. Common variance is assumed to be related to the same factors.

Confidence Interval   An interval of scores within which, with specified confidence, a parameter is expected to lie.

Confidence Level   The confidence $(1 - \alpha)$ that an interval contains a parameter.

Confidence Limits   Two numbers that define the boundaries of a confidence interval.

Constant   A mathematical value that remains the same within a series of operations; for example, regression coefficients *a* and *b* have the same value for all predictions from the same regression line.

Control Group   A baseline group in an experiment to which other groups are compared.

Correlated-Samples Design   An experimental design in which measures from different groups are not independent of each other; also called a dependent-samples design.

Correlation Coefficient   A descriptive statistic calculated on bivariate data; expresses the degree of relationship between the two variables.

Critical Region   The area of the sampling distribution that includes the values of the test statistic that are not due to chance.

Critical Value   The value from a sampling distribution against which a computed statistic is compared to determine whether the null hypothesis may be rejected.

Degrees of Freedom   The number of observations minus the number of necessary relations obtaining among these observations.

Dependent Variable   The observed variable that is expected to be dependent on the independent variable in an experiment.

Descriptive Statistic   A number that expresses some particular characteristic of a set of data. Graphs are sometimes included in this category. (Congratulations to you if you have just looked up this entry after reading footnote 1 in Chapter 1. Very few students make the effort to check out their authors' claims like you just did. You have one of the makings of a scholar.)

Deviation Score   A raw score minus the mean of the distribution that includes the raw score.

Dichotomous Variable   A variable that takes two, and only two, values.

Distribution-Free Statistics   Statistical methods that do not assume any particular population distribution.

Empirical Distribution   An arrangement from highest to lowest of scores from actual observations.

Epistemology   The study or theory of the nature of knowledge.

Error Variance (or Error Term)   Variance due to factors not controlled in the experiment; within-group variance.

Expected Value   The mean value of a random variable over an infinite number of samplings. The expected value of a statistic is the mean of the sampling distribution of the statistic.

Experimental Group   A group that receives a treatment in an experiment and whose dependent-variable scores are compared with those of a control group.

Extraneous Variable   A variable, other than the independent variable, that may affect the dependent variable.

***F* Distribution**   A theoretical sampling distribution of *F* values. There is a different *F* distribution for each combination of degrees of freedom.

***F* Test**   A method of determining the significance of the difference among two or more means or between two variances.

Factor   Independent variable.

Factorial Design   An experimental design using two or more levels of two or more factors; permits an analysis of interaction effects between independent variables.

Frequency   The number of times a score occurs in a distribution.

Frequency Polygon   A graph with quantitative scores on the *X* axis and frequencies on the *Y* axis. Each point on the graph represents a score and the frequency of occurrence of that score.

Goodness of Fit   Degree to which observed data coincide with theoretical expectations.

Grand Mean   The mean of all the scores in an experiment.

Grouped Frequency Distribution An arrangement of scores from highest to lowest in which scores are grouped together into equal-sized ranges called class intervals. The number of scores that occur in each class interval is placed in a column beside the appropriate class interval.

Histogram A graph with quantitative scores on the X axis and frequencies on the Y axis.

Hypothesis A statement about the relationship between two or more variables.

Hypothesis Testing The process of hypothesizing a parameter and comparing (or testing) the parameter with an empirical statistic in order to decide whether the parameter is reasonable.

Independent Events that have nothing to do with each other. The occurrence or variation of one does not affect the occurrence or variation of the other. Two sets of uncorrelated scores are independent of each other.

Independent-Samples Design An experimental design using samples whose dependent-variable scores cannot logically be paired.

Independent Variable The treatment variable; it is selected by the experimenter.

Inferential Statistics A method of reaching conclusions about unmeasurable populations using sample evidence and probability.

Interaction The situation in which the effect of one variable on the dependent variable depends on which level of a second variable is operative.

Interpolation A method for determining a value known to lie between two other values.

Interval Scale A measurement scale in which equal differences between numbers stand for equal differences in the thing measured. The zero point is arbitrarily defined.

J-Curve A severely skewed distribution with the mode at one extreme.

Least Squares Solution Method of fitting a regression line such that the sums of the squared deviations from the straight regression line will be a minimum.

Level One value of the independent variable.

Level of Significance The probability level at which the null hypothesis is rejected.

Line Graph A graph presenting the relationship between two variables.

Lower Limit The bottom of the range of possible values that a score on a quantitative variable can take.

Main Effect The deviation of one or more treatment means from the grand mean.

Mann-Whitney *U* Test A nonparametric method used to determine whether the two independent samples of ranked data came from the same population.

Matched Pairs A correlated-samples design in which individuals are paired before the experiment.

Mean The arithmetic average; the sum of the scores divided by the number of scores.

Mean Square An ANOVA term for the variance; a sum of squares divided by its degrees of freedom.

Median The point that divides a distribution of scores into two equal halves, so that half the scores are above the median and half are below it.

Meta-analysis A technique of reaching conclusions when the data consist of many separate studies done by different investigators.

Mode The score that occurs most frequently in a distribution.

Multiple Comparisons Tests of differences between treatment means or combinations of means following an ANOVA.

Multiple Correlation A correlation method that combines intercorrelations among more than two variables into a single statistic.

Natural Pairs A correlated-samples design in which pairing occurs prior to the experiment.

Nominal Scale A scale of measurement in which numbers are used simply as names and not as quantities.

Nonparametric Statistics   Statistical methods that do not require the estimation of parameters.

Normal Distribution (or Normal Curve)   A theoretical distribution that predicts the frequency of occurrence of chance events.

Normality   The condition of being distributed in the form of the normal curve.

NS   Not statistically significant.

Null Hypothesis   A hypothesis about a population or the relationship between populations.

Observed Frequency   Number of observations actually occurring in a category.

One-Tailed Test of Significance   A statistical test in which the critical region lies in one tail of the distribution.

One-Way ANOVA   A technique that helps determine whether any of several samples are likely to have come from populations with the same mean.

Operational Definition   A definition that specifies a concrete meaning for a variable. The variable is defined in terms of the operations of the experiment; for example, *hunger* may be defined as "24 hours of food deprivation."

Ordinal Scale   A rank-ordered scale of measurement in which equal differences between numbers do not represent equal differences between the things measured.

Ordinate   The vertical or *Y* axis of a graph.

Parameter   Some numerical characteristic of a population.

Partial Correlation   Technique that allows the separation or partialing out of the effects of one variable from the correlation of two other variables.

Point Estimation   Estimating one particular number to be the parameter of a population.

Population   All members of a specified group.

Post Hoc Tests   A category of tests designed to test for significant differences among means after an analysis of variance. The comparisons that are made can be chosen after examining the data.

Proportion   A part of a whole.

Qualitative Variable   A variable that exists in different kinds.

Quantification   The idea that translating a phenomenon into numbers will promote a better understanding of the phenomenon.

Quantitative Variable   A variable that exists in different degrees.

Random Sample   A subset of a population chosen in such a way that all samples of the specified size have an equal probability of being selected.

Range   Upper limit of highest score minus lower limit of lowest score.

Ratio Scale   A scale that has all the characteristics of an interval scale, plus a true zero point.

Raw Score   A score as it is obtained in an experiment.

Rectangular Distribution   A distribution in which all scores have the same frequency.

Regression Coefficients   The values *a* (point where the regression line intersects the *Y* axis) and *b* (slope of the regression line).

Regression Equation   An equation used to predict particular values of *Y* for specific values of *X*.

Regression Line   The "line of best fit" that runs through a scatterplot.

Reliability   The dependability or test-retest consistency of a measure.

Repeated Measures   An experimental design in which more than one dependent-variable measure is taken on each subject.

Sample   A subset of a population.

Sampling Distribution   A theoretical distribution of a statistic based on all possible random samples drawn from the same population; used to determine probabilities.

Scatterplot   The graph of a bivariate frequency distribution.

Simple Frequency Distribution   Scores arranged from highest to lowest, with the frequency of each score placed in a column beside the score.

Skewed Distribution   An asymmetrical distribution. The skew may be positive or negative.

Spearman's r_s   A correlation statistic for two sets of ranked data.

Standard Deviation   A descriptive statistic that indicates the amount of variation within a set of numbers.

Standard Error   The standard deviation of a sampling distribution.

Standard Error of a Difference   The standard deviation of a sampling distribution of differences between means.

Standard Error of Estimate   The standard deviation of the differences between predicted outcomes and actual outcomes.

Standard Score   A score expressed in standard-deviation units.

Statistic   A numerical or nominal characteristic of a sample.

Statistically Significant   A conclusion based on a low probability that the null hypothesis is true; a reliable conclusion.

Stratified Sample   A sample drawn in such a way that it reflects exactly a known characteristic of the population.

Subsample   A subset of a sample.

Sum of Squares   The sum of the squared deviations from the mean; the numerator of the formula for the standard deviation.

***t* Distribution**   Theoretical distribution used to determine the significance of experimental results based on small samples.

***t* Test**   Significance test that uses the *t* distribution.

Theoretical Distribution   Arrangement of hypothesized scores based on mathematical formulas and logic.

Theoretical Frequency   Number of observations expected in a category if the null hypothesis is true; expected frequency.

Treatment   One value of the independent variable.

Truncated Range   The range of the sample is much smaller than the range of its population.

Two-Tailed Test of Significance   Any statistical test in which the critical region is divided into the two tails of the distribution.

Type I Error   Rejection of the null hypothesis when it is true.

Type II Error   Retention of the null hypothesis when it is false.

Univariate Distribution   A frequency distribution of one variable.

Upper Limit   The top of the range of values a score from a quantitative variable can take.

***U* Value**   Statistic used in the Mann-Whitney *U* test.

Variability   Differences.

Variable   Something that exists in more than one amount or in more than one form.

Variance   The square of the standard deviation.

Wilcoxon Matched-Pairs Signed-Ranks *T* Test   A nonparametric method used to determine whether two correlated samples of ranked data came from the same population.

Wilcoxon Rank-Sum Test   A nonparametric test for testing the difference between two independent samples.

Wilcoxon-Wilcox Multiple-Comparisons Test   A nonparametric method for independent samples in which all possible pairs of treatments are compared.

Yates' Correction   A correlation for a 2×2 chi square when expected frequencies are few. (Now obsolete.)

***z* Score**   A score expressed in standard-deviation units; used to compare the relative standing of scores in two different distributions.

***z* Test**   An inferential statistics test based on the normal curve.

Appendix E
GLOSSARY OF FORMULAS

Analysis of Variance

degrees of freedom in one-way ANOVA

$$df_{bg} = K - 1$$
$$df_{wg} = N_{tot} - K$$
$$df_{tot} = N_{tot} - 1$$

degrees of freedom in factorial ANOVA

$$df_A = A - 1$$
$$df_B = B - 1$$
$$df_{AB} = (A - 1)(B - 1)$$
$$df_{wg} = N_{tot} - (A)(B)$$
$$df_{tot} = N_{tot} - 1$$

F value in one-way ANOVA

$$F = \frac{MS_{bg}}{MS_{wg}}$$

F values in factorial ANOVA

$$F_A = \frac{MS_A}{MS_{wg}}$$

$$F_B = \frac{MS_B}{MS_{wg}}$$

$$F_{AB} = \frac{MS_{AB}}{MS_{wg}}$$

mean square

$$MS = \frac{SS}{df}$$

between-groups sum of squares

$$SS_{bg} = \Sigma\left[\frac{(\Sigma X_g)^2}{N_g}\right] - \frac{(\Sigma X_{tot})^2}{N_{tot}}$$

within-groups sum of squares

$$SS_{wg} = \Sigma\left[\Sigma X_g^2 - \frac{(\Sigma X_g)^2}{N_g}\right]$$

total sum of squares

$$SS_{tot} = \Sigma X_{tot}^2 - \frac{(\Sigma X_{tot})^2}{N_{tot}}$$

sum of squares for the interaction effect in factorial ANOVA

$$SS_{AB} = N_g\Sigma[(\bar{X}_{AB} - \bar{X}_A - \bar{X}_B + \bar{X}_{tot})^2]$$

Check:
$$SS_{AB} = SS_{bg} - SS_A - SS_B$$

sum of squares for a main effect in factorial ANOVA

$$SS = \frac{(\Sigma X_1)^2}{N_1} + \frac{(\Sigma X_2)^2}{N_2} + \cdots + \frac{(\Sigma X_K)^2}{N_K} - \frac{(\Sigma X_{tot})^2}{N_{tot}}$$

where 1 and 2 denote levels of a factor and K denotes the last level of a factor

Tukey's HSD

$$HSD = \frac{\bar{X}_1 - \bar{X}_2}{s_{\bar{x}}}$$

where $s_{\bar{x}}$ (for $N_1 = N_2$) $= \sqrt{\dfrac{MS_{wg}}{N}}$

$$s_{\bar{x}} \text{ (for } N_1 \neq N_2) = \sqrt{\frac{MS_{wg}}{2}\left(\frac{1}{N_1} + \frac{1}{N_2}\right)}$$

Chi Square

basic formula

$$\chi^2 = \Sigma\left[\frac{(O - E)^2}{E}\right]$$

degrees of freedom for a chi square table

$df = (R - 1)(C - 1)$
where R = number of rows
C = number of columns

shortcut formula for a 2 × 2 table

$$\chi^2 = \frac{N(AD - BC)^2}{(A + B)(C + D)(A + C)(B + D)}$$

where A, B, C, and D designate the four cells of the table, moving left to right across the top row and then across the bottom row

Confidence Intervals

about a mean $N \geq 30$

$LL = \bar{X} - zs_{\bar{x}}$
$UL = \bar{X} + zs_{\bar{x}}$

about a mean $N < 30$

$LL = \bar{X} - t_\alpha s_{\bar{x}}$
$UL = \bar{X} + t_\alpha s_{\bar{x}}$

about a mean difference (independent samples)

$LL = (\bar{X}_1 - \bar{X}_2) - t_\alpha(s_{\bar{x}_1 - \bar{x}_2})$
$UL = (\bar{X}_1 - \bar{X}_2) + t_\alpha(s_{\bar{x}_1 - \bar{x}_2})$

about a mean difference (correlated samples)

$LL = (\bar{X} - \bar{Y}) - t_\alpha(s_{\bar{D}})$
$UL = (\bar{X} - \bar{Y}) + t_\alpha(s_{\bar{D}})$

Correlation

Pearson product-moment

definition formula

$$r = \frac{\Sigma(z_Y z_X)}{N}$$

computation formulas

$$r = \frac{\dfrac{\Sigma XY}{N} - (\bar{X})(\bar{Y})}{(S_X)(S_Y)}$$

$$r = \frac{N\Sigma XY - (\Sigma X)(\Sigma Y)}{\sqrt{[N\Sigma X^2 - (\Sigma X)^2][N\Sigma Y^2 - (\Sigma Y)^2]}}$$

testing significance from .00 (or use Table A)

$$t = (r)\sqrt{\frac{N-2}{1-r^2}}$$

$$df = N - 2$$

Spearman's r_s

computation formula

$$r_s = 1 - \frac{6\Sigma D^2}{N(N^2 - 1)}$$

Degrees of Freedom

See specific statistical tests.

Deviation Score

$$x = X - \bar{X} \quad \text{or} \quad x = X - \mu$$

Mann-Whitney U Test

value for U

$$U_1 = (N_1)(N_2) + \frac{N_1(N_1 + 1)}{2} - \Sigma R_1$$

where ΣR_1 = sum of the ranks in the N_1 group

testing significance for larger samples $N \geq 21$

$$z = \frac{(U_1 + c) - \mu_U}{\sigma_U}$$

where $c = 0.5$

$$\mu_U = \frac{N_1 N_2}{2}$$

$$\sigma_U = \sqrt{\frac{(N_1)(N_2)(N_1 + N_2 + 1)}{12}}$$

Mean

from a frequency distribution

$$\mu \text{ or } \bar{X} = \frac{\Sigma fX}{N}$$

from raw data

$$\mu \text{ or } \bar{X} = \frac{\Sigma X}{N}$$

of a set of means

$$\bar{\bar{X}} = \frac{\Sigma(N_1\bar{X}_1 + N_2\bar{X}_2 + \cdots + N_K\bar{X}_K)}{\Sigma N}$$

where N_1, N_2, and so on are the number of scores associated with their respective means

Range

range $= X_H - X_L$,
where X_H = upper limit of highest score
X_L = lower limit of lowest score

Regression

for predicting Y from X

$$Y' = r\frac{S_Y}{S_X}(X - \bar{X}) + \bar{Y}$$

for a straight line

$$Y' = a + bX$$
where a = value at the Y intercept
b = slope of the regression line

the Y intercept of a regression line

$$a = \bar{Y} - b\bar{X}$$

the slope of a regression line

$$b = r\frac{S_Y}{S_X}$$

$$b = \frac{N\Sigma XY - (\Sigma X)(\Sigma Y)}{N\Sigma X^2 - (\Sigma X)^2}$$

Standard Deviation of a Population or Sample (for description)

by the raw-score method from ungrouped data

$$\sigma \text{ or } S = \sqrt{\frac{\Sigma X^2 - \dfrac{(\Sigma X)^2}{N}}{N}}$$

by the raw-score method from grouped data

$$\sigma \text{ or } S = \sqrt{\frac{\Sigma fX^2 - \dfrac{(\Sigma fX)^2}{N}}{N}}$$

by the deviation-score method from ungrouped data

$$\sigma \text{ or } S = \sqrt{\frac{\Sigma x^2}{N}}$$

by the deviation-score method from grouped data

$$\sigma \text{ or } S = \sqrt{\frac{\Sigma fx^2}{N}}$$

Standard Deviation of a Sample (for estimation of σ)

by the raw-score method from ungrouped data

$$s = \sqrt{\frac{\Sigma X^2 - \dfrac{(\Sigma X)^2}{N}}{N - 1}}$$

$$s = \sqrt{\frac{N\Sigma X^2 - (\Sigma X)^2}{N(N - 1)}}$$

by the raw-score method from grouped data

$$s = \sqrt{\frac{\Sigma fX^2 - \dfrac{(\Sigma fX)^2}{N}}{N - 1}}$$

$$s = \sqrt{\frac{N\Sigma fX^2 - (\Sigma fX)^2}{N(N - 1)}}$$

by the deviation-score method from ungrouped data

$$s = \sqrt{\frac{\Sigma x^2}{N - 1}}$$

by the deviation-score method from grouped data

$$s = \sqrt{\frac{\Sigma fx^2}{N - 1}}$$

for correlated samples

$$s_D = \sqrt{\frac{\Sigma D^2 - \dfrac{(\Sigma D)^2}{N}}{N - 1}}$$

Standard Error

of the mean
  where the population
  standard deviation is
  known

$$\sigma_{\bar{X}} = \frac{\sigma}{\sqrt{N}}$$

  estimated from a single
  sample

$$s_{\bar{X}} = \frac{s}{\sqrt{N}}$$

of a difference between means
  for large independent
  samples or for equal-N
  small independent
  samples

$$s_{\bar{X}_1 - \bar{X}_2} = \sqrt{s_{\bar{X}_1}^2 + s_{\bar{X}_2}^2}$$

$$= \sqrt{\left(\frac{s_1}{\sqrt{N_1}}\right)^2 + \left(\frac{s_2}{\sqrt{N_2}}\right)^2}$$

$$= \sqrt{\frac{\Sigma X_1^2 - \frac{(\Sigma X_1)^2}{N_1} + \Sigma X_2^2 - \frac{(\Sigma X_2)^2}{N_2}}{N_1(N_2 - 1)}}$$

  for small independent
  samples with unequal
  N's

$$s_{\bar{X}_1 - \bar{X}_2} = \sqrt{\frac{\Sigma X_1^2 - \frac{(\Sigma X_1)^2}{N_1} + \Sigma X_2^2 - \frac{(\Sigma X_2)^2}{N_2}}{N_1 + N_2 - 2}\left(\frac{1}{N_1} + \frac{1}{N_2}\right)}$$

  for correlated samples
  by the direct-difference
  method

$$s_{\bar{D}} = \frac{s_D}{\sqrt{N}}$$

  where $s_D = \sqrt{\dfrac{\Sigma D^2 - \dfrac{(\Sigma D)^2}{N}}{N - 1}}$

  for correlated samples
  when r is known

$$s_{\bar{D}} = \sqrt{s_{\bar{X}}^2 + s_{\bar{Y}}^2 - 2r_{XY}(s_{\bar{X}})(s_{\bar{Y}})}$$

t Test

as a test for whether a sample
  mean came from a population
  with a mean μ

$$t = \frac{\bar{X} - \mu}{s_{\bar{X}}}$$

$$df = N - 1$$

for correlated samples where r
  is known

$$t = \frac{\bar{X} - \bar{Y}}{\sqrt{s_{\bar{X}}^2 + s_{\bar{Y}}^2 - 2r_{XY}(s_{\bar{X}})(s_{\bar{Y}})}}$$

for correlated samples where
  the direct-difference method
  is used

$$t = \frac{\bar{X} - \bar{Y}}{s_{\bar{D}}}$$

$$df = \text{number of pairs minus one}$$

for independent samples

$$t = \frac{\bar{X}_1 - \bar{X}_2}{s_{\bar{X}_1 - \bar{X}_2}}$$

$$df = N_1 + N_2 - 2$$

for testing whether a correlation coefficient is significantly different from .00

$$t = (r)\sqrt{\frac{N-2}{1-r^2}}$$

df = number of pairs minus two

Variance

Use the formulas for the standard deviation. For s^2 and σ^2, square s and σ.

Wilcoxon Matched-Pairs Signed-Ranks T Test

value for T
testing significance for larger samples $N \geq 50$

T = smaller sum of the signed ranks

$$z = \frac{(T + c) - \mu_T}{\sigma_T}$$

where $c = 0.5$

$$\mu_T = \frac{N(N + 1)}{4}$$

$$\sigma_T = \sqrt{\frac{N(N + 1)(2N + 1)}{24}}$$

N = number of pairs

z Score

$$z = \frac{X - \bar{X}}{S}$$

z Tests

as a test for whether a sample mean came from a population with a mean, μ; σ is known

$$z = \frac{\bar{X} - \mu}{\sigma_{\bar{X}}}$$

as a test for whether a sample mean came from a population with a mean, μ; σ_X is estimated by $s_{\bar{X}}$

$$z = \frac{\bar{X} - \mu}{s_{\bar{X}}}$$

as a test of the significance of the difference between two means

$$z = \frac{\bar{X}_1 - \bar{X}_2}{s_{\bar{X}_1 - \bar{X}_2}}$$

Appendix F
ANSWERS TO PROBLEMS

CHAPTER 1

1. **a.** 7.5–8.5   **b.** Qualitative   **c.** 9.995–10.005
   d. Qualitative   **e.** 4.45–4.55   **f.** 2.945–2.955
2. Many paragraphs can qualify as good answers to this question. Your paragraph should include variations of the following definitions:
   Population: an arbitrarily defined set of scores that is of interest to an investigator. *Sample:* some subset of a population. *Statistic:* a numerical characteristic of a sample or subsample. *Parameter:* a numerical characteristic of a population.
3. **a.** Descriptive; this is a statement of a past fact.
   b. Inferential; sample figures are being used to predict a future event.
   c. Descriptive; this is an enumeration of what happened.
   d. Inferential; this is a projection based on past records.
   e. Inferential; this is a generalization about college students based on a study of a specific sample.
   f. Inferential; this is a generalization about the past.
4. Nominal, ordinal, interval, and ratio
5. *Nominal:* different numbers are assigned to different classes of things. *Ordinal:* nominal properties, plus the numbers carry information about greater than and less than. *Interval:* ordinal properties, plus the distance between units is equal. *Ratio:* interval properties, plus there is a true zero point.
6. **a.** Ordinal   **b.** Ratio   **c.** Nominal   **d.** Nominal   **e.** Ratio   **f.** Ordinal

7.

Independent	*Dependent*
a. Amount of previous experience	Problem-solving ability
b. Amount eaten	General feeling
c. Amount eaten	General feeling
d. Amount of anxiety	Time to solve problems
e. Amount of sunlight	Rate of growth
f. Kind of teaching	Amount learned
g. When it rains	When flowers bloom

8. **i. a.** Hypnosis, yes or no
   b. Number of suggestions followed
   c. Set of suggestions given

    d. Response scores of hypnotized people, and those for people just trying to get the best score

    e. The two samples were the two groups of 25.

    f. The mean numbers of suggestions followed (4.8 and 5.1) are statistics.

    g. The mean numbers of suggestions followed by the two populations

    h. Hypnosis (yes or no)

    i. Number of suggestions followed

    j. Barber found that hypnotized people did not follow more suggestions than people who were asked (without hypnosis) to try to get high scores.

ii. a. Whether the question mentioned the barn or not

    b. Answer to the question a week later, "Did you see a barn?"

    c. Subjects saw same film; the week interval; asked same final question

    d. The memory of people who are given false information after an event

    e. Each of the two groups in the study was a sample.

    f. 17 percent and 3 percent

    g. Percent score for people who are given false information after an event

    h. Whether the barn was mentioned in the first question

    i. Percent who said yes to the question, "Did you see a barn?"

    j. People who are given false information after an event may incorporate that false information into later recall of the event.

iii. a. Time shown by the clock on the wall

    b. Weight of crackers consumed

    c. Actual time of day; all volunteers were obese.

    d. Obese males or obese humans

    e. The 60 males who volunteered for this experiment

    f. Mean grams of crackers consumed for any group (i.e., 20, 30, or 40 grams)

    g. Mean grams of crackers consumed by all obese people at dinner time

    h. Gender of the participants

    i. Weight of crackers consumed

    j. Schachter and Gross showed that the weight of crackers that obese male students eat is dependent on what time they think it is rather than on what time it actually is (their biological time).

CHAPTER 2 **1.**

Females			Males		
Height (in.)	Tally marks	f	Height (in.)	Tally marks	f
72	/	1	77	/	1
71		0	76	/	1
70	/	1	75	/	1
69	//	2	74	/	1
68	/	1	73	////	4
67	///	3	72	丗	5
66	丗	5	71	丗//	7
65	丗//	7	70	丗/	6
64	丗丗	10	69	丗///	8
63	丗////	9	68	丗//	7
62	丗/	6	67	//	2
61	///	3	66	//	2
60	/	1	65	///	3
59	/	1	64	/	1
			63		0
			62	/	1

2. The order of candidates is arbitrary. We list them and graph them in the order in which they are given in the problem; any order is correct.

Candidate	f	Candidate	f
Bolivar	13	Lenin	8
Gandhi	19	Mao	11
		Attila	5

3. The range is $63.5 - 4.5 = 59$. You may have tried $i = 3$ and found yourself with an extra interval. That is, $59 \div 3 = 19.67$; therefore, 20 intervals should result. However, the lowest interval must begin with 3 if it is to have as its lower limit a multiple of 3 and include the lowest score, 5. This adds two extra scores to the bottom of the distribution. The range then becomes 61, and $61 \div 3 = 20.33$. An extra interval is necessary to handle the decimal, causing the 10 to 20 intervals convention to be violated. The correct i for these data is 5.

Two solutions using $i = 5$ are acceptable. One solution begins the intervals with multiples of 5. The other solution places multiples of 5 at the midpoints. Both are shown here.

Class interval	Tally marks	f
60–64	/	1
55–59	//	2
50–54	////	4
45–49	////	4
40–44	////// /	6
35–39	////// //	7
30–34	////// ////	9
25–29	////// /	6
20–24	//////	5
15–19	///	3
10–14	//	2
5–9	/	1

Class interval	Tally marks	f
63–67	/	1
58–62	/	1
53–57	///	3
48–52	////	4
43–47	////	4
38–42	////// //	7
33–37	////// //	7
28–32	////// //	7
23–27	////// //	7
18–22	////	4
13–17	///	3
8–12	/	1
3–7	/	1

4.

Number of sentences heard before	Tally marks	f
16	//	2
15		0
14	//	2
13	///	3
12	///	3
11	////// ////	9
10	////// /	6
9	////// //	7
8	////	4
7	//	2
6	/	1

5.

Class interval	Tally marks	*f*
33–35	/	1
30–32	/	1
27–29	//	2
24–26	/	1
21–23	////	4
18–20	乂乂乂乂乂	10
15–17	乂乂乂乂乂	10
12–14	乂乂乂乂乂 /	11
9–11	乂乂乂乂乂乂乂乂 /	16
6–8	乂乂乂乂乂乂乂乂乂乂乂乂 ///	23
3–5	乂乂乂乂乂乂乂乂乂 ////	19
0–2	/	1

6. a. The number 55 is the midpoint of the interval 54–56.

   b. Eight is a frequency number representing eight students.

   c. Two; both students are represented by the score 49, which is the midpoint of the interval 48–50.

   d. This point indicates the number of students whose scores were in the interval 24–26.

7.

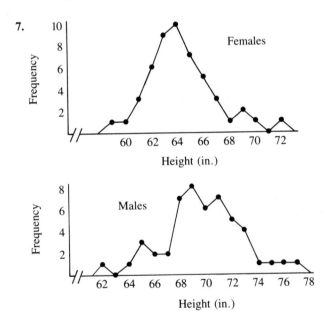

8. They should be graphed as frequency polygons. Note that the midpoint of the class interval is used as the *X*-axis score.

a.

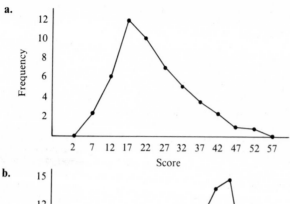

b.

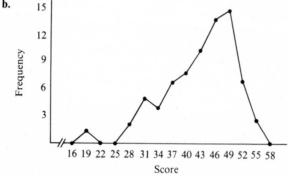

9. Problem **8a** is positively skewed. Problem **8b** is negatively skewed.
10. A bar graph is the proper graph for the qualitative data in Problem 2.

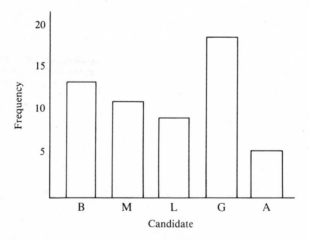

11. Check your sketches against Figures 2.7 and 2.10.
12. Right

13. In a frequency polygon the Y axis is frequency. In a line graph the Y axis is some variable other than frequency.

14. **a.** Positively skewed   **b.** Symmetrical   **c.** Positively skewed
    d. Positively skewed   **e.** Negatively skewed   **f.** Symmetrical

15. **a.** 5   **b.** 12.5, halfway between 12 and 13
    c. 9.5, halfway between 8 and 11

16. Only Distribution C has two modes. They are 14 and 18.

17. Mean = 409/39 = 10.49
    Median. Since there are 39 scores, the halfway point will have 19.5 scores below and 19.5 above. Counting from the bottom of the distribution, there are 14 scores of 9 or less. Thus, you need 5.5 of the 6 scores in the interval of 9.5 to 10.5 (19.5 − 14 = 5.5). 5.5/6 = 0.92. 9.5 + 0.92 = 10.42 = median. Mode = 11.

18. **a.** For these nominal data, only the mode is appropriate. The mode is Gandhi, which occurs on 34 percent of the signs.
    b. Since a precinct was covered and the student's interest was citywide, this is a sample mode, a statistic.
    c. A simple interpretation would be, Since there are more Gandhi yard signs than any other, I expect Gandhi to get the most votes.

19. In both cases the interest is in some larger group (five sections of Introduction to Sociology and all of Midwesternville). Thus, both sets of numbers are samples.

20. **a.** $N = 19$ and half that is 9.5. Counting from the *top,* there are 7 when you include a score of 11. You need 2.5 of the 5 scores in the interval 9.5 to 10.5. 2.5/5 = 0.5, so *subtract* 0.5 from 10.5 to get the median, 10.0.
    b. The easiest and quickest way to deal with 20 scores is to construct a simple frequency distribution. Counting from the top again, there are 6 when you include a score of 6. You need 4 of the 5 scores that lie in the interval 4.5 to 5.5. 4/5 = 0.80, so subtract 0.80 from 5.5 to find the median, 4.70.

21. $\Sigma(X - \bar{X}) = 0$ and $\Sigma(X - \bar{X})^2$ is a minimum.

22. *Mean.* $\bar{X} = \Sigma fX/N = 1725/50 = 34.50$, or $\bar{X} = \Sigma fX/N = 1720/50 = 34.40$. The difference between the two means is minor and is due to the different methods of grouping. The mean computed from the ungrouped scores is 34.43, so neither solution contains much error.
    Median. $N/2 = 50/2 = 25$. The median will have 25 frequencies above it and 25 below it. Count frequencies from the bottom of the first distribution. There are 17 frequencies through the interval 25–29 and 26 frequencies through the interval 30–34. The median, then, must be near the top of the interval 30–34. Subtract the 17 frequencies below 30–34 from 25. 25 − 17 = 8, so 8 more frequencies are needed. There are 9 frequencies in the interval containing the median, so you need 8/9, or 0.89 of the interval. The interval has 5 score points, so 0.89 × 5 = 4.44, the number of score points needed to reach the median. The lower limit of the interval 30–34 is 29.5. Adding 4.44 to 29.5, you get 33.94 as the median.
    In the second distribution, 23 frequencies are below the interval 33–37. Two more are needed to reach the median. Since there are 7 frequencies in the interval, 2/7, or 0.29, of the interval is needed, or 0.29 of 5. Multiply 0.29 × 5 to get 1.43. Add 1.43 points to the lower limit, 32.5, to find the median. Thus, 32.5 + 1.43 = 33.93.
    Mode. In the first distribution, the mode is the midpoint of the interval 30–34, since that interval has the greatest number of frequencies. The mode is 32. In the second distribution, four class intervals in a row have seven frequencies. The mode, in this case, is the point between intervals 28–32 and 33–37, or 32.5.

23. *Mean.* $\bar{X} = \Sigma fX/N = 1125/99 = 11.36$. *Median.* $N/2 = 99/2 = 49.5$. Working from the bottom, there are 43 frequencies below the interval 9–11 and 16 in the interval

9–11. You need 49.5 – 43 = 6.5 of the 16 frequencies. (6.5/16) × 3 = 1.22. The lower limit of the class interval, 8.5, plus 1.22 is 9.72, the median. Working from the top, there are 40 frequencies above the interval 9–11. Of the 16 frequencies in that interval, 9.5 are needed. Thus, you multiply (9.5/16) × 3 to get 1.78, which is subtracted from the upper limit, 11.5, to give the median, 9.72. *Mode* = 7.

24. **a.** The mode is appropriate because the observations are made on a nominal scale.
   b. The median or mode is appropriate because the observations produce an ordinal scale.
   c. The median is appropriate for data with an open-ended category.
   d. The median is the appropriate central value to use. It is conventional to use the median for income data because the distribution is so severely skewed. About half of the frequencies are in the $0–$15,000 range.
   e. The mode is appropriate because these are nominal data.
   f. The mean is appropriate because the data are not severely skewed.
   g. The median is appropriate because the distribution is severely skewed.

25. $\quad 12 \times 74 = 888$
   $\quad 31 \times 69 = 2139$
   $\quad \underline{17 \times 75 = 1275}$
   $\quad \overline{\Sigma \quad 60 \qquad 4302}$ Overall mean: $\dfrac{4302}{60} = 71.7$

26. The distribution in Problem **8a** is positively skewed (mean = 23.80, median = 22.0). The distribution in Problem **8b** is negatively skewed (mean = 43.19, median = 44.62).

27. This is not correct. To find his lifetime batting average he would need to add up his hits for the 3 years and divide by the total number of at-bats. From the description of the problem it appears that his average would be lower than .317.

CHAPTER 3

1. Range = 15.5 – 4.5 = 11

X	x
15	5
13	3
12	2
10	0
8	–2
7	–3
5	–5
$\Sigma X = 70$	$\Sigma x = 0$
μ or $\bar{X} = 10$	

2. Range 17.5 – 0.5 = 17

X	x
17	11
5	–1
1	–5
1	–5
$\Sigma X = 24$	$\Sigma x = 0$
μ or $\bar{X} = 6$	

3. Range = 3.45 – 2.55 = 0.90

X	x
3.4	0.5
3.1	0.2
2.7	–0.2
2.7	–0.2
2.6	–0.3
$\Sigma X = 14.5$	$\Sigma x = 0$
μ or $\bar{X} = 2.9$	

4. Range = 0.455 – 0.295 = 0.16

X	x
0.45	0.10
0.30	–0.05
0.30	–0.05
$\Sigma X = 1.05$	$\Sigma x = 0$
μ or $\bar{X} = 0.35$	

5. σ is used to describe the variability of a population. s is used to estimate σ from a sample of the population. S is used to describe the variability of a sample when you have no desire to estimate σ.

6.

X	x	x^2
7	2	4
6	1	1
5	0	0
2	-3	9
Σ 20	0	14

μ or $\bar{X} = 5$

$$\sigma \text{ or } S = \sqrt{\frac{\Sigma x^2}{N}} = \sqrt{\frac{14}{4}} = \sqrt{3.5} = 1.87$$

7.

X	x	x^2
14	3.8	14.44
11	0.8	0.64
10	-0.2	0.04
8	-2.2	4.84
8	-2.2	4.84
Σ 51	0	24.80

μ or $\bar{X} = 10.2$

$$\sigma \text{ or } S = \sqrt{\frac{\Sigma x^2}{N}} = \sqrt{\frac{24.80}{5}} = \sqrt{4.96}$$
$$= 2.23$$

8.

X	x	x^2
107	2	4
106	1	1
105	0	0
102	-3	9
Σ 420	0	14

μ or $\bar{X} = 105$

$$\sigma \text{ or } S = \sqrt{\frac{\Sigma x^2}{N}} = \sqrt{\frac{14}{4}} = \sqrt{3.5} = 1.87$$

9. Although the numbers in Problem 8 are much larger than those in Problem 6, the standard deviations are the same. Thus, the size of the numbers does not give you any information about the variability of the numbers.

10. Yes; the larger the numbers, the larger the mean.

11.

City	Mean	Standard deviation
San Francisco	56.75°	3.96°
Albuquerque	56.75°	16.24°

Although the mean temperature of the two cities is the same, Albuquerque has a wider variety of temperatures.

12. In eyeballing data for variability, use the range as a quick index.
   a. The second distribution
   b. Equal variability
   c. The first distribution
   d. Equal variability
   e. The second distribution is more variable; however, most of the variability is due to one extreme score, 15.

13. The second distribution (**b**) is more variable than the first.

a.	X	X^2
	6	36
	5	25
	4	16
	3	9
	2	4
Σ	20	90

$$S = \sqrt{\frac{\Sigma X^2 - \dfrac{(\Sigma X)^2}{N}}{N}} = \sqrt{\frac{90 - \dfrac{(20)^2}{5}}{5}} = \sqrt{2.00} = 1.41$$

b.	X	X^2
	6	36
	6	36
	6	36
	2	4
	2	4
Σ	22	116

$$S = \sqrt{\frac{\Sigma X^2 - \dfrac{(\Sigma X)^2}{N}}{N}} = \sqrt{\frac{116 - \dfrac{(22)^2}{5}}{5}} = \sqrt{3.84} = 1.96$$

14. For Problem **13a**, $5/1.41 = 3.55$. For Problem **13b**, $5/1.96 = 2.55$. Yes, these are between 2 and 5.

15. a. $\sigma = \sqrt{\dfrac{\Sigma X^2 - \dfrac{(\Sigma X)^2}{N}}{N}} = \sqrt{\dfrac{262 - \dfrac{(34)^2}{5}}{5}} = 2.48$

b. $\sigma = \sqrt{\dfrac{294 - \dfrac{(38)^2}{5}}{5}} = 1.02$

16. $s = \sqrt{\dfrac{\Sigma X^2 - \dfrac{(\Sigma X)^2}{N}}{N - 1}} = \sqrt{\dfrac{5064 - \dfrac{(304)^2}{21}}{20}} = \sqrt{33.16} = 5.76$

17.

Females				Males			
Height (in.)	*f*	*fX*	*fX²*	*Height (in.)*	*f*	*fX*	*fX²*
72	1	72	5,184	77	1	77	5,929
70	1	70	4,900	76	1	76	5,776
69	2	138	9,522	75	1	75	5,625
68	1	68	4,624	74	1	74	5,476
67	3	201	13,467	73	4	292	21,316
66	5	330	21,780	72	5	360	25,920
65	7	455	29,575	71	7	497	35,287
64	10	640	40,960	70	6	420	29,400
63	9	567	35,721	69	8	552	38,088
62	6	372	23,064	68	7	476	32,368
61	3	183	11,163	67	2	134	8,978
60	1	60	3,600	66	2	132	8,712
59	1	59	3,481	65	3	195	12,675
	Sum	3,215	207,041	64	1	64	4,096
				62	1	62	3,844
					Sum	3,486	243,490

$$s = \sqrt{\frac{207{,}041 - \dfrac{3215^2}{50}}{49}} = 2.54$$

$$s = \sqrt{\frac{243{,}490 - \dfrac{3486^2}{50}}{49}} = 3.02$$

18. s is the more appropriate standard deviation; generalization to the manufacturing process would be expected.

$$\text{Process A:}\quad s = \sqrt{\frac{18 - \dfrac{0^2}{6}}{5}} = 1.90 \qquad \text{Process B:}\quad s = \sqrt{\frac{12 - \dfrac{0^2}{6}}{5}} = 1.55$$

Process B produces the more consistent doodads. Note that both processes have an average error of zero.

19. Before: $s = \sqrt{1.43} = 1.20$, $\bar{X} = 5.0$. After: $s = \sqrt{14.29} = 3.78$, $\bar{X} = 5.0$. It appears that, before studying poetry, students were homogeneous and neutral. After studying poetry for 9 weeks, some students were turned on and some were turned off; they were no longer neutral.

20. a.

X	x	x^2	$z = \dfrac{x}{S}$
10	5	25	1.58
7	2	4	0.63
4	−1	1	−0.32
3	−2	4	−0.63
1	−4	16	−1.26
Σ 25	0	50	0

$$\bar{X} = \frac{25}{5} = 5.00$$

$$S = \sqrt{\frac{50}{5}} = 3.16$$

b. Variance = 10

21. $\Sigma z = 0$. Since $\Sigma x = 0$, it follows that $\Sigma(x/S) = 0$.

22. Zero

23.

Hattie	*Missy*
$z = \dfrac{37 - 39.5}{1.803} = -1.39$	$z = \dfrac{24 - 26.25}{1.479} = -1.52$

Of course, in timed events, the more negative the z score, the better the score. Missy's -1.52 is superior to Hattie's -1.39.

24.

Tobe's apple	*Zeke's orange*
$z = \dfrac{9 - 5}{1} = 4.00$	$z = \dfrac{10 - 6}{1.2} = 3.33$

Tobe's z score is larger so the answer to Hamlet must be a resounding "To be." Notice that each fruit varies from its group mean by the same amount. It is the smaller variability of the apple weights that makes Tobe's fruit a winner.

25.

First test	*Second test*	*Third test*
$z = \dfrac{79 - 67}{4} = 3.00$	$z = \dfrac{125 - 105}{15} = 1.33$	$z = \dfrac{51 - 45}{3} = 2.00$

Milquetoast's performance was poorest on the second test.

CHAPTER 4

1. A bivariate distribution has two variables. The scores of the variables are paired in some logical way.

2. The statement means that variation in either variable is accompanied by predictable variation in the other. Notice that nothing is said here about direction. They may vary in the same direction (positive correlation) or opposite directions (negative correlation).

3. When correlation is positive, X and Y increase and decrease together. When correlation is negative, Y increases as X decreases, and Y decreases as X increases.

4. a. Yes, positive; taller people usually weigh more than shorter people.
   b. No; these scores cannot be correlated, since there is no basis for pairing the scores.
   c. Yes, negative; as temperatures go up, less heat is needed and fuel bills go down.
   d. Yes, positive; people with higher IQs score higher on reading comprehension tests.
   e. No; there is no basis for pairing scores.

5.

	Fathers	*Daughters*
Mean	67.50	62.50
S	3.15	2.22
Sum	405	375
Sum squared	27,397	23,467
ΣXY	25,334	
r		.513

r by the blanched procedure:

$$r = \frac{\dfrac{25{,}334}{6} - (67.5)(62.5)}{(3.15)(2.22)} = \frac{4{,}222.33 - 4{,}218.75}{6.98} = .51$$

By the raw-score procedure:

$$r = \frac{(6)(25{,}334) - (405)(375)}{\sqrt{[(6)(27{,}397) - (405)^2][(6)(23{,}467) - (375)^2]}} = .513$$

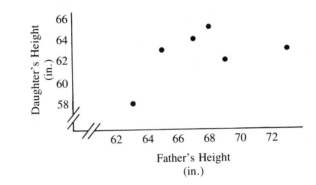

6. $\bar{X} = 46.18$

   $\bar{Y} = 30.00$

$$S_X = \sqrt{\frac{87{,}373 - \dfrac{(1{,}755)^2}{38}}{38}} = \sqrt{166.31} = 12.90$$

$$S_Y = \sqrt{\frac{37{,}592 - \dfrac{(1140)^2}{38}}{38}} = \sqrt{89.26} = 9.45$$

$$r = \frac{\dfrac{55{,}300}{38} - (46.18)(30)}{(12.90)(9.45)} = \frac{1{,}455.26 - 1{,}385.40}{121.91} = \frac{69.86}{121.91} = .57$$

7. $r = \dfrac{(50)(175{,}711) - (202)(41{,}048)}{\sqrt{[(50)(1{,}740) - (202)^2][(50)(35{,}451{,}830) - (41{,}048)^2]}} = .25$

$$r = \frac{\dfrac{175{,}711}{50} - (4.04)(820.96)}{(4.30)(187.25)} = .25$$

8. Coefficient of determination = .32. The two measures of self-esteem have about 32 percent of their variance in common. Although the two measures are to some extent measuring the same traits, they are, in large part, measuring different traits.

9. **a.** $(.40)^2 = .16$, or 16 percent   **b.** $(.10)^2 = .01$, or 1 percent

10.	Cigarette consumption	Death rate
Mean	604.27	205.00
S	367.87	113.72
Sum	6,647	2,255
Sum squared	5,505,173	604,527
ΣXY	1,705,077	
N		11
r		.74

With $r = .74$, these data show that there is a fairly strong relationship between per capita cigarette consumption and male death rate 20 years later.

11. **a.** People who cannot tolerate ambiguity tend to be authoritarian.
   b. Vocational interests tend to remain similar from age 20 to age 40.
   c. Identical twins have very similar IQs.
   d. There is a slight tendency for IQ to be lower as family size increases.
   e. There is a slight tendency for taller men to have higher IQs than shorter men.
   f. The lower a person's income level is, the greater the probability that he or she will be diagnosed as psychotic.

12. **a.** There is some tendency for children with more older siblings to accept less credit or blame for their own successes and failures than children with fewer older siblings.
   b. Nothing; correlation coefficients do not permit you to make cause-and-effect statements.
   c. The coefficient of determination is $.1369 \ (-.37^2)$, so we can say that about 14 percent of the variance in acceptance of responsibility is predictable from knowledge of the number of older siblings; 86 percent is not. This provides very poor ability to predict. However, if we are attempting to devise a theory about acceptance of responsibility, the number of older siblings would be considered important.

13. **a.** Since $.97^2 = .94$, these two tests have 94 percent of their variance in common. They must be measuring essentially the same trait. There is a very strong tendency for persons scoring high on one test to score high on the other.
   b. No; this is a cause-and-effect statement and is not justified on the basis of correlational data.

14. A Pearson r is not a useful statistic for data such as these. The relationship is a curved one. In fact, this is the same relationship you saw in Figure 2.5.

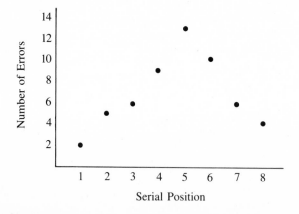

15. a. $b = r\dfrac{S_Y}{S_X} = (.513)\dfrac{2.217}{3.149} = (.513)(.704) = .361$

$a = \bar{Y} - b\bar{X} = 62.5 - (.361)(67.5) = 38.12$

To draw the regression line, we used two points: $(\bar{X}, \bar{Y})$ and $(62, 60.5)$. To get $(62, 60.5)$, we used the formula $Y' = a + bX$. Other X values would work just as well.

b.

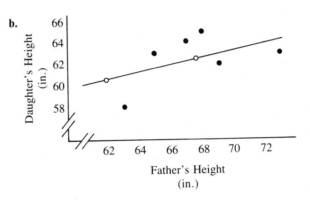

Father's Height
(in.)

16. a. $b = r\dfrac{S_Y}{S_X} = (.57)\dfrac{9.45}{12.90} = (.57)(.73) = .42$

$a = \bar{Y} - b\bar{X} = 30 - (.42)(46.18) = 30 - 19.40 = 10.60$

b. $Y' = a + bX = 10.60 + (.42)(42) = 10.60 + 17.64 = 28.24$

17.

	Advertising	Sales
Mean	4.286	108.57
S	1.030	20.30
Sum	30	760
Sum squared	136	85,400
ΣXY	3360	
r	.703	

a. $Y' = a + bX = 49.23 + 13.85X$

b. One point we used for the regression line on our scatterplot was the two means, $\bar{X} = 4.29$, $\bar{Y} = 108.57$. The other point was $X = 6.00$, $Y = 132.33$. Other X values would work just as well.

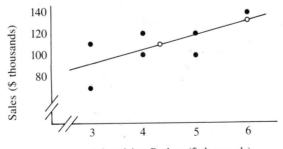

Advertising Budget ($ thousands)

c. $Y' = a + bX$. For $X = 10$, $Y' = 187.69$, or, in terms of sales, $187,690.

d. We warned you that the interpretation of r can be a tricky business. This problem leads you to one of those pitfalls we mentioned. With $r = .70$, you have a fair degree of confidence that the prediction will hold *in general*. However, if your thinking (and your answer) was that by increasing advertising, sales would increase, you fell into the pitfall of inferring a causal relationship from a correlation coefficient. As a matter of fact, some analysts say that sales cause advertising, that advertising budgets are determined by last year's sales. In any case, don't infer more than the data permit. These data allow you to estimate with some confidence what sales were by knowing the amount spent for advertising. In this problem, the percent of sales spent on advertising (just less than 4 percent) is about the same as the percent spent nationally in the United States.

18. $Y' = (.80)\left(\dfrac{15}{16}\right)(65 - 100) + 100 = 73.75 = 74$

19. First, we will present the answers we asked you to find. We designated the year variable as Y.

$$\Sigma X = 55.80 \qquad \Sigma Y = 477 \qquad N = 6$$
$$\Sigma X^2 = 519.02 \qquad \Sigma Y^2 = 37{,}939 \qquad \Sigma XY = 4{,}437.20$$
$$r = .930 \qquad a = -48.38 \qquad b = 13.75$$

For 1 million graduates ($10 \times 100{,}000$),

$$Y' = a + bX = -48.38 + 13.75(10) = 89.12$$

Thus, the predicted first year for 1 million graduates is 1990. This is a simple and straightforward prediction from a regression equation. It will be worthwhile, though, for you to think about the variables that determine the actual year when 1 million Americans graduate from college.

20. We suggest that your essay have at least the following characteristics:

1. An introductory paragraph that orients the reader to what the essay will cover
2. A concluding paragraph that makes a summary statement about descriptive statistics
3. An explanation of the purpose of descriptive statistics
4. Examples of descriptive statistics

CHAPTER 5

1. There are seven cards between the 3 and jack, each with a probability of $4/52 = .077$. So $(7)(.077) = .539$.

2. The probability of drawing a 7 is $4/52$, and there are 52 opportunities to get a 7. Thus, $(52)(4/52) = 4$.

3. There are four cards that are larger than a jack or smaller than a 3. Each has a probability of $4/52$. Thus, $(4)(4/52) = 16/52 = .308$.

4. The probability of a 5 or 6 is $4/52 + 4/52 = 8/52 = .154$. In 78 draws, $(78)(.154) = 12$ cards that are 5's or 6's.

5. a. .7500; adding the probability of one head (.3750) to that of two heads (.3750), you get a figure of .7500.

b. $.1250 + .1250 = .2500$

6. $16(.1250) = 2$ times

7. a. .0668   **b.** The test points are normally distributed.

8. a. .0832   **b.** .4778   **c.** .2912

9. Empirical; the scores are based on observations.
10. $z = (X - \mu)/\sigma$. For IQ = 55, z score = -3.00; for IQ = 110, z score = .67; for IQ = 103, z score = .20; for IQ = 100, z score = .00.
11. A quick check on your answer can be made by comparing it with the proportion having IQs of 120 or higher, .0918. The proportion with IQs of 70 or lower will be less than the proportion with IQs of 120 or higher. Is your calculated proportion less? Following is a picture of a normal distribution in which the proportion of the population with IQs of 70 or lower is shaded. For IQ = 70, $z = (70 - 100)/15 = -30/15 = -2.00$. The proportion beyond $z = 2.00$ is .0228 (column C). Thus, .0228 or 2.28 percent of the population would be expected to be in special education classes.

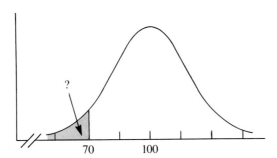

12. $(.0228)(4000) = 91.2$, or 91 students
13. $z = (110 - 100)/15 = 10/15 = .67$. The proportion beyond $z = .67$ is .2514, which is the proportion of people with IQs of 110 or higher.
14. **a.** $.2514 \times 250 = 62.85$ or 63 students   **b.** $250 - 62.85 = 187.15$ or 187 students
    c. $1/2 \times 250 = 125$. We hope you were able to get this one immediately by thinking about the symmetrical nature of the normal distribution.
15. The z score that includes a proportion of .02 is 2.06. If $z = 2.06$, $X = 100 + (2.06)(15) = 100 + 30.9 = 130.9$. In fact, Mensa requires an IQ of 130 on tests that have a standard deviation of 15.
16. **a.** The z score you need is 1.65. (1.64 will include more than the tallest 5 percent.)
    $1.65 = (X - 64.3)/2.5$; $X = 64.3 + 4.1 = 68.4$ inches.
    b. $z = (58 - 64.3)/2.5 = -6.3/2.5 = -2.52$. The proportion excluded is .0059.
17. **a.** $z = (60 - 69.7)/3.0 = -9.7/3.0 = -3.23$. The proportion excluded is .0006.
    b. $z = (62 - 69.7)/3.0 = -7.7/3.0 = -2.57$. The proportion taller than Napoleon is .9949 (.5000 + .4949).
18. **a.** $z = \dfrac{3.20 - 3.11}{.05} = 1.80$, proportion = .0359
    b. The z score corresponding to a proportion of .1000 is 1.28. Thus,

$$X = 3.11 + (1.28)(.05)$$
$$= 3.11 + .06$$
$$= 3.17$$
$$X = 3.11 - .06$$
$$= 3.05$$

Thus, the middle 80 percent of the pennies weigh between 3.05 and 3.17 grams.

19.

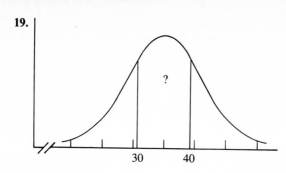

a. $z = \dfrac{30 - 35}{6} = -.83$, proportion $= .2967$. $z = \dfrac{40 - 35}{6} = .83$, proportion $= .2967$.

(2)(.2967) = .5934 = the proportion of students with scores between 30 and 40

b. The probability is also .5934.

20. No, because the distance between 30 and 40 straddles the mean, where scores that occur frequently are found. The distance between 20 and 30 is all in one tail of the curve, where scores are achieved less frequently. If you missed this problem, it was probably because you failed to draw a normal curve and write in the mean and the scores in question.

21.

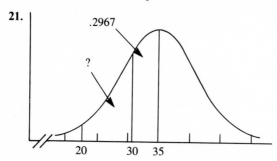

$$z = \frac{20 - 35}{6} = -2.50,\ \text{proportion} = .4938$$

$$z = \frac{30 - 35}{6} = -.83,\ \text{proportion} = .2967$$

.4938 − .2967 = .1971. We have found the proportion of students with scores between 35 and 20 and subtracted from it the proportion with scores between 35 and 30. There are also other correct ways to set up this problem.

22. 800 people × .1971 = 157.7 = 158 people

23.

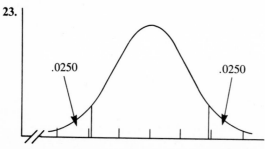

The z score corresponding to a proportion of .0250 is 1.96 (column C). Thus,

$$X = \mu + (z)(\sigma)$$
$$= 100 + (1.96)(15)$$
$$= 129.4 = 129$$

and

$$X = 100 + (-1.96)(15)$$
$$= 70.6 = 71$$

IQ scores of 129 or higher or 71 or lower are achieved by 5 percent of the population.

24. Since .05 of the curve lies outside the limits of the scores 71 and 129, the probability is .05. The probability is .025 that the randomly selected person has an IQ of 129 or higher and .025 that it is 71 or lower.

25. $z = (13.00 - 11.00)/2.68 = 2/2.68 = .75$. The proportion of those with z scores greater than .75 is .2266, so just less than one-fourth of the secretaries would be expected to be transferred if the firm manager implemented the decision.

26. $z = \dfrac{.85 - .99}{.17} = \dfrac{-.14}{.17} = -.82$. The proportion associated with $z = -.82$ is .2061 (column C). $(.2061)(185,822) = 38,298$ truck drivers. The usual procedure would be to report this result as "about 38,000 truck drivers," since the normal curve corresponds only approximately to the empirical curve.

27. This problem has two parts: determining the number of trees over 8 inches DBH on 1 acre and multiplying by 100:

$$z = \dfrac{8 - 13.68}{4.83} = -1.18, \text{ proportion} = .5000 + .3810 = .8810$$

$.8810(199) = 175.319$ trees on 1 acre

$175.319(100) = 17,532$ trees on 100 acres (about 17,500 trees)

28. $z = (300 - 268)/14 = 32/14 = 2.29$. The proportion of gestation periods expected to last 10 months or longer is .0110. In modern obstetrical care, pregnancies are normally brought to term (by drugs or surgery) earlier than 300 days.

29. $\mu = \dfrac{3600}{100} = 36.0$ inches

$$\sigma = \sqrt{\dfrac{130,000 - \dfrac{(3600)^2}{100}}{100}} = 2.00 \text{ inches}$$

a. $z = \dfrac{32 - 36}{2} = -2.00$, proportion $= .5000 + .4772 = .9772$

b. $z = \dfrac{39 - 36}{2} = 1.50$, proportion $= .5000 + .4332 = .9332$

$.9332(300) = 279.96 = 280$ hobbits

c. $z = \dfrac{44 - 36}{2} = 4.00$, proportion $= .00003$ or, practically speaking, zero

CHAPTER 6

1. A statistic is a numerical characteristic of a sample.
2. A parameter is a numerical characteristic of a population.
3. μ is the mean of a population, and $\bar{X}$ is the mean of a sample.

4. *Population:* light bulbs that were produced the previous month and shipped to stores. *Parameter:* the proportion of recently produced bulbs that do not work. *Sample:* the bulbs taken from the shelves at the 25 stores. *Statistic:* the proportion of the sample that did not work, the numerical value of which is .005.

5. $\bar{X} = \dfrac{\Sigma X}{N} = \dfrac{245}{7} = 35.0$

6. In this problem it is not the self-esteem scores in your sample that are "right" or "wrong," but the method you used to get them. The proper method is to haphazardly find a starting place in the table of random numbers, record in sequence five two-digit numbers from the table (ignoring numbers over 24 and duplications), and write down the five self-esteem scores whose identification number you got from the table.

7. Proceed in the same way as for Problem 6, but obtain ten two-digit numbers from the table of random numbers.

8. If you begin every random sample at the same place, the same scores will be chosen for every sample. This violates the principle that every sample has an equal probability of being selected.

9. Again, the method you used is the important thing. You should have given each score an identifying number (01 to 38), begun at some chance starting place in Table B, and selected 12 numbers. The numbers identify 12 scores, which then constitute a random sample.

10. Yes, those with greater educational accomplishments are more likely to return the questionnaire. Such a biased sample will overestimate the accomplishments of the population.

11. The sample will be biased. Some of the general population will not even read the article and will thus have a zero probability of being selected. Of those who do read it, some will have no opinion and will not return the ballot. Of those who read it and have an opinion, many will not take the time and trouble to mark the ballot and return it. The opinions of those who do not return the ballot have a zero probability of being in the sample. In short, the results of such a sample cannot, with confidence, be generalized to a larger population.

12. A biased sample is one in which there is a systematic over- or under-representation of certain members of the population. In a representative sample, there is no such systematic bias.

13. Yes; a mean computed from a biased sample can be arithmetically "correct" but will not be representative of a population.

14. **a.** Biased, although the bias is not considered serious. It is biased because every sample does not have an equal chance to be selected. For example, two brothers named Kotki, adjacent to each other on the list, could not both be in one sample. Thus, samples with both Kotki brothers are not possible.

    b. Biased, because freshmen, sophomores, and seniors, although they are members of the population, have no chance to be selected for the sample.

    c. Random, assuming the names in the box were thoroughly mixed.

    d. Random; although small, this sample fits the definition of a random sample.

    e. Stratified

    f. Biased, because some members of the population have completed the English course and, therefore, have no chance to be chosen for the sample.

15. Biased; only bulbs in the area of the factory were included in the sample. Bulbs that had to travel farther might have been found to have more need for the new packaging.

16.

N	σ = 1	σ = 2	σ = 4	σ = 8
1	1	2	4	8
4	0.50	1	2	4
16	0.25	0.50	1	2
64	0.125	0.25	0.50	1

17. As N increases by a factor of 4, $\sigma_{\bar{x}}$ is reduced by one-half. More simply but less accurately, as N increases, $\sigma_{\bar{x}}$ decreases.

18. 16 times

19. When $N = 1$

20. If this standard error (whose name is the standard error of estimate) was very small, it would mean that you would be confident that the actual Y would be close to the predicted Y. A very large standard error would indicate that you could not put much confidence in a predicted Y, that the actual Y would be subject to lots of other influences besides X.

21. a. "REPORTER LAMBASTED BY EDITOR FOR BIASED SAMPLING." The subhead would read, "N too small also." Children in other schools and children in that school who were not in rooms with windows next to the sidewalk were excluded from the sample.

b. $z = \dfrac{23.3 - 25.5}{3.1/\sqrt{12}} = -2.46$

$p = .0069$ that the strike was having no effect on attendance

22. $s = \sqrt{\dfrac{40{,}464 - \dfrac{1188^2}{36}}{35}} = 6.00$

$s_{\bar{x}} = \dfrac{6.00}{\sqrt{36}} = 1.00$

$\bar{X} = \dfrac{1188}{36} = 33.00$

$z = \dfrac{33 - 30}{1.00} = 3.00$

$p = .0013$

The difference between the trained clients and the national norms does not seem to be a chance difference. It seems more reasonable to attribute the improvement to the 8-week training course.

23. $s_{\bar{x}} = \dfrac{10{,}000}{\sqrt{38}} = 1622$. Therefore, $z = \dfrac{36{,}900 - 38{,}500}{1622} = -.99$; $p = .1611$. Thus, the junior's suspicion that the $36,900 sample mean might be just a chance fluctuation from a true mean of $38,500 has some foundation. Such a sample mean (or one even smaller) would be expected in about 16 percent of a large number of random samples if the true campus mean was $38,500.

24. $s_{\bar{x}} = \dfrac{s}{\sqrt{N}} = \dfrac{10{,}000}{\sqrt{400}} = 500$, $z = \dfrac{38{,}000 - 38{,}500}{500} = -1.00$; $p = .1587$. Thus, in this hypothetical case, the increased sample size, which produced a new mean, did not reduce the uncertainty associated with the decision.

25. $LL = \bar{X} - z(s_{\bar{x}}); \qquad UL = \bar{X} + z(s_{\bar{x}})$

a. $LL = 36 - (1.96)\left(\dfrac{2}{\sqrt{100}}\right) = 35.61$

$UL = 36 + (1.96)\left(\dfrac{2}{\sqrt{100}}\right) = 36.39$

b. $LL = 36 - (1.96)\left(\dfrac{10}{\sqrt{100}}\right) = 34.04$

$UL = 36 + (1.96)\left(\dfrac{10}{\sqrt{100}}\right) = 37.96$

c. $LL = 36 - (1.96)\left(\dfrac{10}{\sqrt{1000}}\right) = 35.38$

$UL = 36 + (1.96)\left(\dfrac{10}{\sqrt{1000}}\right) = 36.62$

Notice that a fivefold increase in s (**b** versus **a**) causes a fivefold increase in the size of the confidence interval. A tenfold increase in N (**c** versus **b**) causes a threefold decrease in the confidence interval.

26. $LL = 36,900 - (2.58)\left(\dfrac{10,000}{\sqrt{38}}\right) = \$32,715$

$UL = 36,900 + (2.58)\left(\dfrac{10,000}{\sqrt{38}}\right) = \$41,085$

Our two economics students can be 99 percent confident that the interval \$32,715 to \$41,085 captures the mean family income for their campus.

27. $LL = 49.5 - (1.96)\left(\dfrac{9}{\sqrt{36}}\right) = 46.56$

$UL = 49.5 + (1.96)\left(\dfrac{9}{\sqrt{36}}\right) = 52.44$

The investigators can be 95 percent confident that the interval 46.56 to 52.44 contains the mean standardized test score for children whose parents supplied them with answers to their homework.

28. The size of the standard deviation of each sample. If you happen to get a sample that produces a large standard deviation, a large standard error will result and the confidence interval will be wide.

29. a. 19

b. More narrow; since N was increased by four, $s_{\bar{x}}$ would be reduced to half of its size. Therefore, the lines would be about half as long.

30. a. 18

b. More narrow; the z score is smaller.

31. From a specified population, you draw many samples of the same size, find the median for each sample, and arrange those sample medians into a frequency distribution. This frequency distribution of medians is called a sampling distribution of the median.

32. Standard error of the median

CHAPTER 7

1. Your outline should include the following points:

a. Two samples are drawn from one population.

b. The samples are treated the same except for one thing.
c. The samples are measured, and the difference found is attributed to chance or to the treatment difference.

2. Random assignment of subjects to groups
3. Your outline should include the following points:
   a. Recognize two logical possibilities:
      H_0: The treatment had no effect.
      H_1: The treatment had an effect.
   b. Assume H_0 to be correct.
   c. Compare the actual difference found with those in the sampling distribution of mean differences, which is based on the assumption that H_0 is true. If the difference found has a very low probability, reject H_0 and accept H_1. If the difference found has a high probability, suspend judgment, claiming that your experiment did not allow you to choose between H_0 and H_1.

4. Null hypothesis
5. The null hypothesis states that there is no difference between the populations from which the samples come. Stated another way, the samples come from the same population.
6. *Level of significance* is the arbitrary cutoff point between considering a difference "due to chance" or "not due to chance."
7. Events in the critical region have a probability less than the level of significance.
8. .05
9. $p < .01$ refers to the probability of obtaining such a difference in samples if the two populations are identical. That is, if only chance is operating and the populations are the same, differences as large or larger than the one found occur less than 1 time in 100.
10. 2.33
11. a. Independent variable—experience with electrical switch; dependent variable—time to solve problem
    b. H_0: Experience with the light switch does not affect the time required to solve the problem.

    c. $$s_{\bar{X}_1 - \bar{X}_2} = \sqrt{\left(\frac{2.13}{\sqrt{43}}\right)^2 + \left(\frac{2.31}{\sqrt{41}}\right)^2}$$
    $$= \sqrt{.1055 + .1301} = \sqrt{.2357} = .4854$$
    $$z = \frac{7.40 - 5.05}{.4854} = 4.84, p < .0001$$

    d. Since the experienced group took longer, you should conclude that the previous experience with the switch *retarded* the subject's ability to recognize the solution.
12. The z-score test requires large samples (more than 30). These samples do not meet that requirement. Chapters 8 and 12 describe techniques for analyzing small-sample experiments.
13. Independent variable—percent of cortex removed; dependent variable—number of errors

    $$s_{0 \text{ percent}} = \sqrt{\frac{\Sigma X^2 - \dfrac{(\Sigma X)^2}{N}}{N - 1}} = 4.00$$

    $$s_{20 \text{ percent}} = \sqrt{\frac{\Sigma X^2 - \dfrac{(\Sigma X)^2}{N}}{N - 1}} = 4.00$$

$$s_{\bar{X}_1 - \bar{X}_2} = \sqrt{\left(\frac{4.00}{\sqrt{40}}\right)^2 + \left(\frac{4.00}{\sqrt{40}}\right)^2} = 0.894$$

$$z = \frac{\bar{X}_0 - \bar{X}_{20}}{s_{\bar{X}_1 - \bar{X}_2}} = \frac{5.2 - 6.3}{0.894} = -1.23$$

Since $1.23 < 1.96$, retain the null hypothesis and conclude that a 20 percent loss of cortex does not reduce significantly a rat's memory for a simple maze.

14. $z = \dfrac{\bar{X}_1 - \bar{X}_2}{s_{\bar{X}_1 - \bar{X}_2}} = \dfrac{60 - 36}{\sqrt{\left(\dfrac{12}{\sqrt{45}}\right)^2 + \left(\dfrac{6}{\sqrt{52}}\right)^2}} = \dfrac{24}{1.97} = 12.18$

The meaning of very large z scores is quite clear; they indicate a very, very small probability. Thus, conclude that children whose illness begins before age 3 exhibit symptoms longer than do those whose illness begins after age 6.

15. Head shaking is in order. This "experiment" is so poorly designed that a comparison of mean scores is meaningless. The most glaring problem is that the two groups were not tested on the same dependent variable; one took a multiple choice test and the other took an essay test. Also, there is no reason to think that the two groups were equivalent to begin with, since students were allowed free choice between the classes (free choice will probably lead to biased samples). In addition, there were two professors, and one may be a more effective teacher than the other. If so, any difference in the means may be due to the professors rather than to the two methods. The professors in this problem are an example of an extraneous variable. The main reason we included this tricky question was to remind you of the importance of a sound experimental design. Without it, statistics cannot produce meaningful answers.

16. A Type I error is a rejection of the null hypothesis when it is in fact true.

17. A Type II error is retaining the null hypothesis when it is in fact false.

18. No, a Type I error can be made only when the null hypothesis is true.

19. α is the probability of a Type I error. The level of significance is the cutoff point between "due to chance" and "not due to chance," and the experimenter chooses a particular α for this cutoff point.

20. The probability of a Type I error decreases from .05 to .01, and the probability of a Type II error increases.

21. First of all, you might point out that if the populations are those particular freshman classes, no statistics are necessary; you have the population data and there is no sampling error. State U. is one-tenth of a point higher than The U. If the question is not about those freshman classes but about the two schools, and the two freshman classes can be treated as representative samples, a two-tailed test is called for, since superiority of either school would be of interest.

$$s_{\bar{X}_1 - \bar{X}_2} = \sqrt{\left(\frac{s_1}{\sqrt{N_1}}\right)^2 + \left(\frac{s_2}{\sqrt{N_2}}\right)^2} = 0.0474$$

$$z = \frac{\bar{X}_1 - \bar{X}_2}{s_{\bar{X}_1 - \bar{X}_2}} = \frac{21.4 - 21.5}{0.0474} = -2.11, \, p < .05$$

Students at State U. have statistically significant higher ACT admission scores than do students at The U. You may have noted how small the difference is, only one-tenth of a point. We discuss this in a later section in the text.

22. $z = 2.33$

23. **a.** One-tailed test
    b. Since the new Brand Z was slower than Brand Y on hand, no test is necessary.

24. Independent variable—experience with aerobics program; dependent variable—blood pressure

$$z = \cfrac{125 - 116}{\sqrt{\left(\cfrac{15}{\sqrt{36}}\right)^2 + \left(\cfrac{19}{\sqrt{36}}\right)^2}} = \frac{9}{\sqrt{16.28}} = 2.23, p < .05$$

Veterans of the noontime aerobics program had significantly lower blood pressure than newcomers did.

25. Our own words are, "Significant differences are not due to chance. Important differences are ones that change our understanding about something."

26. Our list is (1) sample size, (2) alpha level, (3) the actual difference between the population means, and (4) the preciseness of measuring the dependent variable.

27. Regardless of the material you choose to include, your essay should have good introductory and concluding paragraphs. As for content, the purpose of inferential statistics should receive prominent attention along with the logic of inferential statistics. Sampling should be covered. Your essay should explain the null hypothesis and the alternative hypothesis and describe sampling distributions as the means for determining the probability of obtaining the observed results *if the null hypothesis is true.* Level of significance and possible decisions about the population should be discussed.

CHAPTER 8

1. **a.** Normal, t **b.** t, normal   **c.** t
2. Larger than
3. $df = N - 1$ **a.** $25 - 1 = 24$ **b.** $4 - 1 = 3$ **c.** $42 - 1 = 41$
4. W. S. Gosset, who wrote under the pseudonym "Student," invented the t distribution so that he could assess probabilities for small samples.

5. $t = \dfrac{\bar{X} - \mu}{s_{\bar{X}}}$ **a.** 1.00   **b.** -0.33 **c.** 4.00   **d.** 0.50

   e. -4.51; notice that the mean difference on **5e** is small; on **5d** it is large. The difference in the t values shows the importance of $s_{\bar{X}}$.

6. **a.** $df = 14$; $t = 1.96$; $p > .05$. Retain H_0.
   b. $df = 6$; $t = -2.10$; $p > .05$. Retain H_0.
   c. $df = 23$; $t = -2.10$; $.01 < p < .05$. Reject H_0.
   d. $df = 20$; $t = 2.845$; $p = .01$. Reject H_0.
   e. $df = 4$; $t = -4.90$; $p < .01$. Reject H_0.

7. The probabilities refer to the event of getting such a sample mean with a random sample from a population with $\mu = 81.00$.

8. Correlated samples. With twins divided between the two groups, there is a logical pairing (natural pairs).

9. Correlated samples. The depression score after treatment is compared to the depression score before treatment (repeated measures).

10. Correlated samples. This is a before-and-after experiment. For each group, the amount of aggression before the screen was lifted is paired with the amount after the screen was lowered (repeated measures).

11. Correlated samples. Each rat that is deprived of sleep is paired with a rat that must exercise just as much (whenever the platform turns)(matched pairs, yoked control design).

12. Independent samples. The dean randomly assigned individuals to one of the two groups.

13. Correlated samples. The IQ of the firstborn is paired with the IQ of his or her sibling (natural pairs).

This paragraph is really not about statistics, and you may skip it if you wish. Were you somewhat more anxious about your decisions on Problems 10, 11, or 13 than on Problems 8 and 9? If so, and if this anxiety was based on the expectation that surely it was time for an answer to be "independent samples" and if you based your answer on this expectation rather than on the problem, you were exhibiting a *response bias*. A response bias is when a current response is made on the basis of previous responses rather than on the basis of the current stimulus. Such response biases often lead to a correct answer in textbooks, and you may learn to make some decisions (those you are not sure about) on the basis of irrelevant cues (such as what your response was on the last question). To the extent it rewards your response biases, a textbook is doing you a disservice. So, be forewarned; recognize response bias and resist it.

14.

	Laboratory	Car
N	4	5
ΣX	52	30
ΣX^2	722	210
$\bar{X}$	13	6

$$t = \frac{13 - 6}{\sqrt{\left(\dfrac{722 - \dfrac{52^2}{4} + 210 - \dfrac{30^2}{5}}{4 + 5 - 2}\right)\left(\dfrac{1}{4} + \dfrac{1}{5}\right)}} = \frac{7}{2.21} = 3.17 \qquad df = 7$$

Note that you must use the longer formula for $s_{\bar{X}_1 - \bar{X}_2}$ because the N's are not equal. Since the obtained $|t|$ is greater than 2.37 $[t_{.05}(7\ df) = 2.37]$, the probability that chance produced such a difference is less than .05 $(p < .05)$, so reject the null hypothesis. Since the car group made fewer errors, conclude that immediate, concrete experience facilitates memory. A two-tailed test is appropriate, since the interest was in "the effect of immediate, concrete experience," which could be positive or negative.

15.

	"Winners"	"Losers"
$\bar{X}$	3.46	3.52
s	0.35	0.30
N	16	16

$$t = \frac{\bar{X}_1 - \bar{X}_2}{\sqrt{\left(\dfrac{s_{\bar{X}_1}}{\sqrt{N_1}}\right)^2 + \left(\dfrac{s_{\bar{X}_2}}{\sqrt{N_2}}\right)^2}} = \frac{3.46 - 3.52}{\sqrt{\left(\dfrac{0.35}{\sqrt{16}}\right)^2 + \left(\dfrac{0.30}{\sqrt{16}}\right)^2}} = \frac{-0.060}{0.115} = -.52$$

$t_{.05}(30\ df) = 2.04$ (two-tailed test)

There is no evidence that being in the sophomore honors course has a significant effect on grade point average for those students who qualified for the sophomore honors course.

16.

	New package	Old package
ΣX	45.9	52.8
ΣX^2	238.73	354.08
$\bar{X}$	5.1	6.6
N	9	8
s	0.76	0.89

$$t = \frac{5.1 - 6.6}{\sqrt{\left(\frac{10.24}{15}\right)(0.24)}} = \frac{-1.5}{0.40} = -3.75$$

$t_{.005}(15 \ df) = 2.95$ (one-tailed test). Thus, $p < .005$; therefore, reject the null hypothesis and conclude that the test procedure can be worked more quickly with the new package. A one-tailed test is appropriate here because the only interest is whether the new package is better than the one on hand. (See the section "One-Tailed and Two-Tailed Tests" and Problem 23 in Chapter 7.)

17. a. $s_D = \sqrt{\dfrac{\Sigma D^2 - \dfrac{(\Sigma D)^2}{N}}{N - 1}}$

s_D is the standard deviation of the distribution of differences between correlated scores.

b. $D = X - Y$. D is the difference between two correlated scores.

c. $s_{\bar{D}} = s_D/\sqrt{N}$. $s_{\bar{D}}$ is the standard error of the difference between means for a correlated set of scores.

d. t is the name of a theoretical probability distribution. So far in this chapter you have used it to determine the probability that two samples came from populations with the same mean. As you will see, t has other uses, too.

e. $\bar{Y} = \Sigma Y/N$. $\bar{Y}$ is the mean of a set of scores that is correlated with another set.

18.

	X	Y	D	D^2
	16	18	-2	4
	10	11	-1	1
	17	19	-2	4
	4	6	-2	4
	9	10	-1	1
	12	14	-2	4
Σ	68	78	-10	18
Mean	11.3	13.0		

$$s_D = \sqrt{\frac{\Sigma D^2 - \dfrac{(\Sigma D)^2}{N}}{N - 1}} = \sqrt{\frac{18 - \dfrac{(-10)^2}{6}}{5}} = \sqrt{0.2667} = 0.516$$

$$s_{\bar{D}} = \frac{s_D}{\sqrt{N}} = \frac{0.516}{\sqrt{6}} = 0.211$$

$$t = \frac{11.333 - 13.000}{0.211} = \frac{-1.667}{0.211} = -7.91$$

$t_{.001}(5\ df) = 6.86$. Therefore, $p < .001$. Notice that the small difference between means (1.67) is highly significant even though the data consist of only six pairs of scores. This illustrates the power that a large correlation can have in reducing the standard error. The conclusion reached by the experimenters was that frustration (produced by seeing others treated better) leads to aggression.

19. On this problem you cannot use the direct-difference method, since you do not have the raw data to work with. This leaves you with the definition formula, $t = (\bar{X} - \bar{Y})/\sqrt{s_{\bar{X}}^2 + s_{\bar{Y}}^2 - 2r_{XY}(s_{\bar{X}})(s_{\bar{Y}})}$, which requires a correlation coefficient. Fortunately, you have the data necessary to calculate r. (The initial clue for many students is the term ΣXY.)

Females	Males
$\bar{X} = \$12,782.25$	$\bar{Y} = \$15,733.25$
$s_X = 2,358.99$	$s_Y = 3,951.27$
$s_{\bar{X}} = 589.75$	$s_{\bar{Y}} = 987.82$
	$r = .693$

$$t = \frac{12,782.25 - 15,733.25}{\sqrt{(589.75)^2 + (987.82)^2 - (2)(.693)(589.75)(987.82)}} = -4.11$$

Since $t_{.001}$ (15 df) = 4.07, the null hypothesis can be rejected at the .001 level. Here is the conclusion about these data written for the court.

> Chance is not a likely explanation for the \$2951 annual difference in favor of men. Chance would be expected to produce such a difference less than one time in a thousand. In addition, the difference cannot be attributed to education, since both groups were equally educated. Likewise, the two groups were equal in their work experience at the rehabilitation center. One explanation that has not been eliminated is discrimination, based on sex.[1]

20. This is a correlated-samples study.

$$\Sigma D = -.19 \qquad \Sigma D^2 = .0101 \qquad N = 11 \qquad \bar{X} - \bar{Y} = \frac{\Sigma D}{N} = -.0173$$

$$s_D = \sqrt{\frac{\Sigma D^2 - \frac{(\Sigma D)^2}{N}}{N - 1}} = \sqrt{0.00068} = 0.026$$

$$s_{\bar{D}} = \frac{0.026}{\sqrt{11}} = 0.00787$$

$$t = \frac{-0.0173}{0.00787} = -2.19$$

$t_{.05}$ (10 df) = 2.23. The obtained $|t|$ does not reach statistical significance. (This is a case in which a small sample led to a Type II error. On the basis of larger samples, auditory RT has been found to be faster than visual RT—approximately 0.14 second for auditory RT and 0.18 second for visual RT, using practiced subjects.)

21. We are most hopeful that you did not treat these data as correlated samples. If you did, you exhibited a response bias that led you astray. If you treated these data as the

---

[1] These data are 1979 salary data submitted in Hartman and Hobgood v. Hot Springs Rehabilitation Center, HS-76-23-C.

independent samples they are, you are on your (mental) toes. So much for our role as dispensers of verbal reinforcement; here is the data analysis.

Recency	Primacy
$\Sigma X_1 = 86$	$\Sigma X_2 = 104$
$\Sigma X_1^2 = 934$	$\Sigma X_2^2 = 1312$
$N_1 = 9$	$N_2 = 9$

$$s_{\bar{X}_1 - \bar{X}_2} = \sqrt{\frac{\Sigma X_1^2 - \dfrac{(\Sigma X_1)^2}{N} + \Sigma X_2^2 - \dfrac{(\Sigma X_2)^2}{N}}{N(N-1)}} = 1.75$$

$$t = \frac{\bar{X}_1 - \bar{X}_2}{s_{\bar{X}_1 - \bar{X}_2}} = \frac{9.56 - 11.56}{1.75} = 1.14$$

$t_{.05}(16\ df) = 2.12$. Therefore, retain the null hypothesis. Our conclusion was, "This experiment does not provide evidence that primacy or recency is more powerful than the other."

22. $\text{LL} = (\bar{X}_1 - \bar{X}_2) - t(s_{\bar{X}_1 - \bar{X}_2}) = 1.5 - (2.13)(0.04) = 1.41$
    $\text{UL} = (\bar{X}_1 - \bar{X}_2) + t(s_{\bar{X}_1 - \bar{X}_2}) = 1.5 + (2.13)(0.04) = 1.59$

23. The decision now is to stay with the old package. The difference in time per problem—between 1.41 and 1.59 minutes—is not enough to justify the higher price.

24. Buy the new package, without a doubt (or at least with only a very tiny doubt).

25. $\text{LL} = (\bar{X} - \bar{Y}) - t(s_{\bar{D}}) = -0.0173 - (3.17)(0.00787) = -0.0422$
    $\text{UL} = (\bar{X} - \bar{Y}) + t(s_{\bar{D}}) = -0.0173 + (3.17)(0.00787) = 0.00768$
    Notice that, since the interval includes .00, the null hypothesis cannot be rejected at the .01 level.

26. $\Sigma D = -15$; $\Sigma D^2 = 41$

$$s_D = \sqrt{\frac{\Sigma D^2 - \dfrac{(\Sigma D)^2}{N}}{N-1}} = \sqrt{\frac{41 - \dfrac{(-15)^2}{8}}{7}} = 1.36$$

$$s_{\bar{D}} = \frac{s_D}{\sqrt{N}} = \frac{1.36}{\sqrt{8}} = 0.48$$

$t_{.10}(7\ df) = 1.895$

$\text{LL} = (\bar{X} - \bar{Y}) - t(s_{\bar{D}}) = -1.88 - (1.895)(0.48) = -2.79$
$\text{UL} = (X - \bar{Y}) + t(s_{\bar{D}}) = -1.88 + (1.895)(0.48) = -0.97$

27. With 90 percent confidence, you can state that the real (parametric) effect of sleep is to reduce the number of CVCs forgotten from a ten-item list by 0.97 to 2.79 items.

28. The formulas for the limits are $\text{LL} = \bar{X} - t(s_{\bar{X}})$ and $\text{UL} = \bar{X} + t(s_{\bar{X}})$. With several confidence intervals to calculate, it is convenient to concentrate on the $-t(s_{\bar{X}})$ and the $+t(s_{\bar{X}})$ parts of the formulas for the limits. In the table (p.360), these two parts have been combined into $\pm t(s_{\bar{X}})$. In the graph, the values for $\pm t(s_{\bar{X}})$ were simply added to the mean for the trial. For example, on Trial 1, the confidence interval extends from approximately 19 to 23, as seen on the Y axis in the figure; $t_{.05}(16\ df) = 2.12$.

	Trial					
	1	2	3	4	5	6
$\bar{X}$	21	27	30	25	30	36
s	3.3	2.7	3.0	3.6	5.1	4.5
N	17	17	17	17	17	17
$s_{\bar{x}}$	0.80	0.65	0.73	0.87	1.24	1.09
$\pm (t)(s_{\bar{x}})$	± 1.70	± 1.39	± 1.54	± 1.85	± 2.62	± 2.31

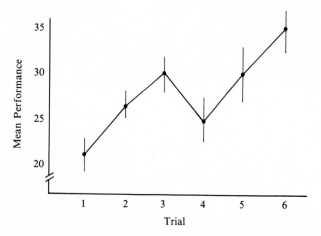

We think you will agree: the confidence intervals display the reliability of the dip in the curve.

29. Assumption 3; the scores are not independent. When one person contributes two or more scores to one sample, those scores are not independent, and Assumption 3 has been violated.

30. $t = (r)\sqrt{\dfrac{N-2}{1-r^2}}$

 a. $t = 2.24$; $df = 8$. Since $t_{.05}\,(8\ df) = 2.31$, $r = .62$ is not significantly different from .00.

 b. $t = -2.12$; $df = 120$. Since $t_{.05}\,(120\ df) = 1.98$, $r = -.19$ is significantly different from .00.

 c. $t = 2.08$; $df = 13$; $t_{.05}\,(13\ df) = 2.16$. Retain the null hypothesis.

 d. $t = -2.85$; $df = 62$; $t_{.05}\,(60\ df) = 2.00$. Reject the null hypothesis.

31. See Table A and read the next to last paragraph of the chapter.

CHAPTER 9

1. *Example 2:* The independent variable is the schedule of reinforcement; there are four levels. The dependent variable is persistence. The null hypothesis is that the mean persistence is the same for the four populations, which is to say that the schedule of reinforcement has no effect on persistence.

 Example 3: The independent variable is degree of modernization; there are three levels. The dependent variable is suicide rate. The null hypothesis is that suicide rates are not dependent on degree of modernization, that the mean suicide rates for countries with low, medium, and high degrees of modernization are the same. In formal terms, the null hypothesis is H_0: $\mu_{\text{low}} = \mu_{\text{med}} = \mu_{\text{high}}$.

Example 4: The independent variable is the injection; there are three levels. The dependent variable is memory. The null hypothesis is that the three populations of memory scores are the same (the effect of the three injections will be the same).

2. These are correlated-samples designs, and the ANOVA technique described here is appropriate for only independent samples. ANOVA can be used on correlated-samples data using techniques described in Howell (1987, chap. 14) and Guilford and Fruchter (1978, chap. 13).

3. Larger

4. One estimate is obtained from variability among the sample means, and another estimate is obtained by pooling the sample variances.

5. *F* is a ratio of the two estimates of the population variance.

6. **a.** Retain the null hypothesis and conclude that the data do not provide evidence that social classes differ in their attitude toward religion.

   b. Reject the null hypothesis and conclude that there is a relationship between degree of modernization and suicide rate.

7.

	X_1	X_2	X_3	X_4
ΣX	28	26	25	25
ΣX^2	176	162	151	147
N	5	5	5	5

$$SS_{tot} = 636 - \frac{104^2}{20} = 636 - 540.80 = 95.20$$

$$SS_{bg} = \frac{28^2}{5} + \frac{26^2}{5} + \frac{25^2}{5} + \frac{25^2}{5} - \frac{104^2}{20}$$

$$= 156.80 + 135.20 + 125.00 + 125.00 - 540.80$$

$$= 542.00 - 540.80 = 1.20$$

$$SS_{wg} = \left(176 - \frac{28^2}{5}\right) + \left(162 - \frac{26^2}{5}\right) + \left(151 - \frac{25^2}{5}\right) + \left(147 - \frac{25^2}{5}\right)$$

$$= 19.20 + 26.80 + 26.00 + 22.00 = 94.00$$

Check: $95.20 = 1.20 + 94.00$

8.

	Degree of modernization		
	Low	*Medium*	*High*
ΣX	24	48	84
ΣX^2	154	614	1522
$\bar{X}$	6.0	12.0	16.8

$$SS_{tot} = 2290 - \frac{(156)^2}{13} = 418.00$$

$$SS_{mod} = \frac{(24)^2}{4} + \frac{(48)^2}{4} + \frac{(84)^2}{5} - \frac{(156)^2}{13} = 259.20$$

$$SS_{wg} = \left(154 - \frac{(24)^2}{4}\right) + \left(614 - \frac{(48)^2}{4}\right) + \left(1522 - \frac{(84)^2}{5}\right) = 158.80$$

Check: $418.00 = 259.20 + 158.80$

Incidentally, the rate per 100,000 is about 18 in the United States and about 17 in Canada.

9.

	X_1	X_2	X_3
ΣX	137	88	82
ΣX^2	2015	836	746
$\bar{X}$	13.70	8.80	8.20
N	10	10	10

$$SS_{tot} = 3597 - \frac{307^2}{30} = 3597 - 3141.63 = 455.37$$

$$SS_{methods} = \frac{137^2}{10} + \frac{88^2}{10} + \frac{82^2}{10} - \frac{307^2}{30}$$

$$= \frac{18,769}{10} + \frac{7744}{10} + \frac{6724}{10} - 3141.63 = 182.07$$

$$SS_{wg} = \left(2015 - \frac{137^2}{10}\right) + \left(836 - \frac{88^2}{10}\right) + \left(746 - \frac{82^2}{10}\right)$$

$$= 138.10 + 61.60 + 73.60 = 273.30$$

Check: $455.37 = 182.07 + 273.30$

10. Data from Problem 7:

Source	df	SS	MS	F	p
Between groups	3	1.20	0.40	0.07	>.05
Within groups	16	94.00	5.88		
Total	19	95.20			

$F_{.05}$ (3, 16 df) = 3.24. Since 0.07 does not reach this critical value, retain the null hypothesis and conclude that these four groups could have come from the same population.

11. Durkheim's modernization and suicide data; Problem 8. For df, use

$$df_{tot} = N_{tot} - 1 = 13 - 1 = 12$$
$$df_{bg} = K - 1 = 3 - 1 = 2$$
$$df_{wg} = N_{tot} - K = 13 - 3 = 10$$

Source	df	SS	MS	F	p
Between groups	2	259.20	129.60	8.16	<.01
Within groups	10	158.80	15.88		
Total	12	418.00			

Since $F_{.01}$ (2, 10) = 7.56, reject the null hypothesis and conclude that there is a relationship between suicide rate and a country's degree of modernization.

12. Data from Problem 9. The independent variable here is methods, which is reflected in the source column.

Source	df	SS	MS	F	p
Between methods	2	182.07	91.04	9.00	<.01
Within groups	27	273.30	10.12		
Total	29	455.37			

$F_{.01}$ (2, 27 *df*) = 5.49. Since 9.00 > 5.49, reject the null hypothesis at the .01 level and conclude that the teaching method had an effect on the number of errors on a comprehensive examination. By the end of this chapter, you will be able to make comparisons among the pairs of means to determine which method is best (or worst).

13. Six groups. The critical values of F are based on 5, 70 *df* and are 2.35 and 3.29 at the .05 and .01 levels, respectively. If $\alpha = .01$, retain the null hypothesis; if $\alpha = .05$, reject the null hypothesis.

14. *A priori* tests are those planned in advance. They are tests based on the logic of the design of an experiment. *Post hoc* tests are chosen after examining the data. Recall that the suicide rates for the different degrees of modernization were: low—6.0; medium—12.0; high—16.8.

15. Comparing countries with low and medium degrees of modernization:

$$\text{HSD} = \frac{12.0 - 6.0}{\sqrt{\dfrac{15.88}{4}}} = \frac{6.0}{1.99} = 3.02 \qquad \text{HSD}_{.05} = 3.88 \qquad \text{NS}$$

Comparing countries with medium and high degrees of modernization:

$$\text{HSD} = \frac{16.8 - 12.0}{\sqrt{\dfrac{15.88}{2}\left(\dfrac{1}{4} + \dfrac{1}{5}\right)}} = \frac{4.8}{1.890} = 2.54 \qquad \text{HSD}_{.05} = 3.88 \qquad \text{NS}$$

Comparing countries with low and high degrees of modernization:

$$\text{HSD} = \frac{16.8 - 6.0}{1.886} = \frac{10.8}{1.886} = 5.73 \qquad \text{HSD}_{.01} = 5.27 \qquad p < .01$$

Suicide rates are significantly higher in countries with a high degree of modernization than they are in countries with a low degree of modernization. Although the other differences were not significant, the trend is in the direction predicted by Durkheim's hypothesis.

16. We calculated HSDs for each of the six pairwise differences. For some of these HSDs the equal N formula was appropriate, and for some the unequal N formula was necessary. The two values of $s_{\bar{x}}$ are

$$s_{\bar{x}} = \sqrt{\frac{MS_{wg}}{N}} = \sqrt{\frac{3.93}{5}} = 0.887; \quad s_{\bar{x}} = \sqrt{\frac{MS_{wg}}{2}\left(\frac{1}{N_1} + \frac{1}{N_2}\right)} = \sqrt{\frac{3.93}{2}\left(\frac{1}{5} + \frac{1}{7}\right)} = 0.821$$

The table shows all six HSD values. The significance level of each comparison is indicated with one or two asterisks. $\text{HSD}_{.05} = 4.00$, $\text{HSD}_{.01} = 5.09$.

	crf	FR2	FR4
FR2	4.63*		
FR4	7.44**	3.41	
FR8	9.47**	5.60**	2.03

* *p* < .05
** *p* < .01

The trend of the means is consistent; the higher the ratio of response to reinforcement during learning, the greater the persistence during extinction. The crf schedule produced the least persistence, significantly less than any other schedule. The FR2 schedule produced less persistence than the FR4 schedule (NS) and significantly less than the FR8 schedule (*p* < .01).

17. Normally distributed dependent variable, homogeneity of variance, and random sampling

18.

	Normals	*Schizophrenics*	*Depressives*
ΣX	178	138	72
ΣX^2	4162	2796	890
$\bar{X}$	22.25	19.71	12.00
N	8	7	6

Source	df	SS	MS	F	p
Between categories	2	376.31	188.16	11.18	<.01
Within groups	18	302.93	16.83		
Total	20	679.24			

$F_{.01}$ (2, 18) = 6.01. Since the ANOVA showed that the populations are not equal in responsiveness to the Rorschach Test, pairwise comparisons are appropriate. After we examined the means, we decided to test the difference between the depressives and the schizophrenics. If that difference proved significant, we could conclude that the normal-depressive difference would be significant, too (greater difference between means and larger samples). Thus,

$$\text{HSD} = \frac{19.71 - 12.00}{\sqrt{\dfrac{16.83}{2}\left(\dfrac{1}{7}+\dfrac{1}{6}\right)}} = \frac{7.71}{1.614} = 4.78 \qquad \text{HSD}_{.01} = 4.70 \qquad p < .01$$

Since the depressives are significantly less responsive than the schizophrenics, they will also be significantly less than the normals ($p < .01$). For the comparison of the normals and the schizophrenics,

$$\text{HSD} = \frac{22.25 - 19.71}{\sqrt{\dfrac{16.83}{2}\left(\dfrac{1}{8}+\dfrac{1}{7}\right)}} = \frac{2.54}{1.501} = 1.69 \qquad \text{HSD}_{.05} = 3.61 \qquad \text{NS}$$

The R scores on the Rorschach Test do not distinguish between normals and schizophrenics.

19.

Source	df	SS	MS	F	p
Between injections	2	235.64	117.82	7.98	<.01
Within groups	23	339.40	14.76		
Total	25	575.04			

$F_{.01}$ (2, 23 df) = 5.66. Conclude that the effect of injections was seen 7 days later in the number of trials necessary to relearn the avoidance task. Note that strychnine *improved* the memory scores.

20.

Source	df	SS	MS	F	p
Between varieties	2	44.40	22.20	12.13	<.01
Within groups	12	22.00	1.83		
Total	14	66.40			

$F_{.01}$ (2, 12 df) = 6.93. Comparing Varieties 3 and 1 (the intermediate difference), you get

$$\text{HSD} = \frac{9.4 - 7.0}{\sqrt{\dfrac{1.83}{5}}} = \frac{2.4}{0.605} = 3.97 \qquad \text{HSD}_{.05} = 3.77 \qquad p < .05$$

Variety 3 is significantly more productive than Variety 1. Since Variety 2 produced even more than Variety 3, you can conclude also that Variety 2 is significantly better than Variety 1. Comparing Variety 2 to Variety 3, you get

$$\text{HSD} = \frac{11.2 - 9.4}{0.605} = \frac{1.8}{0.605} = 2.98 \qquad \text{HSD}_{.05} = 3.77 \qquad \text{NS}$$

Variety 2 is not significantly more productive than Variety 3. Summarizing, Variety 1 is the worst of the three, but Variety 2 is not significantly better than Variety 3.

CHAPTER 10

1. a. 3 × 2   **b.** 4 × 3   **c.** 2 × 2   **d.** 4 × 3 × 3. This is a factorial design in which there are *three* independent variables. This text does not cover the analysis of three-way ANOVAs. If you are interested in such complex designs, see footnote 1 in Chapter 10.

2. A main effect is the effect that the different levels of one factor have on the dependent variable.

3. An interaction occurs when the effect of one factor depends on the level of another factor.

4. Main effect

5. Main effect

6. i. a.

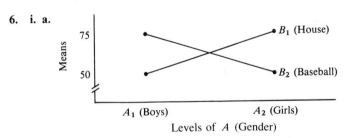

b. There appears to be an interaction.

c. Attitude toward playing the games depends on gender. Boys would rather play baseball and girls would rather play house.

d. There appear to be no main effects. Means for games are house = 62.5 and baseball = 62.5. Means for gender are boys = 62.5 and girls = 62.5.

e. Later in this chapter, you will learn that when there is a significant interaction, the interpretation of main effects depends on the interaction. Although all four means are exactly the same, there is a difference in attitudes toward the two games, but this difference depends on the gender of the child.

ii. a.

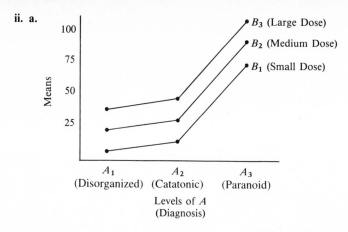

b. There appears to be no interaction.

c. The effect of the drug dosage does not depend on the diagnosis of the patient.

d. There appear to be main effects for both factors.

e. For all schizophrenics, the larger the dose, the higher the "happiness score" (Factor B, main effect). Paranoids have higher "happiness" scores than the others at all dose levels (Factor A, main effect).

iii. a.

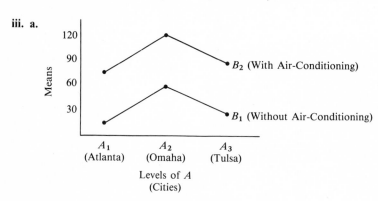

b. There appears to be no interaction.

c. Whether people buy cars with or without air-conditioning does not depend on which of the three cities they live in.

d. There appear to be main effects for both factors.

e. It appears that Overseas Motors sells more cars in Omaha than in Tulsa or Atlanta (Factor A). More cars are sold with than without air-conditioning (Factor B).

iv. a.

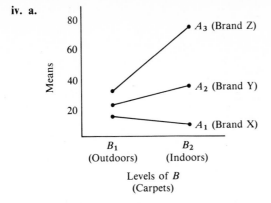

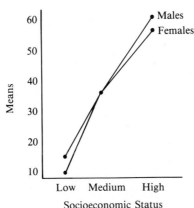

b. There appears to be an interaction.

c. The effect of placing carpet indoors or outdoors depends on the brand. Brand X seems to last a little longer outdoors than indoors, but the reverse is true for Brand Y. Brand Z lasts considerably longer indoors.

d, e. Again, when the interaction is significant, main effects must be interpreted in the light of the interaction effect. Clearly, the effect of placing the carpet indoors or outdoors causes differences in carpet life, but these differences depend on the brand. The brands of carpet differ in life, but these differences depend on whether the carpet is indoors or outdoors.

v. a.

b. There appears to be no interaction. Although the lines in the graph are not parallel—they even cross—the departure from parallel is slight and probably should be attributed to sampling variation.

c. Gender differences in attitudes toward lowering taxes on investments do not depend on socioeconomic status.

d. There appears to be a main effect for socioeconomic status but not for gender.

e. Males and females do not differ in their attitudes toward lowering taxes on investments (Factor A). The higher the socioeconomic status, the more positive are attitudes toward lowering taxes on investments.

7. a. The number of scores in each group (cell) must be the same.

b. The samples in each cell must be independent.

c. The levels of both factors must be chosen by the experimenter and not left to chance.

d. The samples are drawn from normally distributed populations.

e. The variances of the populations are equal.

f. The samples are drawn randomly from the populations.

8. a. 2×3 **b.** Diets and sex

c. $SS_{tot} = 553 - \dfrac{95^2}{18} = 553 - 501.39 = 51.61$

$$SS_{bg} = \frac{(17)^2}{3} + \frac{(11)^2}{3} + \frac{(24)^2}{3} + \frac{(14)^2}{3} + \frac{(17)^2}{3} + \frac{(12)^2}{3} - \frac{(95)^2}{18}$$

$$= 538.33 - 501.39$$

$$= 36.94$$

$$SS_{diets} = \frac{(31)^2}{6} + \frac{(28)^2}{6} + \frac{(36)^2}{6} - \frac{(95)^2}{18} = 506.83 - 501.39 = 5.44$$

$$SS_{sex} = \frac{(52)^2}{9} + \frac{(43)^2}{9} - \frac{(95)^2}{18} = 505.89 - 501.39 = 4.50$$

$$SS_{AB} = 3[(5.6667 - 5.1667 - 5.7778 + 5.2778)^2$$
$$+ (4.6667 - 5.1667 - 4.7778 + 5.2778)^2$$
$$+ (3.6667 - 4.6667 - 5.7778 + 5.2778)^2$$
$$+ (5.6667 - 4.6667 - 4.7778 + 5.2778)^2$$
$$+ (8.00 - 6.00 - 5.7778 + 5.2778)^2$$
$$+ (4.00 - 6.00 - 4.7778 + 5.2778)^2] = 27.00$$

Check: $SS_{AB} = 36.94 - 5.44 - 4.50 = 27.00$

$$SS_{wg} = \left[101 - \frac{(17)^2}{3}\right] + \left[41 - \frac{(11)^2}{3}\right] + \left[194 - \frac{(24)^2}{3}\right]$$

$$+ \left[68 - \frac{(14)^2}{3}\right] + \left[99 - \frac{(17)^2}{3}\right] + \left[50 - \frac{(12)^2}{3}\right]$$

$$= 4.67 + 0.67 + 2.00 + 2.67 + 2.67 + 2.00 = 14.68$$

Check: $51.61 = 36.94 + 14.67$ (The difference between answers using the two methods is due to rounding.)

9. a. 2×3 **b.** Teacher response and gender

c. Errors on a comprehensive examination in arithmetic

d. $SS_{tot} = 13,281 - \dfrac{845^2}{60} = 1380.58$

$$SS_{bg} = \frac{(169)^2}{10} + \frac{(136)^2}{10} + \frac{(118)^2}{10} + \frac{(146)^2}{10} + \frac{(167)^2}{10} + \frac{(109)^2}{10} - \frac{(845)^2}{60} = 306.28$$

$$SS_{response} = \frac{(315)^2}{20} + \frac{(303)^2}{20} + \frac{(227)^2}{20} - \frac{(845)^2}{60} = 227.73$$

$$SS_{gender} = \frac{(423)^2}{30} + \frac{(422)^2}{30} - \frac{(845)^2}{60} = 0.02$$

$$SS_{AB} = 10[(16.90 - 15.75 - 14.10 + 14.0833)^2$$
$$+ (14.60 - 15.75 - 14.0667 + 14.0833)^2$$
$$+ (13.60 - 15.15 - 14.10 + 14.0833)^2$$
$$+ (16.70 - 15.15 - 14.0667 + 14.0833)^2$$
$$+ (11.80 - 11.35 - 14.10 + 14.0833)^2$$
$$+ (10.90 - 11.35 - 14.0667 + 14.0833)^2]$$
$$= 78.53$$

Check: $SS_{AB} = 306.28 - 227.73 - 0.02 = 78.53$

$$SS_{wg} = \left(3129 - \frac{(169)^2}{10}\right) + \left(2042 - \frac{(136)^2}{10}\right) + \left(1620 - \frac{(118)^2}{10}\right)$$
$$+ \left(2244 - \frac{(146)^2}{10}\right) + \left(2955 - \frac{(167)^2}{10}\right) + \left(1291 - \frac{(109)^2}{10}\right) = 1074.30$$

Check: $1380.58 = 306.28 + 1074.30$

10. $SS_{tot} = 1405 - \dfrac{145^2}{20} = 353.75$

$$SS_{bg} = \frac{25^2}{5} + \frac{57^2}{5} + \frac{43^2}{5} + \frac{20^2}{5} - \frac{145^2}{20} = 173.35$$

$$SS_{learn} = \frac{68^2}{10} + \frac{77^2}{10} - \frac{145^2}{20} = 4.05$$

$$SS_{recall} = \frac{63^2}{10} + \frac{82^2}{10} - \frac{145^2}{20} = 18.05$$

$$SS_{AB} = 5[(8.60 - 6.80 - 6.30 + 7.25)^2 + (5.00 - 6.80 - 8.20 + 7.25)^2$$
$$+ (4.00 - 7.70 - 6.30 + 7.25)^2 + (11.40 - 7.70 - 8.20 + 7.25)^2]$$
$$= 151.25$$

Check: $SS_{AB} = 173.35 - 4.05 - 18.05 = 151.25$

$$SS_{wg} = \left(400 - \frac{43^2}{5}\right) + \left(100 - \frac{20^2}{5}\right) + \left(155 - \frac{25^2}{5}\right) + \left(750 - \frac{57^2}{5}\right) = 180.40$$

Check: $353.75 = 173.35 + 180.40$

11. There is not an equal number of scores in each cell.

12.

Source	df	SS	MS	F	p
A (diets)	2	5.44	2.72	2.23	>.05
B (sex)	1	4.50	4.50	3.69	>.05
AB	2	27.00	13.50	11.07	<.01
Within groups	12	14.67	1.22		
Total	17	51.61			

$F_{.05}$ (2, 12 df) = 3.88; $F_{.01}$ (2, 12 df) = 6.93; $F_{.05}$ (1, 12 df) = 4.75

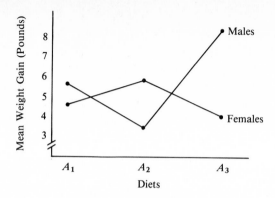

Interpretation: The null hypothesis $\mu_{A_1} = \mu_{A_2} = \mu_{A_3}$ is retained. There are no significant differences among the means of the puppies receiving different diets. The null hypothesis $\mu_{B_1} = \mu_{B_2}$ is retained. Mean weight gains of male and female puppies did not differ significantly. There was a significant interaction between diet and sex. Males and females were affected differently by the three diets. Whereas Diet 2 was best for female puppies, Diet 3 was best for males.

13.

Source	df	SS	MS	F	p
A (response)	2	227.73	113.87	5.72	<.01
B (gender)	1	0.02	0.02	0.00	>.05
AB	2	78.53	39.27	1.97	>.05
Within groups	54	1,074.30	19.89		
Total	59	1,380.58			

$F_{.05}(2, 50\ df) = 3.18;\ F_{.01}(2, 50\ df) = 5.06$

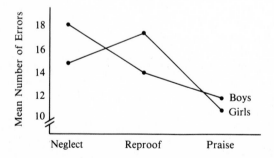

Interpretation: Only the teacher-response variable produced results that were significantly different from chance ($p < .01$). The null hypothesis $\mu_{neglect} = \mu_{reproof} = \mu_{praise}$ is rejected. The three methods of response produced different numbers of errors. Neither the gender main effect nor the interaction was statistically significant.

14.

Source	df	SS	MS	F	p
Learning	1	4.05	4.05	0.36	>.05
Recall	1	18.05	18.05	1.60	>.05
AB	1	151.25	151.25	13.41	<.01
Within groups	16	180.40	11.28		
Total	19	353.75			

$F_{.05} (1, 16 \, df) = 4.49; F_{.01} (1, 16 \, df) = 8.53$

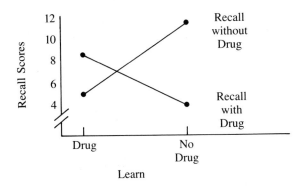

Interpretation: Only the interaction is significant. The conclusion is that when conditions of learning and recall are the same, memory is good. When conditions of learning and recall are different, memory is poorer.

15. a. Independent variables: (1) treatment the subject received and (2) whether the experimenter who administered the humor test had treated the subject or not. Dependent variable: number of captions produced. $p(A) < .01$, $p(B) < .05$, $p(AB) < .05$.

The significant interaction means that the effect of the experimenter *depended* on whether the subject had been insulted or had been treated neutrally. If the subject had been insulted, then taking the humor test from the person who had done the insulting caused a larger number of captions to be written—larger, that is, than if a different experimenter had administered the test. If the subject had not been insulted, there was little difference between the two kinds of experimenters.

The main effect of treatments was significant. On the average, more captions were produced when subjects had been insulted than when they had been treated neutrally. However, this average effect appears to be due primarily to those taking the test from the same experimenter who did the insulting (Cell $A_1 B_1$).

The main effect of the experimenters was significant, with more captions being produced when the experimenter who administered the test was the same as the one who administered the treatment. Again, this main effect seems to be heavily influenced by Cell $A_1 B_1$.

b. Independent variables: (1) name and (2) occupation of father. Dependent variable: score assigned to the theme. $p(A) > .05$, $p(B) > .05$, $p(AB) < .01$.

The significant interaction means that the effect of the name depends on the occupation of the father. The effect of the name David is good if his father is a research chemist (Cell $A_1 B_1$) and bad if his father is unemployed (Cell $A_1 B_2$). The effect of the name Elmer is good if his father is unemployed (Cell $A_2 B_2$) and bad if his father is a research chemist (Cell $A_2 B_1$).

The main effects are not significant. The average "David" theme is about equal to the average "Elmer" theme. Similarly, the themes do not differ significantly according to whether the father was unemployed or a research chemist.

Notice that these main effects, if interpreted by themselves, would be misleading; that is, *names do have an effect,* but the effect depends on the occupation of the father. Main effects, however, are only sensitive to average differences.

c. Independent variables: (1) gender of speaker and (2) job experience of speaker. Dependent variable: number of students looking at the speaker. $p(A) < .05$, $p(B) < .01$, $p(AB) > .05$.

Since the interaction is not significant, the interpretations of the main effects are straightforward. The gender of the speaker was significant; students paid more attention to men than to women. The experience of the speaker was significant; students paid more attention to those who had been on the job for longer than 2 years than to those who had been on the job less than 6 months.

16. No, because the interaction in that study was significant.

17.

Source	df	SS	MS	F	p
A (rewards)	3	199.31	66.44	3.90	$<.05$
B (classes)	1	2.25	2.25	<1	$>.05$
AB	3	0.13	0.04	<1	$>.05$
Within groups	56	953.25	17.02		
Total	63	1154.94			

Up to this point the interpretation is that freshmen and seniors do not differ significantly and there is no significant interaction between rewards and class standing. There is a significant difference among the rewards; that is, rewards have an effect on attitudes toward the police. Pairwise comparisons are called for.

The size of a mean difference that will be significant at the .05 level with $N = 16$ is 3.91.

$$(\text{HSD}_{.05})(s_{\bar{x}}) = (3.79)\sqrt{\frac{17.02}{16}} = 3.91$$

To facilitate an overall view, we arranged all six mean differences into a table, marking with an asterisk the one that is larger than the critical value of 3.91.

	$10	$5	$1	50¢
$10		1.31	3.19	4.62*
$5			1.88	3.31
$1				1.43

The less students were paid to defend the police, the more positive their attitude toward the police. Those paid 50¢ had significantly more positive attitudes toward the police than those paid $10, $p < .05$.

18. $SS_{tot} = 817 - \dfrac{(129)^2}{24} = 123.625$

$SS_{bg} = \dfrac{(22)^2}{3} + \dfrac{(9)^2}{3} + \dfrac{(23)^2}{3} + \dfrac{(15)^2}{3} + \dfrac{(11)^2}{3} + \dfrac{(26)^2}{3} + \dfrac{(9)^2}{3}$

$+ \dfrac{(14)^2}{3} - \dfrac{(129)^2}{24} = 104.29$

$SS_{gender} = \dfrac{(69)^2}{12} + \dfrac{(60)^2}{12} - \dfrac{(129)^2}{24} = 3.375$

$SS_{pictures} = \dfrac{(33)^2}{6} + \dfrac{(35)^2}{6} + \dfrac{(32)^2}{6} + \dfrac{(29)^2}{6} - \dfrac{(129)^2}{24} = 3.125$

$SS_{G \times P} = 3[(7.333 - 5.5 - 5.75 + 5.375)^2 + (3.667 - 5.50 - 5.0 + 5.375)^2$

$+ (3.0 - 5.833 - 5.75 + 5.375)^2 + (8.667 - 5.833 - 5.0 + 5.375)^2$

$+ (7.667 - 5.333 - 5.75 + 5.375)^2 + (3.0 - 5.333 - 5.0 + 5.375)^2$

$+ (5.0 - 4.833 - 5.75 + 5.375)^2 + (4.667 - 4.833 - 5.0 + 5.375)^2]$

$= 3(32.60) = 97.79$

Check: $SS_{bg} = 3.375 + 3.125 + 97.79 = 104.29$

$SS_{wg} = \left(166 - \dfrac{(22)^2}{3}\right) + \left(41 - \dfrac{(11)^2}{3}\right) + \left(29 - \dfrac{(9)^2}{3}\right) + \left(226 - \dfrac{(26)^2}{3}\right)$

$+ \left(179 - \dfrac{(23)^2}{3}\right) + \left(29 - \dfrac{(9)^2}{3}\right) + \left(77 - \dfrac{(15)^2}{3}\right) + \left(70 - \dfrac{(14)^2}{3}\right) = 19.33$

Check: $123.62 = 104.29 + 19.33$

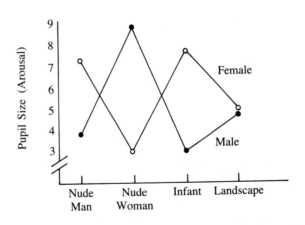

Source	df	MS	F	p
Gender	1	3.375	2.79	NS
Pictures	3	1.042	0.86	NS
Gender × pictures	3	32.597	26.98	<.01
Within groups	16	1.208		
Total	23			

$F_{.05}\,(1, 16) = 4.49; \ F_{.01}\,(3, 16) = 5.29$

Only the interaction is significant. Based on the graph, males were aroused by nude women and not by nude men or infants. Females, however, were aroused by nude men and infants and not by nude women. Responses toward the landscape were about the same for both genders. Tukey's HSDs are not appropriate because the main effects were not significant. (Even if a main effect had been significant, the significant interaction would prevent HSDs from being appropriate.)

CHAPTER 11

1. $\chi^2 = \Sigma \left[\dfrac{(O-E)^2}{E} \right] = \dfrac{(316-300)^2}{300} + \dfrac{(84-100)^2}{100}$

   $= 0.8533 + 2.5600$

   $= 3.4133; \quad df = 1$

   $\chi^2_{.05} (1\,df) = 3.84$. Retain the null hypothesis and conclude that these data are consistent with a 3:1 hypothesis.

2. The expected frequencies are 20 correct and 40 incorrect: $(1/3)(60) = 20$ and $(2/3)(60) = 40$.

   $\chi^2 = \dfrac{(32-20)^2}{20} + \dfrac{(28-40)^2}{40} = 10.80 \qquad df = 1, p < .01$

   Thus, these three emotions can be distinguished from each other. Subjects did not respond in a chance fashion.

   There is an interesting sequel to this experiment. Subsequent researchers did not find the simple, clear-cut results that Watson reported. One experimenter (Sherman, 1927) found that if only the infant's reactions were observed, there was a great deal of disagreement. However, if the observers also knew the stimuli (dropping, stroking, and so forth), they agreed with each other. This seems to be a case in which Watson's design did not permit him to separate the effect of the infant's reaction (the independent variable) from the effect of knowing what caused the reaction (an extraneous variable).

3.

	Received program	Control	Total
Police record	114	101	215
No police record	211	224	435
Total	325	325	650

The expected frequencies are:

$\dfrac{(215)(325)}{650} = 107.5; \quad \dfrac{(215)(325)}{650} = 107.5; \quad \dfrac{(435)(325)}{650} = 217.5; \quad \dfrac{(435)(325)}{650} = 217.5$

By the shortcut method:

$\chi^2 = \dfrac{650[(114)(224) - (101)(211)]^2}{(215)(435)(325)(325)} = 1.17$

$\chi^2_{.05} (1\,df) = 3.84$. Therefore, retain the null hypothesis and conclude that the two groups did not differ significantly in the number who had police records. The Cambridge-Somerville Youth Study, discussed in many sociology texts, was first reported by Powers and Witmer (1951).

4. $\chi^2 = \dfrac{100[(9)(34) - (26)(31)]^2}{(35)(65)(40)(60)} = 4.579$

Calculations of the expected frequencies are as follows:

$\dfrac{(35)(40)}{100} = 14 \qquad \dfrac{(35)(60)}{100} = 21$

$\dfrac{(65)(40)}{100} = 26 \qquad \dfrac{(65)(60)}{100} = 39$

O	E	$O - E$	$(O - E)^2$	$\dfrac{(O - E)^2}{E}$
9	14	-5	25	1.786
26	21	5	25	1.191
31	26	5	25	0.962
34	39	-5	25	0.641
				$\chi^2 = 4.58$
				$df = 1$

$\chi^2_{.05}$ (1 df) = 3.84. Reject the hypothesis that group size and joining are independent. Conclude that passersby are more likely to join a group of five than a group of two.

5. a.

Recapture site

		Issaquah	East Fork
	Issaquah	46	0
Capture site	East Fork	8	19

$\chi^2 = \dfrac{73[(46)(19) - (8)(0)]^2}{(46)(27)(54)(19)} = 43.76$

$\chi^2_{.001}$ (1 df) = 10.83. Reject the null hypothesis (which is that the second choice is independent of the first) and conclude that choices are very consistent; salmon tend to choose the same stream each time.

b. $\chi^2 = \dfrac{70[(39)(3) - (16)(12)]^2}{(51)(19)(55)(15)} = 0.49$

$\chi^2_{.05}$ (1 df) = 3.84. Hasler's hypothesis is supported; fish with plugged nasal openings do not make a consistent choice of streams, but those that get olfactory cues consistently choose the same stream. For further confirmation and a summary of Hasler's work, see Hasler, Scholz, and Horrall (1978).

6. The work that needs to be done on the data is to divide each set of applicants into two independent categories, hired and not hired. The following table results:

	White	*Black*	*Total*
Hired	390	18	408
Not hired	3810	832	4642
Total	4200	850	5050

$$\chi^2 = \frac{5050[(390)(832) - (3810)(18)]^2}{(4200)(850)(408)(4642)} = \frac{3.3070 \times 10^{14}}{6.7514 \times 10^{12}} = 48.91$$

Given that $\chi^2_{.001}$ (1 df) = 10.83, the null hypothesis can be rejected. Since 9.29 percent of the white applicants were hired compared to 2.12 percent of the black applicants, the statistical consultant and the company in question concluded that discrimination had occurred.

7. To get the expected frequencies, multiply the percentages given by Professor Stickler by the 340 students. Then enter these expected frequencies in the usual table.

O	E	$O - E$	$(O - E)^2$	$\dfrac{(O - E)^2}{E}$
20	23.8	−3.8	14.44	0.61
74	81.6	−7.6	57.76	0.71
120	129.2	−9.2	84.64	0.66
88	81.6	6.4	40.96	0.50
38	23.8	14.2	201.64	8.47
				$\chi^2 = 10.95$
				$df = 5 - 1 = 4$

(In this problem, the only restriction on the theoretical frequencies is that $\Sigma E = \Sigma O$.) $\chi^2_{.05}$ (4 df) = 9.49. Therefore, reject the contention of the professor that his grades conform to "the curve." By examining the data, you can also reject the contention of the colleague that the professor is too soft. The primary reason the data do not fit the curve is that there were too many "flunks."

8. Goodness of fit

9. Each of the six sides of a die is equally likely. Since the total number of throws was 1200, the expected value for any one of them is $\frac{1}{6}$ times 1200, or 200. Thus, retain the null hypothesis. The results of the evening do not differ significantly from the "unbiased dice" model.

O	E	$O - E$	$(O - E)^2$	$\dfrac{(O - E)^2}{E}$
195	200	−5	25	0.13
200	200	0	0	0.00
220	200	20	400	2.00
215	200	15	225	1.13
190	200	−10	100	0.50
180	200	−20	400	2.00
Σ 1200	1200			$\chi^2 = 5.76$
$\chi^2_{.05}$ (5 df) = 11.07				

10. Goodness of fit

11.

Younger students		Older students		Faculty	
O	E	O	E	O	E
29	23.54	6	11.64	11	10.82
31	26.10	10	12.90	10	12.00
20	23.03	13	11.38	12	10.59
7	14.33	14	7.08	7	6.59

$\chi^2 = 17.26$

$df = 3 \times 2 = 6$

$\chi^2_{.01} (6\ df) = 16.81$

Thus, reject the null hypothesis at the .01 level. A person's opinion is not independent of group membership. By examining the separate results of each $(O - E)^2/E$, you can see that the principal contributions to the final χ^2 value were from the opinion "Intervene with Military Force." The younger and older students differed on this, with the older students being pro and the younger students being con. Our political science student might conclude that age appears to be an important variable in predicting opinions.

12. Independence

13. Expected frequencies:

$(\frac{1}{4})(130) = 32.5; \qquad (\frac{3}{4})(130) = 97.5$

O	E	O − E	$(O - E)^2$	$\dfrac{(O - E)^2}{E}$
23	32.5	−9.5	90.25	2.777
107	97.5	9.5	90.25	0.9256
				$\chi^2 = 3.703$
				$df = 2 - 1 = 1$

$\chi^2_{.05} (1\ df) = 3.84$. Thus, retain the null hypothesis and conclude that these data are consistent with the 1:3 ratio predicted by the theory.

14. The difference between the tests for goodness of fit and for independence is in how the expected frequencies are obtained. In the goodness-of-fit test, expected frequencies are predicted by a theory, whereas in the independence test, they are obtained from the data.

15. You should be gentle with your friend but explain that the observations in his study are not independent and cannot be analyzed with χ^2. Explain that, since one person is making five observations, the observations are not independent; the choice of one female candidate may cause the subject to pick a male candidate next (or vice versa).

16.

		Candidates		
		Hill	Dale	Σ
	Yes	57	31	88
Signaled turn	No	11	2	13
	Σ	68	33	101

$$\chi^2 = \frac{101[(57)(2) - (31)(11)]^2}{(88)(13)(68)(33)} = 2.03$$

$\chi^2_{.05}$ (1 df) = 3.84. With respect to failure to signal a left turn, there was not a statistically significant difference between cars with stickers for Hill and cars with stickers for Dale.

17. We hope you would say something like, "Gee, I've been able to analyze data like these since I learned the t test. I will need to know the standard deviations for those means, though." If you attempted to analyze these data using χ^2, you erred (which, according to Alexander Pope, is human). If you erred, forgive yourself and reread pages 234 and 250.

18.

Houses

Candidates		Brick	Frame	Σ
	Hill	17	59	76
	Dale	88	51	139
	Σ	105	110	215

$$\chi^2 = \frac{N(AD - BC)^2}{(A + B)(C + D)(A + C)(B + D)}$$

$$= \frac{215(867 - 5192)^2}{(76)(139)(105)(110)} = 32.96; df = 1$$

$\chi^2_{.001}$ (1 df) = 10.83; $p < .001$. 215 houses with yard signs were classified as brick (affluent) or one-story frame (less affluent). Hill's signs were more often found at frame houses, and Dale's signs were more often found at brick houses. This difference was so statistically significant that it is not likely that it was due to chance.

19. $df = (R - 1)(C - 1)$, except for a table with only one row.
    a. 3   **b.** 12   **c.** 3   **d.** 10

CHAPTER 12

1. Sum of the ranks of those given estrogen = 76
   Sum of ranks of the control animals = 44

   Smaller $U = (7)(8) + \dfrac{(7)(8)}{2} - 76 = 8$

   Upon looking in Table H under $N_1 = 7$, $N_2 = 8$, for a two-tailed test with $\alpha = .05$, you see that a U value of 10 or less is required to reject the null hypothesis. Since the obtained $U = 8$, you should reject the null hypothesis. By examining the data and noting that the estrogen-injected animals were the lowest ranking ones, you can conclude that estrogen causes rats to be less dominant.

2. The sum of the ranks are 574.5 and 245.5. The smaller U is 109.5 (the larger is 274.5).

   $$U = (24)(16) + \frac{(24)(25)}{2} - 574.5 = 109.5$$

   $$\mu_U = \frac{(24)(16)}{2} = 192$$

   $$\sigma_U = \sqrt{\frac{(24)(16)(41)}{12}} = 36.22$$

$$z = \frac{(109.5 + 0.5) - 192}{36.22} = -2.26$$

With such a z value you may reject the null hypothesis that the distributions are the same. By examining the average ranks (23.9 for present-day birds and 15.4 for 10-years-ago birds), you should conclude that present-day birds have significantly fewer brain parasites. (If you gave the 0 birds a rank of 2, the average ranks are 17.1 for present-day birds and 25.6 for 10-years-ago birds. The same conclusion is reached, however.)

3. With $\alpha = .01$, there are *no* possible results that would allow H_0 to be rejected. We must find more cars.

4. $\Sigma R_Y = 13$, $\Sigma R_Z = 32$.

$$\text{Smaller } U = (4)(5) + \frac{(5)(6)}{2} - 32 = 3$$

A U of 1 or less is required to reject H_0 (two-tailed test). Since the obtained $U = 3$, you must conclude that the quietness test did not produce evidence that Y cars are quieter than Z cars.

5. $\mu_T = \dfrac{N(N + 1)}{4} = \dfrac{112(113)}{4} = 3164$

$$\sigma_T = \sqrt{\frac{N(N + 1)(2N + 1)}{24}} = \sqrt{\frac{112(113)(225)}{24}} = 344.46$$

$$z = \frac{(4077 + 0.5) - 3164}{344.46} = 2.65$$

Since 2.65 is greater than 1.96, reject H_0 and conclude that incomes were significantly different after the program. Because you do not have the actual data, you cannot tell whether the incomes were higher or lower than before.

6.

Worker	Without rests	With rests	D	Signed ranks
1	2240	2421	181	2
2	2069	2260	191	4
3	2132	2333	201	5
4	2095	2314	219	6
5	2162	2297	135	1
6	2203	2389	186	3

Check: $21 + 0 = 21$

$\dfrac{(6)(7)}{2} = 21$

Σ positive $= 21$

Σ negative $= 0$

$T = 0$

$N = 6$

The critical value of T at the .05 level for a two-tailed test is 0. Since 0 (obtained) is equal to or less than 0 (table), reject the null hypothesis and conclude that the output with rests is greater than the output without rests. Here is the story behind this study.

From 1927 to 1932 the Western Electric Company conducted a study on a group of six workers at their Hawthorne Works near Chicago. The final conclusion reached by this study has come to be called the Hawthorne effect. In the study, six workers who assembled telephone relays were separated from the rest of the workers. A variety of changes in their daily routine followed, one at a time, with the following results. Five-minute rest periods morning and afternoon increased output, 10-minute rest periods increased output, company-provided snacks during rest periods increased output, and quitting 30 minutes early increased output. Going back to the original no-rest schedule increased output again, as did the reintroduction of the rest periods. Finally, the management concluded that it was the special attention paid to the workers, rather than the rest periods, that increased output. Thus, the Hawthorne effect is an improvement in performance that is due just to being in an experiment (getting special attention) rather than to any specific manipulation in the experiment. For a summary of this study, see Mayo (1946).

7. The appropriate test for this study is one for two independent samples. A Mann-Whitney U test will work. If high scores are given high rank (that is, 39 ranks 1):

$$\Sigma \text{(Canadians)} = 208$$
$$\Sigma \text{(US)} = 257$$

(If low scores are given high ranks, the sums are reversed.)

$$U = (15)(15) + \frac{(15)(16)}{2} - 257 = 88$$

By consulting the second page of Table H (boldface type), you will find that a U value of 64 is required in order to reject the null hypothesis ($\alpha = .05$, two-tailed test). The lower of the two calculated U values for this problem is 88. (The U value using the Canadian sum is 137.) Since 88 is larger than the tabled value of 64, the null hypothesis must be retained. Thus, our political science student must conclude that he has no evidence that Canadians and people from the United States have different attitudes toward the regulation of business.

Once again you must be cautious in writing the conclusion when the null hypothesis is retained. You have not demonstrated that the two groups are the same; you have shown only that the groups were not significantly different, which is a rather unsatisfactory way to leave things.

8.

Student	Before	After	D	Signed rank
1	18	4	14	13
2	14	14	0	1.5
3	20	10	10	9
4	6	9	−3	−4
5	15	10	5	6
6	17	5	12	11
7	29	16	13	12
8	5	4	1	3
9	8	8	0	−1.5
10	10	4	6	7
11	26	15	11	10
12	17	9	8	8
13	14	10	4	5
14	12	12	0	Eliminated

Check: $85.5 + 5.5 = 91$ Σ positive $= 85.5$

$\dfrac{(13)(14)}{2} = 91$ Σ negative $= -5.5$

$T = 5.5$

$N = 13$

Since the tabled value for T for a two-tailed test with $\alpha = .01$ is 9, you may reject H_0 and conclude that the after distribution is from a different population than the before distribution. Since, except for one person, the number of misconceptions stayed the same or decreased, you may conclude that the course *reduced* the number of misconceptions.

We would like to remind you here of the distinction we made in Chapter 7 between statistically significant and important. There is a statistically significant decrease in the number of misconceptions, but a professor might be quite dismayed at the number of misconceptions that remain.

9. Σ positive $= 81.5$ 　　　　Check: $81.5 + 54.5 = 136$

Σ negative $= -54.5$ 　　　　$\dfrac{(16)(17)}{2} = 136$

$T = 54.5$

$N = 16$

A $T \leqslant 29$ is required at the .05 level. Thus, there is no significant difference in the weight 10 months later. Put in the most positive language, there is no evidence that the participants tended to gain back the weight lost during the workshop.

10. Arrange the data into the usual summary table:

Groups

		1	2	3	4	5
	2	85				
	3	31	54			
Groups	4	18	67	13		
	5	103	188*	134	121	
	6	31	116	62	49	72

$*p < .05$

For $K = 6$, $N = 8$, differences of 159.6 and 188.4 are required to reject the null hypothesis at the .05 and .01 levels, respectively. Thus, the only significant difference is between the means of Groups 2 and 5 at the .05 level. (With data such as these, the novice investigator may be tempted to report "almost significant at the .01 level." Resist that temptation.)

11.

Authoritarian		Democratic		Laissez-faire	
X	Rank	X	Rank	X	Rank
77	5	90	10.5	50	1
86	9	92	12	62	2
90	10.5	100	16	69	3
97	13	105	21	76	4
100	16	107	22	79	6
102	19	108	23	82	7
120	27.5	110	24	84	8
121	29	118	26	99	14
128	32	125	30	100	16
130	33	131	34	101	18
135	37	132	35.5	103	20
137	38	132	35.5	114	25
141	39.5	146	41	120	27.5
147	42	156	44	126	31
153	43	161	45	141	39.5
Σ	$\overline{393.5}$		$\overline{419.5}$		$\overline{222.0}$

	Authoritarian (393.5)	Democratic (419.5)	Laissez-faire (222)
Democratic (419.5)	26		
Laissez-faire (222)	171.5*	197.5*	

*p < .05

For $K = 3$, $N = 15$, a difference in the sum of the ranks of 168.8 is required to reject H_0 at the .05 level. Therefore, both the authoritarian leadership and the democratic leadership resulted in higher personal satisfaction scores than did laissez-faire leadership, but the authoritarian and democratic types of leadership did not significantly differ from each other.

12. The Mann-Whitney U test can be performed on ranked data from two independent samples with equal or unequal N's. The Wilcoxon matched-pairs signed-ranks test can be performed on ranked data from two correlated samples. The Wilcoxon-Wilcox multiple-comparisons test makes all possible comparisons among K independent equal-N groups of ranked data.

13.

Year	Rank in marriages	Rank in grain
1930	1	3
1931	2	1
1927	3	4
1925	4	2
1926	5	5
1923	6	6
1922	7	7
1919	8.5	11
1924	8.5	9
1928	10	9
1929	11	9
1918	12	12.5
1921	13	12.5
1920	14	14

This is a case of tied ranks, so we will compute a Pearson r on the ranks.

$$r = \frac{N\Sigma XY - (\Sigma X)(\Sigma Y)}{\sqrt{[N\Sigma X^2 - (\Sigma X)^2][N\Sigma Y^2 - (\Sigma Y)^2]}}$$

$$= \frac{14(1002.5) - (105)(105)}{\sqrt{[(14)(1014.5) - (105)^2][(14)(1012.5) - (105)^2]}} = .95$$

Consulting Table A for the significance of correlation coefficients, we find that for 12 df, a coefficient of .78 is significant at the .001 level. Our sentence of explanation would be: There is a highly significant relationship between the number of marriages and the value of the grain crop; as one goes up the other does too.

14. $r_s = 1 - \dfrac{6\Sigma D^2}{N(N^2 - 1)} = 1 - \dfrac{6(308)}{16(16^2 - 1)} = .55$

From Table L, a coefficient of .51 is required for significance. Thus, an $r_s = .55$ is significantly different from .00.

15.

Candidates	Locke	Kant	D	D^2
A	7	8	−1	1
B	10	10	0	0
C	3	5	−2	4
D	9	9	0	0
E	1	1	0	0
F	8	7	1	1
G	5	3	2	4
H	2	4	−2	4
I	6	6	0	0
J	4	2	2	4

$$\Sigma D^2 = 18$$

$$r_s = 1 - \frac{6\,\Sigma D^2}{N(N^2 - 1)} = 1 - \frac{6(18)}{10(99)} = 1 - .109 = .891$$

The two professors seem to be in pretty close agreement on the selection criteria.

16. Unfortunately, with only four pairs of scores there is no possible way to reject the hypothesis that the population correlation is .00 (see Table L). Advise your friend that more data must be obtained before any inference can be made about the population.

17. a. Wilcoxon matched-pairs signed-ranks test; this is a before-and-after study.

  b. Wilcoxon matched-pairs signed-ranks test; pairs are formed by family.

  c. Spearman's r_s; the degree of relationship is desired.

  d. Wilcoxon-Wilcox multiple-comparisons test; the experiment has three independent groups.

  e. Wilcoxon matched-pairs signed-ranks test; again, this is a before-and-after design.

18.

	Symbol of statistic	Appropriate for what design?	Calculated statistics must be (larger, smaller) than the tabled statistic to reject H_0
Mann-Whitney	U	Two independent samples	Smaller
Wilcoxon matched-pairs signed-ranks	T	Two correlated samples	Smaller
Wilcoxon-Wilcox multiple-comparisons	None	More than two independent samples, equal N's	Larger

CHAPTER 13   **Set A**

1. A mean of a set of means is needed, and to find it you need to know the number of dollars invested in each division.

2. Wilcoxon matched-pairs signed-ranks test. You have good evidence that the population of dependent scores (reaction time) is not normally distributed.

3. The one-way ANOVA technique you covered in this book requires independent groups and these measures are correlated. A repeated-measures ANOVA for four groups would be appropriate.

4. A median is appropriate, since the scores are skewed.

5. Assume that these measures are normally distributed and use a normal curve to find the proportion.

6. Either an independent-measures t test or a Mann-Whitney U test will do.
7. A 95 or 99 percent confidence interval about the mean reading achievement of those 50 sixth-graders will answer the question.
8. χ^2 goodness of fit
9. Mode
10. Neither will do. The relationship described is nonlinear—one that first increases and then decreases. Neither r nor r_s is appropriate. A statistic, *eta*, is appropriate for this.
11. Correlated-samples t test or Wilcoxon matched-pairs signed-ranks test
12. A regression equation will provide for the predictions and a correlation coefficient will indicate how accurate the predictions will be.
13. A 2×2 factorial ANOVA
14. The student's wondering may be translated into a question of whether a correlation of .20 is a statistically significant one. Use a t test to find whether $r = .20$ for that class is significantly different from $r = .00$. Or, look in Table A in Appendix C.
15. This is a χ^2 problem but it cannot be worked using the techniques in this text because these before-and-after data are correlated, not independent. Intermediate texts describe appropriate techniques for correlated χ^2 problems.
16. Finding the probability that the sample mean came from a population in which $\mu = 1000$ (a t test or a z test) will answer the question.
17. Pearson product-moment correlation coefficient
18. One-way analysis of variance
19. A χ^2 test of independence will determine whether a decision to report shoplifting is influenced by gender and by the dress of the shoplifter.
20. A line graph with serial position on the X axis and number of errors on the Y axis will illustrate this relationship.
21. A Mann-Whitney U test would be preferred, since the dependent variable seems to be skewed. The question is whether there is a relationship between diet and cancer. (Note that if the two groups differ, two interpretations are possible. For example, if the incidence is higher among the red meat cultures, it might be because of the red meat or because of the lack of cereals.)
22. An r_s will give the degree of relationship for these two ranked variables.
23. For each species of fish, a set of z scores may be calculated. The fish with the largest z score should be declared the overall winner.

Set B

24. The critical values are 377.6 at the .05 level and 457.6 at the .01 level. You can conclude that Herbicide B is significantly better than A, C, or D at the .01 level and that D is better than C at the .05 level.
25. The screening process is clearly worthwhile in making each batch profitable. Of course, the cost of screening will have to be taken into account.
26. $t_{.001} (40\ df) = 3.55$; therefore, reject the null hypothesis and conclude that extended practice *improved* performance. Since this is just the opposite of the theory's prediction, we may conclude that the theory is not adequate to explain the serial position effect.
27. Since $\chi^2_{.05} (2\ df) = 5.99$, the null hypothesis for this goodness-of-fit test is retained. The data do fit the theory; the theory is adequate.
28. The overall F is significant, so we can conclude that the differences should not be attributed to chance. The low dose is significantly better than the placebo $(HSD_{.01} = 4.70)$ and significantly better than the high dose (which has a smaller mean than the placebo). The placebo and high dose are not significantly different.
29. The critical value of U for a two-tailed test with $\alpha = .01$ is 70. You may conclude that the difference between the two methods is statistically significant at the .01 level. You cannot, from the information supplied, tell which of the two methods is superior.

30. Obtained $t = 4.58$; $t_{.001}$ (16 df) = 4.02; attitudes of college students and people in business toward the 18 groups are similar. There is little likelihood that the similarity of attitudes found in the sample is due to chance.

31. $t_{.05}$ (60 df) = 2.00; $F_{.01}$ (30, 40) = 2.11; the poetry unit appears to have no significant effect on mean attitudes toward poetry. There is a very significant effect on the variability of the attitudes of those who studied poetry. It appears that the poetry unit turned some students on and some students off, thus causing a great deal of variability in the attitude scores.

32. $F_{.05}(1, 44) = 4.06$; $F_{.01}(1, 44) = 7.24$; only the interaction is significant. A graph, as always, will help in the interpretation. See the graph here. Although neither of the main effects is statistically significant, both are important. The question of whether to present one or both sides to get the most attitude change depends on the level of education of the audience. If the members of a group have less than a high school education, present one side. If they have some college education, present both sides.

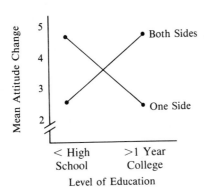

33. Since the tabled value of $F_{.05}$ (2, 60) = 3.15, the null hypothesis is retained. There is no significant difference in the verbal ability scores of third-graders who are left- or right-handed or mixed.

34. Since $\chi^2_{.02}$ (2 df) = 7.82, the null hypothesis for this test of independence is rejected at the .02 level. Alcoholism and toilet-training are not independent, they are related. Experiments such as this do not allow you to make cause-and-effect statements, however.

APPENDIX B—ARITHMETIC AND ALGEBRA REVIEW

PART 1
(PRETEST)

1. About 16	**2.** About 16	**3.** About 16	**4.** 6.1
5. 0.4	**6.** 10.3	**7.** 11.424	**8.** 27.141
9. 19.85	**10.** 2.065	**11.** 49.972	**12.** 0.0567
13. 3.02	**14.** 17.11	**15.** 1.375	**16.** 0.83
17. 0.1875	**18.** 0.056	**19.** 0.60	**20.** 0.10
21. 1.50	**22.** 3.11	**23.** 6	**24.** −22
25. −7	**26.** −6	**27.** 25	**28.** −24
29. 3.33	**30.** −5.25	**31.** 33%	**32.** 22.68
33. .19	**34.** 60 birds	**35.** 5	**36.** 4
37. 6, 10	**38.** 4, 22	**39.** 16	**40.** 6.25
41. 0.1225	**42.** 3.05	**43.** 0.96	**44.** 0.30
45. 4.20	**46.** 4.00	**47.** 4.50	**48.** 0.45

49. 15.25   **50.** 8.00   **51.** 16.25   **52.** 1.02
53. 12.00   **54.** 13.00   **55.** 11.00   **56.** 2.50
57. −2.33

PART 2
(REVIEW)

1. a. A sum is the answer to an addition problem.
   b. A quotient is the answer to a division problem.
   c. A product is the answer to a multiplication problem.
   d. A difference is the answer to a subtraction problem.

2. a. About 10   **b.** About 7   **c.** About 100   **d.** About 0.2
   e. About 0.16   **f.** About 60   **g.** Just less than 1   **h.** About 0.1

3. a. 14   **b.** 126   **c.** 9   **d.** 0   **e.** 128
   f. 13   **g.** 12   **h.** 13   **i.** 9

4. a. 6.33   **b.** 13.00   **c.** 0.05   **d.** 0.97   **e.** 2.61
   f. 0.34   **g.** 0.00   **h.** 0.02   **i.** 0.99

5.
$$\begin{array}{r} 0.001 \\ 10.000 \\ 3.652 \\ 2.500 \\ \hline 16.153 \end{array}$$

6.
$$\begin{array}{r} 14.20 \\ -7.31 \\ \hline 6.89 \end{array}$$

7.
$$\begin{array}{r} 1.26 \\ \times 0.04 \\ \hline 0.0504 \end{array}$$

8.
$$\begin{array}{r} 143.300 \\ 16.920 \\ 2.307 \\ 8.100 \\ \hline 170.627 \end{array}$$

9.
$$\begin{array}{r} 76.5 \\ 0.04\,\overline{)3.06} \\ 28 \\ \hline 26 \\ 24 \\ \hline 20 \\ 20 \\ \hline \end{array}$$

10.
$$\begin{array}{r} 2.04 \\ 11.75\,\overline{)24.00} \\ 23\,50 \\ \hline 5000 \\ 4700 \\ \hline 300 \end{array}$$

11.
$$\begin{array}{r} 152.12 \\ -127.40 \\ \hline 24.72 \end{array}$$

12.
$$\begin{array}{r} 0.07 \\ \times 0.5 \\ \hline 0.035 \end{array}$$

13. $\dfrac{9}{10} + \dfrac{1}{2} + \dfrac{2}{5} = 0.90 + 0.50 + 0.40 = 1.80$

14. $\dfrac{9}{20} \div \dfrac{19}{20} = 0.45 \div 0.95 = 0.47$

15. $\left(\dfrac{1}{3}\right)\left(\dfrac{5}{6}\right) = 0.333 \times 0.833 = 0.28$

16. $\dfrac{4}{5} - \dfrac{1}{6} = 0.80 - 0.167 = 0.63$

17. $\dfrac{1}{3} \div \dfrac{5}{6} = 0.333 \div 0.833 = 0.40$

18. $\dfrac{3}{4} \times \dfrac{5}{6} = 0.75 \times 0.833 = 0.62$

19. $18 \div \dfrac{1}{3} = 18 \div 0.333 = 54$

20. a. $(24) + (-28) = -4$ **b.** -23
   c. $(-8) + (11) = 3$ **d.** $(-15) + (8) = -7$

21. −40   **22.** 24
23. 48   **24.** −33
25. $(-18) - (-9) = -9$ **26.** $14 \div (-6) = -2.33$
27. $12 - (-3) = 15$ **28.** $(-6) - (-7) = 1$
29. $(-9) \div (-3) = 3$ **30.** $(-10) \div 5 = -2$
31. $4 \div (-12) = -0.33$ **32.** $(-7) - 5 = -12$
33. $6 \div 13 = .46$ **34.** $.46 \times 100 = 46$ percent
35. $18 \div 25 = .72$ **36.** $85 \div 115 = .74 \times 100 = 74$ percent
37. 25.92   **38.** $|-31| = 31$
39. $|21 - 25| = |-4| = 4$ **40.** $12 \pm (2)(5) = 12 \pm 10 = 2, 22$
41. $\pm(5)(6) + 10 = \pm 30 + 10 = -20, 40$ **42.** $\pm(2)(2) - 6 = \pm 4 - 6 = -10, -2$

43. $(2.5)^2 = 2.5 \times 2.5 = 6.25$

44. $9^2 = 9 \times 9 = 81$

45. $0.3 \times 0.3 = 0.09$

46. $(1/4)^2 = (0.25)(0.25) = 0.0625$

47. a. 25.00   **b.** 2.50   **c.** 0.79   **d.** 0.25   **e.** 4.10
   f. 0.0548   **g.** 426.00   **h.** 0.50   **i.** 4.74   **j.** 5.14

48. $\dfrac{(4-2)^2 + (0-2)^2}{6} = \dfrac{2^2 + (-2)^2}{6} = \dfrac{4+4}{6} = \dfrac{8}{6} = 1.33$

49. $\dfrac{(12-8)^2 + (8-8)^2 + (5-8)^2 + (7-8)^2}{4-1} = \dfrac{4^2 + 0^2 + (-3)^2 + (-1)^2}{3}$

$= \dfrac{16 + 0 + 9 + 1}{3} = \dfrac{26}{3} = 8.67$

50. $\left(\dfrac{5+6}{3+2-2}\right)\left(\dfrac{1}{3} + \dfrac{1}{2}\right) = \left(\dfrac{11}{3}\right)(0.33 + 0.500) = (3.667)(0.833) = 3.05$

51. $\left(\dfrac{13+18}{6+8-2}\right)\left(\dfrac{1}{6} + \dfrac{1}{8}\right) = \left(\dfrac{31}{12}\right)(0.167 + 0.125) = (2.583)(0.292) = 0.75$

52. $\dfrac{8[(6-2)^2 - 5]}{(3)(2)(4)} = \dfrac{8[4^2 - 5]}{24} = \dfrac{8[16-5]}{24} = \dfrac{8[11]}{24} = \dfrac{88}{24} = 3.67$

53. $\dfrac{[(8-2)(5-1)]^2}{5(10-7)} = \dfrac{[(6)(4)]^2}{5(3)} = \dfrac{[24]^2}{15} = \dfrac{576}{15} = 38.40$

54. $\dfrac{6}{1/2} + \dfrac{8}{1/3} = (6 \div 0.5) + (8 \div 0.333) = 12 + 24 = 36$

55. $\left(\dfrac{9}{2/3}\right)^2 + \left(\dfrac{8}{3/4}\right)^2 = (9 \div 0.667)^2 + (8 \div 0.75)^2$

$= 13.5^2 + 10.667^2 = 182.25 + 113.785 = 296.03$

56. $\dfrac{10 - (6^2/9)}{8} = \dfrac{10 - (36/9)}{8} = \dfrac{10-4}{8} = \dfrac{6}{8} = 0.75$

57. $\dfrac{104 - (12^2/6)}{5} = \dfrac{104 - (144/6)}{5} = \dfrac{104 - 24}{5} = \dfrac{80}{5} = 16.00$

58. $\dfrac{x-4}{2} = 2.58,\ x - 4 = 5.16,\ x = 9.16$

59. $\dfrac{x-21}{6.1} = 1.04,\ x - 21 = 6.344,\ x = 27.34$

60. $x = \dfrac{14-11}{2.5} = \dfrac{3}{2.5} = 1.20$

61. $x = \dfrac{36-41}{8.2} = \dfrac{-5}{8.2} = -0.61$

INDEX

GLOSSARY OF SYMBOLS

Greek Letter Symbols

α	The probability of a Type I error
β	The probability of a Type II error
μ	The mean of a population
Σ	The sum; an instruction to add
σ	Standard deviation of a population
σ^2	Variance of a population
$\sigma_{\bar{X}}$	Standard error of the mean for a population
χ^2	The chi square statistic

Mathematical and Latin Letter Symbols

∞	Infinity
$>$	More than
$<$	Less than
a	Point where
b	The slope of
D	The differenc
$\bar{D}$	The mean of
df	Degrees of
E	In chi squa
$E(\bar{X})$	The expect
F	A ratio of population
f	Frequency
H_0	The null h zero
H_1	Any hypot
HSD	Tukey's h ANOVA
i	The interv
K	The numb
LL	Lower lim
MS	Mean squa
N	The numb